北京历史文化名城 北京皇城保护规划

Conservation plan for the Historic City of Beijing and Imperial City of Beijing

北京市规划委员会 编　　EDIT: Beijing Municipal City Planning Commission

中国建筑工业出版社　China Architecture & Building Press

北京历史文化名城北京皇城保护规划
Conservation plan for the Historic City of Beijing and Imperial City of Beijing

图书在版编目(CIP)数据

北京历史文化名城北京皇城保护规划／北京市规划委员会编．
－北京：中国建筑工业出版社，2004
ISBN 7-112-06237-3

Ⅰ．北…Ⅱ．北…Ⅲ．①古城－保护－北京市②古城－城市规划－北京市　Ⅳ．TU984.21

中国版本图书馆CIP数据核字(2003)第112923号

顾问委员会：	刘 淇	汪光焘	王岐山	单霁翔	刘敬民			
Consulting Board:	Liu Qi	Wang Guangtao	Wang Qishan	Shan Jixiang	Liu Jingmin			

主　　编：陈　刚　梅宁华　朱嘉广
Chief Editors: Chen Gang　Mei Ninghua　Zhu Jiaguang

副 主 编：黄　艳
Associate Chief Editor: Huang Yan

编委会：陈　刚　梅宁华　朱嘉广　黄　艳　姚　莹　邱　跃　张　维　曹跃进　孔繁峙　杜立群　范耀邦
Editorial Board: Chen Gang　Mei Ninghua　Zhu Jiaguang　Huang Yan　Yao Ying　Qiu Yue　Zhang Wei　Cao Yuejin　Kong Fanzhi　Du Liqun　Fan Yaobang

执行主编：温宗勇
Executive Chief Editor: Wen Zongyong

执行副主编：苏　晶　宋晓龙
Executive Associate Chief Editors: Su Jing　Song Xiaolong

编　　辑：赵　琪　郑　皓　桂　琳　廖正昕　吕海虹　王丹江　王玉伟　黄　威
Editors: Zhao Qi　Zheng Hao　Gui Lin　Liao Zhengxin　Lu Haihong　Wang Danjiang　Wang Yuwei　Huang Wei

特约编辑：李晓鸿
Contributing Editor: Li Xiaohong

责任编辑：黄居正　李迪悃　尹怀梅
Editors: Huang Juzheng　Li Dikun　Yin Huaimei

英文翻译：王秋海
Language Translator: Wang Qiuhai

英文校译：温宗勇　张　利　刘　勃　倪　峰
Language Proofreaders: Wen Zongyong　Zhang Li　Liu Bo　Ni Feng

封面设计：蒙安童
Cover: Meng Antong

装帧设计：水晶石数字传媒　蒙安童　郭丽伟　陈卓夫
Layout: Crystal Digital Media　Meng Antong　Guo Liwei　Chen Zhuofu

北京历史文化名城北京皇城保护规划
北京市规划委员会　编
＊
中国建筑工业出版社出版、发行(北京西郊百万庄)
新华书店经销
北京水晶石数字传媒制版
北京雅昌彩色印刷有限公司印刷
＊
开本：965×1270毫米 1/12　印张：16 1/3
2004年1月第一版　2004年1月第一次印刷
印数：1—3,000册　定价：258.00元
ISBN 7-112-06237-3
TU·5499 (12251)

版权所有　翻印必究
如有印装质量问题，可寄本社退换
(邮政编码：100037)
本社网址：http://www.china-abp.com.cn
网上书店：http://www.china-building.com.cn

目 录

序 5

前言 7

第一部分：《北京历史文化名城保护规划》

《北京历史文化名城保护规划》文本 10
《北京历史文化名城保护规划》说明 20
《北京历史文化名城保护规划》规划图纸 44

附件：

政协北京市第九届委员会第四次会议党派团体提案 81
政协委员建议抓紧编制《北京历史文化名城保护规划》 81
关于《北京历史文化名城保护规划》编制工作的报告 82
政协委员对《北京历史文化名城保护规划》编制工作的意见 .. 83
《北京历史文化名城保护规划》专家论证会专家意见要点 84
北京市人民政府常务会议纪要（第48期） 85
首都规划建设委员会第21次全体会议纪要 86
建设部关于《北京历史文化名城保护规划》审查的函 86
建设部关于对《北京历史文化名城保护规划》审核意见的函 .. 86
关于报审《北京历史文化名城保护规划》的请示 87
关于报请批复《北京历史文化名城保护规划》的请示 88
北京市人民政府关于《北京历史文化名城保护规划》的批复 .. 88
北京市人民政府关于实施《北京历史文化名城保护规划》的决定 ... 89
北京市城市总体规划（部分） 90
《北京历史文化名城保护规划》编制工作大事记 91

第二部分：《北京皇城保护规划》

《北京皇城保护规划》文本 94
《北京皇城保护规划》说明 100
皇城历史文化保护区文物保护单位名单 140
《北京皇城保护规划》规划图纸 156

附件：

《北京皇城保护规划》专家评审会评审意见 185
北京市人民政府市长办公会议纪要 187
首都规划建设委员会第22次全体会议纪要 187
关于报请批复《北京皇城保护规划》的请示 188
北京市人民政府关于《北京皇城保护规划》的批复 188
有关历史文化名城保护的国内、国际法案摘编 189

后记 196

Contents

Preface 5

Foreword 7

Part One: Conservation Plan for the Historic City of Beijing
    The Text of "Conservation Plan for the Historic City of Beijing" 10
    Introduction to the "Conservation Plan for the Historic City of Beijing" 20
    Drawings of Conservation Plan for the Historic City of Beijing 44

Appendixes:
    Proposal submitted by the 4th session of the 9th committee of
    Beijing Political Consultative Conference 81
    Members of Political Consultative Conference (PCC) propose to make an early
    Conservation Plan for the Historic City of Beijing 81
    Report on the drafting of Conservation Plan for the Historic City of Beijing 82
    Comments by PCC members on the Conservation Plan
    for the Historic City of Beijing 83
    Experts' suggestions on Conservation Plan for the Historic City of Beijing 84
    Minutes (no.48) of meeting of the Standing Committee of
    Beijing Municipal Government 85
    Summary of 21st plenary session of Capital Planning and Construction Commission 86
    Appraisal Reply of Ministry of Construction pertaining Conservation Plan
    for the Historic City of Beijing 86
    Appraisal Reply of Ministry of Construction pertaining Conservation Plan
    for the Historic City of Beijing 86
    Submission of Conservation Plan for the Historic City of Beijing for examination 87
    Submission of Conservation Plan for the Historic City of Beijing for approval 88
    Beijing Government's approval of Conservation Plan for the Historic City of Beijing 88
    Beijing Government's decision to implement Conservation Plan
    for the Historic City of Beijing 89
    Master Plan for Beijing Construction (part) 90
    Chronicle of major events in the working out of Conservation Plan
    for the Historic City of Beijing 91

Part Two: Conservation Plan for the Imperial City of Beijing
    The Text of "Conservation Plan for the Imperial City of Beijing" 93
    Introduction to the "Conservation Plan for the Imperial City of Beijing" 100
    List of Relic Units under protection in Imperial City Historic Cultural Protection Area 140
    Drawings of Conservation Plan for the Imperial City of Beijing 156

Appendixes:
    Experts' comments on Conservation Plan for the Imperial City of Beijing 185
    Summary of mayor's working conference of Beijing Municipal Government 187
    Summary of 22nd plenary session of Capital Planning and Construction Commission 187
    Submission of Conservation Plan for the Imperial City of Beijing for approval 188
    Beijing Government's Approval of Conservation Plan for the Imperial City of Beijing 188
    Extracts of Chinese and international law and regulations
    on conservation of historic cities 189

Postscript 196

序

北京是世界著名的古都，历史悠久，文化深厚。它有着3000多年的城市发展史和850年的建都史，悠长的岁月中积累了丰富的地上、地下文物资源，形成了具有鲜明中华民族传统文化特色的城市格局和古都风貌。这不仅是北京宝贵的财富，也是我们国家、民族乃至世界的宝贵财富。

党中央、国务院非常重视北京的规划建设。1993年国务院就提出"通过不懈的努力，将北京建成经济繁荣、社会安定和各项公共服务设施及环境达到世界一流水平的历史文化名城和现代化国际城市。"2000年9月，胡锦涛总书记、温家宝总理对北京的历史文化名城保护工作做出重要指示，指出保护好历史文化遗产，是继承和发扬中华民族优秀文化传统，也是北京可持续发展路线的重要组成部分。同时，保护好历史文化名城是我们必须履行的重大历史责任。要处理好保护古都风貌与改善人民生活，加快城市现代化的关系，关键在于坚持以人为本的原则，树立全面、协调、可持续的发展观，用科学的发展观指导城市的建设和管理。要在旧城改造、改善市民居住条件和环境的同时，十分注意保护好文物和历史文化遗产，保护好古都风貌，努力把北京建设成为既拥有现代化新姿又保留历史文化名城特色的世界一流的国际化大都市。

经过建国50多年来的实践和探索，我们认识到，要在城市现代化建设过程中，有效地保护好城市的传统格局和风貌，必须要坚持规划先行，从系统地编制保护规划入手。为此，我们编制了《北京旧城25片历史文化保护区保护规划》、《北京历史文化名城保护规划》、《北京皇城保护规划》等重要规划，确定了文物保护单位、历史文化保护区和历史文化名城整体保护三个层次的保护规划体系。

目前，北京正处于举办一届历史上最出色的奥运会和向国际化大都市迈进的关键时期。我们要认真贯彻中央的指示精神，本着对历史负责、对人民负责的积极态度，不断完善规划，要广泛汲取专家学者、人民代表的意见，采用科学规划的方式来确定保护范围和保护途径。同时，要严格按规划办事，将规划的内容向社会公开，广泛接受群众的监督和评议。我相信，通过社会各界富有创造性的扎实工作，一定能留给后人一个既有历史文化名城特色又具现代化水平的新北京。

北京市市委书记 刘淇

Preface

Beijing is a famous ancient capital city with a long history and rich culture. With an urban history of 3,000 years and a history of capital city of 850 years, Beijing has accumulated an abundance of cultural heritage both above and under the ground, developing itself into an urban layout with distinct ancient features and Chinese traditional culture. This heritage is not only the treasure of Beijing, but also of China, the Chinese people as well as the world.

The Party central committee and the State Council has been attaching great importance to the developmental plan of Beijing. As early as 1993, the State Council stipulated that "through unremitting efforts, we will build Beijing into a historic modern metropolis with a prosperous economy, social stability and various services and environment that are up to the top-grade standard of the world." In September of 2003, General Secretary Hu Jintao and Premier Wen Jiabao also made important remarks about the preservation of this historic city, saying conserving historic heritage is to carry forward and foster the best of cultural tradition of Chinese nation, and therefore, an essential component of sustainable development strategy of Beijing. Meanwhile, to preserve historic city is also an important historical obligation we must undertake. In so doing, the relation between protecting the features of ancient capital and improving the living standard of the people and quickening the pace of modernization must be tackled in a satisfactory manner. And the key to this issue is to care for the people, assume the strategy of comprehensive, coordinated and sustainable development, and guide the urban development and management with scientific approaches. Priority should be given to the protection of historical and cultural heritage and ancient features of the age-old capital in the effort of urban redevelopment and improvement of people's life and environment, so as to turn Beijing into a first-rate international city with a modern style as well as characteristics of a historic city.

Through more than 50 years of practice and experience, we have realized that effective protection of traditional urban layout in the midst of modernization drive must be preceded by planning and systematic drawing up of conservation plans. As a result, we have mapped out The Conservation Plan of the 25 Conservation Districts of Historic Sites in Old City of Beijing, The Conservation Plan of Historic City of Beijing and The Conservation Plan of the Imperial City of Beijing, which establish a comprehensive conservation system with three areas: the identifications of cultural heritage, conservation districts of historic sites and historic city.

At the moment, Beijing is entering a crucial period of holding an Olympic Game which will become the best one in the history, and being transformed into a international metropolis. So, we must carry out the instructions of the Party central committee, have a sense of responsibility for history and the people, further perfect the plans by soliciting the comments and opinions of experts and representatives of the people, and determine the conservation scope and methods through scientific planning. In the meantime, we must proceed according to the plans in a strict way, inform the public of the contents of the plans so as to subject ourselves to the supervision and evaluation of the masses. It is my conviction that by virtue of solid and creative work of people from all walks of life, we can certainly turn over to our posterity a new Beijing that both retains the features of a historic city and boasts the high standard of modernization in the future.

Secretary of Beijing Municipal Party Committee　Liu Qi

前 言

北京是世界著名的文化古都，是国务院首批公布的国家级历史文化名城。

在古都北京，红墙黄瓦的皇城宫府、青砖灰瓦的百姓之家，积淀着北京乃至整个中华民族的政治、人文底蕴，释放着不朽的东方文化魅力。这是先人留给我们的优秀文化遗产，是中华民族城市文明的重要标志，是最宝贵的、不可再生的资源。尽管在名城保护工作中，认识在不断提高，经验在不断积累，然而，随着城市的快速发展，北京，也像世界上的许多古城一样，历史文化风貌的保护面临着巨大的冲击和挑战。当一幢幢高楼大厦拔地而起的时候，当一条条胡同、一个个四合院在"拆"字声中化作记忆的时候，对北京旧城的"建设性"破坏已不容忽视。探寻一条保护历史文化与实现城市发展相结合的新路子，构筑统一和谐的整体，延续城市的历史文脉，已成为刻不容缓的紧迫工作。

为切实保护北京历史文化名城，体现"人文奥运"精神，在市委、市政府的统一领导下，从2001年4月至2002年10月，北京市规划委员会会同北京市文物局、北京市城市规划设计研究院组织编制完成了《北京历史文化名城保护规划》。该项规划是对《北京城市总体规划》的延续和深化，是对历史上北京历史文化名城保护经验的总结和发展，是北京建国以来第一次在市域范围内对历史文化名城保护进行系统而完整的思考和规划。在此基础上，又进一步组织编制完成了《北京皇城保护规划》，进而深化并丰富了历史文化名城保护的内涵。

这两项规划已分别于2002年10月、2003年4月由市政府正式批复，成为保护北京历史文化名城的重要依据。现在，我们将其汇编成书，向社会公开出版发行，希冀全社会各界人士关注、参与名城保护事业，同时也期望我们的工作对名城保护的实践和理论研究工作提供一定参考价值。

我们愿与社会各界携手，本着对历史负责，对人民负责的态度，促进名城的有效保护、合理建设和科学利用。使被誉为"在地球表面上人类最伟大的个体工程"的北京城留存为人类文明永久的丰碑。

2003年9月

Foreword

Beijing is a world renowned historic capital city which is on the list of the first group of State-level historic cities by the State Council.

The imperial palaces and mansions of red walls and yellow tiles, and the houses gray brick and black tile of the ordinary people are the political and cultural reflections of the Chinese people and everlasting charm of the orient. They are also the invaluable cultural heritage left to us by our ancestors, the important urban landmark of the Chinese nation as well as the most precious unrenewable resources. Although our awareness has been raised and experience gained in the conservation process of historic cities, the historical feature of Beijing, like many other ancient cities of the world, are facing severe challenge brought about by the rapid urban development. When buildings after buildings are being constructed, one hutong after another being torn down along with the courtyard houses, the "constructive" damage to the old Beijing can no longer be ignored. It has become an urgent task to explore a new approach to combine the protection of the historical culture with the urban development, so as to strike a harmonious balance and retain the historical context.

With a view to conserving the historic city of Beijing, and demonstrating the spirit of "humanistic Olympiad", the Beijing Municipal City Planning Commission, in cooperation with the Beijing Bureau of Cultural Heritage and Beijing Municipal City Planning Institute, have organized and completed the conservation plan for the Historic City of Beijing under the guidance of the Beijing Party Committee and the municipal government from April 2001 to October 2002. The Plan is a continuation and extension of the Master Plan for Beijing, an accumulation and improvement of the conservation experience of Beijing to the past, and a first systematic and comprehensive planning and reflection on preserving a historic city in a city-wide area. On the basis of this plan, another project entitled the Conservation Plan of the Imperial City of Beijing was also finished, thus further enriching the experience of the historic city conservation.

The two plan, approved respectively in October 2002 and April 2003 by the municipal government, have become the important foundation for the conservation of historic city of Beijing. We are now put these plans into one book to be published with the hope of drawing the attention of people from all walks of life so that they can join in this cause. Meanwhile, it is also our hope that our work would serve as some reference to the theoretical research and practice of the conservation of historic city.

We would like to cooperate with people of all circles to effectively preserve, reasonably construct and utilize the historic city with a sense of responsibility for the people and history, so that to make Beijing, which is regarded as " the greatest individual project of mankind on earth", an everlasting monument in the civilization of human being.

September 2003

北京历史文化名城保护规划

北京市规划委员会
北京市文物局
北京市城市规划设计研究院

統一思想，廣爲宣傳北京名城保護的重要性

認真貫徹中央對北京城市規劃建設和管理工作的一系列重要指示，立足于把北京建設成爲第一流的現代化國際城市和歷史文化名城的高度，充分認識實施《北京歷史文化名城保護規劃》的重要性，廣爲宣傳，在全社會形成"熱愛名城，保護名城"的共識。

進一步落實《北京城市總體規劃》關于"兩個戰略轉移"的方針

進一步落實《北京城市總體規劃》的要求，積極推進城市建設重點逐步從市區向遠郊區縣轉移，市區建設從外延擴展向調整改造轉移的步伐，疏解舊城區人口和功能，落實與城四區危改相應地段的經濟適用房項目，爲保護歷史文化名城創造良好條件。

在《北京歷史文化名城保護總體規劃》審批后，應盡快深化落實

"十五"期間，組織編制落實北京歷史文化名城保護規劃的具體規劃。主要包括：第一批歷史文化保護區保護規劃，舊城內道路規劃的調整與定綫，舊城建築高度控制的調整規劃，第五批、第六批文物保護單位保護範圍和建設控制地帶的劃定等。

加快制定北京歷史文化名城保護的相關法規

由市文物、規劃、建設等部門共同開展立法的前期調研工作，結合北京實際，借鑒外省市經驗，廣泛聽取社會各界意見，制定《北京市歷史文化名城保護條例》，報市人大審議。制定《北京市歷史文化保護區管理規定》，作爲法律依據在城市管理工作中嚴格執行。制定故宮、頤和園、長城、天壇等世界文化遺産的專項保護法規，明確管理責任，保證保護資金，規範建設和管理行爲。

舊城內危改不宜采用開發的方式進行

舊城內危改區建設的好壞對北京舊城的整體風貌影響很大，舊城危改區住宅建設項目不宜采用房地産開發方式。危改中要注意保護原有主要胡同、大樹、古樹和具有一定歷史價值的建築。

深入研究舊城房産産權改革的相關法規政策并開展試點

積極推進舊城歷史文化保護區內房屋産權的改革，并研究制定相關實施政策，各區都要開展歷史文化保護區規劃實施試點，積累經驗。

多渠道籌集資金，初步建立資金保障機制

根據"國家保護爲主，動員全社會參與"的主導思想，積極開展試點工作，探索文物保護單位和歷史文化保護區保護、維修、整治、利用的有效途徑，建立資金保障機制。除政府財政撥款外，還要通過開發旅游經濟，鼓勵使用單位自籌資金、社會募捐等多種渠道，建立保護基金。

加強對北京傳統風貌和文化特色的繼承和發揚的

《北京历史文化名城保护规划》文本

1. 总则

1.1 为深化《北京城市总体规划（1991年～2010年）》，正确处理发展与保护的关系，加强北京历史文化名城的保护，特制定本规划。

1.2 在本次保护规划范围内进行各项建设活动的一切单位和个人，均应按《中华人民共和国城市规划法》的有关规定，执行本规划。

1.3 本规划未涉及的控制指标和管理规定，应遵循国家及北京市的相关法规、规定。

1.4 本规划经北京市人民政府批准后，由市城市规划行政主管部门负责执行；如有重大调整，必须经北京市人民政府批准。

2. 规划依据

2.1 《中华人民共和国城市规划法》(1989年12月)

2.2 《中华人民共和国文物保护法》(1982年11月通过，1991年6月修改)

2.3 《中华人民共和国文物保护法实施细则》(1992年5月)

2.4 《北京市文物保护管理条例》(1987年6月)

2.5 《北京市文物保护单位范围及建设控制地带管理规定》(1994年)

2.6 《北京城市总体规划(1991～2010年)》

2.7 国务院关于《北京城市总体规划》的批复(1993年10月)

2.8 《北京旧城历史文化保护区保护和控制范围规划》(1999年4月)

2.9 《北京市区中心地区控制性详细规划》(1999年9月)

2.10 《北京25片历史文化保护区保护规划》(2001年3月)

3. 规划指导思想及思路

3.1 规划指导思想

3.1.1 坚持北京的政治中心、文化中心和世界著名古都的性质。

3.1.2 正确处理历史文化名城保护与城市现代化建设的关系。

3.1.3 重点搞好旧城保护，最大限度地保护北京历史文化名城。

3.1.4 贯彻"以人为本"的思想，使历史文化名城在保护中得以持续发展。

3.2 规划基本思路

3.2.1 规划基本思路为：三个层次和一个重点。

3.2.2 保护的"三个层次"是：文物的保护、历史文化保护区的保护、历史文化名城的保护。"一个重点"是：旧城区。

4. 文物保护单位的保护

4.1 文物保护的层次和内容

4.1.1 北京拥有世界文化遗产5处：故宫、长城、周口店北京猿人遗址、颐和园、天坛；国家级重点文物保护单位60处；市级文物保护单位234处；区、县级文物保护单位501处；区、县级文物暂保单位237处；普查登记在册文物2521处；共计3553处。

4.1.2 旧城范围内(四至以二环路界定，占地62.5km²)有世界文化遗产2处：故宫、天坛；国家级重点文物保护单位34处；市级文物保护单位134处；区级文物保护单位115处；区级文物暂保单位39处；普查在册文物478处；总计800处。

4.1.3 规划市区范围内(四至是东以定福庄，南以南苑，西以石景山，北以清河界定，占地1040km²)除旧城范围以外，有世界文化遗产1处：颐和园；国家级重点文物保护单位16处；市级文物保护单位49处；区级文物保护单位59处；区级文物暂保单位181处；普查在册文物310处；总计615处。

4.1.4 市域范围内（辖10个远郊区县，总面积达16800km²）除规划市区及旧城范围外，有世界文化遗产2处：长城、周口店北京猿人遗址；国家级重点文物保护单位10处；市级文物保护单位51处；区、县级文物保护单位327处；区、县级文物暂保单位17处；普查在册文物1733处；总计2138处。

4.2 文物保护规划

4.2.1 在现有5处世界文化遗产(故宫、长城、周口店北京猿人遗址、颐和园、天坛)的基础上，继续积极申报。

4.2.2 继续公布市级文物保护单位名单，使市级以上文物保护单位达到300处以上。各区、县要继续公布区、县级文物保护单位及区、县级文物暂保单位。同时，近期完成第六批国家级重点文物保护单位（北京部分）的论证及申报工作。

4.2.3 继续完成第五批、第六批市级文物保护单位保护范围和建设控制地带的划定工作；逐步完善区、县级文物保护单位保护范围和建设控制地带的划定工作。

4.2.4 各级文物保护单位的修缮与保护，必须将文物单体建筑修缮与环境整治和改善相结合，景点保护与街、区成片保护相结合。

4.2.5 必须加强保护重要的近现代建筑，逐步提出一批具有较高历史文化价值的近现代建筑名单，并划定保护范围和建设控制地带。

4.2.6 公布第四批地下文物埋藏区，使全市地下文物埋藏区达到50处。

4.2.7 必须重视并加强城市考古工作，特别注意对辽南京、金中都、元大都城市遗址、遗迹的考古调查、勘探和发掘。

5. 历史文化保护区的保护

5.1 历史文化保护区的含义

5.1.1 历史文化保护区是具有某一历史时期的传统风貌、民族地方特色的街区、建筑群、小镇、村寨等，是历史文化名城的重要组成部分。

5.1.2 历史文化保护区中的危房，允许在符合保护规划要求的前提下，逐步进行改造和更新，并不断提高城市基础设施的现代化水平。

5.2 旧城第一批25片历史文化保护区的保护规划

5.2.1 《北京旧城25片历史文化保护区保护规划》已于2002年2月由市政府批准，必须严格执行。

5.2.2 北京旧城25片历史文化保护区中有14片分布在旧皇城内：南、北长街，西什库大街，南、北池子，东华门大街，景山东、西、后、前街，地安门内大街，文津街，五四大街，陟山门街；有7片分布在旧皇城外的内城：西四北头条至八条，东四三条至八条，南锣鼓巷地区，什刹海地区，国子监地区，阜城门内大街，东交民巷；有4片分布在外城：大栅栏，东、西琉璃厂，鲜鱼口地区。

The Text of "Conservation Plan for the Historic City of Beijing"

1. General Principles

1.1 The plan is formulated with a view of better implementing the "Master Plan of Beijing (1991-2010)", correctly dealing with the relationship between development and preservation and enhancing the conservation of the historic city of Beijing.

1.2 All organizations and individuals involved in construction activities within the territory prescribed by this plan must abide by this plan according to the relevant regulations of "The Urban Planing Code of the People's Republic China".

1.3 With regard to the guiding numbers and management regulations not specified in this plan, relevant rules and regulations of the State and Beijing Municipal government must be followed.

1.4 Upon approval by the Beijing Municipal government, this plan will be implemented by agencies in charge of urban planning administrative work and, any important adjustments must be subjected to the approval of the Beijing Municipal government.

2. Basis of the Plan

2.1 "Urban Planning Code of the People's Republic of China" (December, 1989)

2.2 "Conservation Code of Historical Relics of the People's Republic of China" (passed in November 1982, revised in June, 1991).

2.3 "Rules and Regulations for Implementation of the Conservation Code of Historical Relics of the People's Republic of China" (May, 1992)

2.4 "Ordinance of Historical Relics Conservation in Beijing" (June, 1987)

2.5 "Scope and Construction Control Regulations for Cultural Relics Entities of Preservation in Beijing" (1994)

2.6 "Master Plan of Beijing (1991-2010)"

2.7 The Clearance by the State Council for the "Master Plan of Beijing" (October, 1993)

2.8 "Conservation and Control Scope Plan for the Conservation Districts of Historic Sites in Old City of Beijing" (April, 1999)

2.9 "Detailed Guiding Plan of Central Beijing" (September, 1999)

2.10 "Conservation Plan for the 25 Conservation Districts of Historic Sites in Beijing" (March, 2001)

3. Plan Guidelines and Concepts

3.1 Guidelines of the Plan

3.1.1 Maintain the nature of Being as a political and cultural center, as well as one of the world's ancient capitals

3.1.2 Correctly address the relationship between the conservation of the historic city and construction of a modern metropolis.

3.1.3 Priority is given to the preservation of the old city, and strive for the conservation of Beijing historic city.

3.1.4 Promote the man-based concept, so that the historic city can be further developed through conservation.

3.2 Key ideas for the Plan

3.2.1 The basic idea for the plan is: three levels and one priority.

3.2.2 The three levels of conservation are: conservation of historic relics, conservation of the conservation districts of historic sites, and conservation of historic city. One priority is: the old city.

4. Conservation of the Entities of Historic Relics

4.1 Levels and Contents of Conservation of Historic Relics

4.1.1 Beijing boasts 5 world cultural heritages: The Palace Museum, the Great Wall, Zhoukoudian, site of the fossils of Peking Man, Summer Palace, and the Temple of Heaven; 60 key historic relic sites on the conservation list of the State; 234 entities of historic relics under the conservation of Beijing municipal government; 501 by counties and districts; 237 temporarily under the conservation of counties and districts, and 2,521 registered entities of historic relics in a general survey, totaling 3,553.

4.1.2 Within the old urban area (defined by the Second Ring Road with an area of 62.5 square kilometers) are 2 world cultural heritages: Forbidden City and Temple of Heaven; 34 key sites under the State conservation; 134 sites under the conservation of the municipal government; 115 sites under the conservation of districts; 39 sites temporarily protected by districts, and 478 registered sites in general survey, totaling 800.

4.1.3 In the planned area excluding the old city (borders Dingfuzhuang in the east, Nanyuan in the south, Shijingshan in the west and Qinghe in the north, with an area of 1,040 square kilometers) are 1 world cultural heritage: the Summer Palace; 16 key sites on the State conservation list; 49 sites under the conservation of the municipal government; 59 sites protected at the district-level; 181 sites under the temporary conservation of districts; 310 registered sites in general survey, totaling 615.

4.1.4 Within the citywide area excluding the planned area and old city (embracing 10 outlying counties with an area of 16,800 square kilometers) are 2 world cultural heritages: the Great Wall and Zhoukoudian, site of the fossils of Peking Man; 10 key sites under the State-level conservation; 51 sites under the city-level conservation; 327 sites protected by counties and districts; 17 sites temporarily protected by districts and counties, and 1,733 registered sites in general survey, totaling 2,138.

4.2 Plan for Conservation of Cultural Relics

4.2.1 Apart from the 5 world cultural heritages (Forbidden City, Great Wall, Zhoukoudian Peking Man Site, Summer Palace and Temple of Heaven), more heritages of this caliber will be submitted for approval.

4.2.2 Continue making public the list of cultural relics sites under the conservation of the municipal government with a view of having 300 such sites on the list. Various districts and counties should continue publicizing the historic sites protected by districts and counties and those protected by them on a temporary basis. In the meantime, complete in the near future the evaluation and application of the sixth group of key historical sites (those in Beijing area) under the conservation of the State.

4.2.3 Complete the delimitation of conservation sites and construction-controlled regions for the fifth and sixth groups of historical sites under the conservation of the municipal government. Gradually perfect the delimitation of conservation sites and construction-controlled regions for the historical sites under the conservation of districts and counties.

4.2.4 The renovation and conservation of historical sites of various levels must proceed through the integrity of the renovation of individual building with the improvement of its environment, and the integrity of conservation of places of interest with the conservation of streets and surrounding areas.

4.2.5 The conservation of key modern and contemporary architecture should be strengthened. A list of modern and contemporary buildings with fairly high historical and cultural values should be drawn up, and the conservation range and construction-controlled area designated.

4.2.6 Publicize the burial locations of the fourth group of underground historical relics, in order to make the number of such locations in Beijing reach 50.

4.2.7 Attach importance to and enhance archeological work in the city, especially archeological and excavating work of the ruins and relics of Nanjing of the Liao Dynasty, Zhongdu of the Jin Dynasty and Dadu of the Yuan Dynasty.

5. Conservation of the Conservation Districts of Historic Sites

5.1 Definition of conservation districts of historic sites.

5.1.1 Conservation districts of historic sites refer to the streets, architectural complex, small towns and villages with traditional features and national or local characteristics of a certain historical period. They are an important part of the historic city.

5.1.2 The ramshackle buildings in the conservation districts can be allowed to be renovated and repaired gradually in compliance with the requirements of the conservation plan. The modernization of the urban infrastructure should also be improved constantly.

5.2 Conservation Plan for the First Group of 25 Conservation Districts of Historic Sites

5.2.1 The "Conservation Plan for the 25 Conservation Districts of Historic Sites in the Old City of Beijing" has been approved by the municipal government in February, 2002, and must be carried out in a strict manner.

5.2.2 Out of the 25 areas, 14 are located within the old imperial city: Nan and Bei Changjie Streets, Xihuamen Street, Nanchizi and Beichizi Streets,

5.2.3 北京旧城25片历史文化保护区总占地面积1038hm²，约占旧城总用地的17%。其中重点保护区占地面积649hm²，建设控制区占地面积389hm²。

5.2.4 《北京旧城25片历史文化保护区保护规划》强调必须以"院落"为基本单位进行保护与更新，危房的改造和更新不得破坏原有院落布局和胡同肌理，须遵照执行。

5.2.5 《北京旧城25片历史文化保护区保护规划》对保护区内的建筑保护和更新分为6类进行规划管理：文物类建筑、保护类建筑、改善类建筑、保留类建筑、更新类建筑、整饰类建筑，须遵照执行。

5.2.6 《北京旧城25片历史文化保护区保护规划》对保护区内的用地性质变更、人口疏解、道路调整、市政设施改善、环境绿化保护等方面提出了具体的原则、对策和措施，须遵照执行。

5.3 第二批历史文化保护区名单

5.3.1 在旧城第一批25片历史文化保护区基础上确定北京第二批历史文化保护区名单。其中，在旧城内继续补充历史风貌较完整、历史遗存较集中和对旧城整体保护有较大影响的街区进行保护；在旧城外确定一批文物古迹比较集中，能较完整地体现一定历史时期传统风貌和地方特色的街区或村镇，使其得到有效保护。

5.3.2 旧城内第二批5片历史文化保护区为：皇城、北锣鼓巷、张自忠路北、张自忠路南、法源寺。

5.3.3 旧城外确定10片历史文化保护区为：海淀区西郊清代皇家园林，丰台区卢沟桥宛平城，石景山区模式口，门头沟区三家店、川底下村，延庆县岔道城、榆林堡，密云区古北口老城、遥桥峪和小口城堡，顺义区焦庄户。

5.4 第二批历史文化保护区特色

5.4.1 应保护旧城内第二批历史文化保护区特色。

（1）皇城历史文化保护区：是北京旧城整体保护的重点区域，内含紫禁城、太庙、社稷坛、北海、中南海及14片第一批历史文化保护区，面积约6.8km²。

（2）北锣鼓巷历史文化保护区：位于东城区，南至鼓楼东大街，北至车辇店、净土胡同，东至安定门内大街，西至赵府街，总面积约为46hm²。该地区与什刹海、南锣鼓巷、国子监等三个历史文化保护区相邻，是皇城的重要背景，也是保护旧城整体风貌和沿中轴线对称格局不可缺少的地段。

（3）张自忠路北历史文化保护区：位于东城区，南至张自忠路，北至香饵胡同，东至东四北大街，西至交道口南大街，总面积约为42hm²。该街区有和敬公主府、段祺瑞执政府旧址、孙中山逝世纪念地等多处市、区级文物保护单位。

（4）张自忠路南历史文化保护区：位于东城区，南至钱粮胡同，北至张自忠路，东至东四北大街，西至美术馆后街，总面积约为42hm²。该区域处于皇城与东四三条至八条保护区之间，现有胡同格局完整，有马辉堂花园等文物保护单位。

（5）法源寺历史文化保护区：位于宣武区，南至南横西街，北至法源寺后街，东至菜市口南大街，西至教子胡同，总面积约20hm²。该街区内有法源寺、湖南会馆、绍兴会馆等文物保护单位，街区整体风貌保存较好。

5.4.2 旧城内新增的第二批历史文化保护区占地面积约为249hm²，约占旧城总面积的4%。

5.4.3 旧城外历史文化保护区的特色。

（1）西郊清代皇家园林历史文化保护区：位于海淀区，包括颐和园、圆明园、香山静宜园、玉泉山静明园等，即清代的"三山五园"地区，是我国现存皇家园林的精华。

（2）卢沟桥宛平城历史文化保护区：位于丰台区，卢沟桥、宛平城是国家和市级文物保护单位，也是震惊中外的"卢沟桥事变"发生地，具有重要的历史和革命纪念意义。

（3）模式口历史文化保护区：位于石景山区西北部，金顶山路与京门公路之间，为京西古道。在模式口大街以北，传统村落的风貌保存较好，并有承恩寺、田义墓、法海寺等文保单位。

（4）三家店历史文化保护区：位于门头沟区永定河北岸，三家店村中现存多处文物，与煤业发展有关的建筑群、会馆等成为此地独特的景观，具有浓厚的京西地方特色。

（5）川底下村历史文化保护区：是门头沟区斋堂镇的一个自然村，房屋依山而建，村中现保存着许多明清时期的四合院民居，其建筑艺术相当精湛，风貌相当完整。

（6）榆林堡历史文化保护区：位于延庆县康庄镇西南，元、明、清时期是京北交通线上的重要驿站之一，其平面呈"凸"字形。

（7）岔道城历史文化保护区：位于延庆县八达岭镇，是北京通往西北的重要军事据点和驿站，紧邻八达岭长城，至今原有城墙、城门尚在。

（8）古北口老城历史文化保护区：位于密云区古北口镇的东北部，自古以来为兵家必争之地。现存药王庙戏楼、财神庙、古关址等文物和南北大街，风貌较完整。

（9）遥桥峪城堡、小口城堡历史文化保护区：遥桥峪城堡位于密云区新城子乡东部，建于明万历二十六年（公元1599年），此堡呈方形，南面正中一座城门，至今保存完好。小口城堡位于密云区新城子乡北部，距遥桥峪城堡约4km，是明代戍边营堡，城墙"北圆南方"，保存完好。

（10）焦庄户历史文化保护区：焦庄户地道战遗址属顺义区龙湾屯镇焦庄户村，1943年，当地党组织和群众，利用地道和日寇周旋作战，创造了抗战时期闻名的"地道战"，被誉为"人民第一堡垒"。

5.4.4 北京第一批、第二批历史文化保护区合计共有40片。其中，旧城内有30片，总占地面积约1278hm²，占旧城总面积的21%。

5.4.5 旧城第一批、第二批历史文化保护区和文物保护单位保护范围及其建设控制地带的总面积为2617hm²，约占旧城总面积的42%。

5.4.6 第二批历史文化保护区名单经市政府批准后，必须尽快组织编制各保护区的保护规划。

6. 旧城整体格局的保护

必须从整体上考虑北京旧城的保护，具体体现在历史河湖水系、传统中轴线、皇城、旧城"凸"字形城廓、道路及街巷胡同、建筑高度、城市景观线、街道对景、建筑色彩、古树名木十个层面的内容。

7. 历史河湖水系的保护

7.1 规划目标

重点保护与北京城市历史沿革密切相关的河湖水系，部分恢复具有重要历史价值的河湖水面，使市区河湖形成一个完整的系统。

Donghuameng Street, Jingshandong Street, Jingshanxi Street, Jingshanhou Street and Jingshanqian Street, Di'anmen Street, Wenjin Street, Wusi Street, and Zhishanmen Street; 7 areas are scattered outside the imperial city but within the Outer City Wall: the First to Eighth Lanes of Xisi Stree, The Third to Eighth Lanes of Dongsi Street, Nanluoguxiang area, Shichahai area, the old Imperial College area, Fuchengmennei Street, Dongjiaominxiang; another 4 areas are located in the Outer City: Dashilan, Dongliulichang and Xiliulichang, Xianyukou area.

5.2.3 Occupying an area of 1.38 hectare, the 25 conservation districts account for 17% of the total area of the old city. The key conserved area takes up 649 hectares while the construction-controlled part occupies an area of 389 hectares.

5.2.4 The "Plan for the 25 Conservation Districts of Historic Sites in the Old City of Beijing" stresses that conservation and renovation must be done with "courtyards" as basic modules. Renovation and repair of unsafe buildings cannot be undertaken at the expense of damaging the original layout of courtyards and the urban fabrics of Hutongs.

5.2.5 The Plan divides the buildings in the conservation districts into 6 categories for management and conservation: historical monuments, conserved buildings, buildings needing improvement, buildings that should be retained, buildings needing renovation and buildings needing polishing. This rule must be implemented.

5.2.6 The detailed principles and measures stipulated in the Plan in regard to the nature of land use, population reduction, road adjustment, improvement of urban facilities and environmental conservation within the conservation districts must be strictly implemented.

5.3 The List of the Second Group of Conservation Districts of Historic Sites

5.3.1 To determine the second list of conservation districts of historic sites in Beijing on the basis of the 25 conservation districts in the old city. Meanwhile in the old city, continue to identify streets that have fairly complete historical features, concentrated historical sites and big impact on the conservation of the old city. Outside the old city, try to determine a number of streets or towns with cluster of cultural relics for conservation. These places should reflect in a comprehensive way the traditional features and local flavor of certain historical periods.

5.3.2 The 5 conservation districts of the second group within the old city are: imperial city, Beiluoguxiang, Beizhangzizhong Road, Nanzhangzizhong Road and Fayuan Temple.

5.3.3 The 10 conservation districts outside of the old city are: Qing Dynasty Imperial Garden in the suburbs of Haidian District, in Lugou Bridge Wanping City in Fengtai District, Moshikou in Shijingshan District, Sanjiadian in Mentougou District, Chuandixia Village, Chadaokou in Yanqing County, Yulinbao, at Gubeikou in Old City Miyun County, Yaoqiaoyu and Xiaokou Castle, and Jiaozhuanghu Village in Shunyi District.

5.4 Features of the Second Group of Conservation Districts of Historic Sites

5.4.1 The features of the second group of conservation districts in the old city must be protected.

(1) The conservation districts of the imperial city: the key area under the comprehensive conservation in the old city, including Forbidden City, the Imperial Ancestral Temple, the Altars to the Gods of Earth and Grain, Beihai Zhongnanhai, and first group of 14 conservation districts with an area of 6.8 square kilometers.

(2) Beiluoguxiang conservation district: located in the Dongcheng District with an area of 46 hectares, it stretches to the East Drum Tower Street in the south, Cheniandian and Jingtu Hutong in the north, Andingmennei Street in the east, and Zhaofu Street in the west. An important background of the Imperial City, it borders Shichahai, Nanluoguxiang and the Imperial College, thus an indispensable section in protecting the overall features of the old city and in forming a symmetrical layout along the axis.

(3) Conservation district north of Zhangzizhong Road: located in Dongcheng District with an area of 42 hectares, it stretches to the Zhangzizhong Road in the south, Xiang'er Hutog in the north, Dongsibeijie Street in the east, and Jiaodaokounanjie Street in the west. The place boasts many cultural relics under the city- and district-level governments, such as the Mansion of Princess Hejing, Site of Duanqirui Government, and the Memorial of Sun Ya-san.

(4) Conservation district south of Zhangzizhong Road: located in Dongcheng District with an area of 42 hectares, it stretches Qianliang Hutong in the south, Zhangzizhong Road in the north, Dongsibeijie Street in the east and Meishuguanhoujie Street in the west. Sandwiched between the Imperial City and the First and Eighth Lanes of Dongsi, it has a complete hutong layout, and cultural relics such as Mahuitang Garden.

(5) Fayuan Temple conservation district: located in Xuanwu District with an area of 20 hectares, it stretches to the West Nanheng Street in the south, Fayuansihou Street in the north, South Caishikou Street in the east, and Jiaozi Hutong in the west. It has Fayuan Temple, Hunan and Shaoxing Guild Halls and other historic sites with a fairly preserved street layout.

5.4.2 The newly-added second group of conservation districts covers an area of 249 hectares, accounting for 4% of the total area of old city.

5.4.3 Features of the conservation districts outside of the Old City

(1) The Qing Dynasty Imperial Garden Conservation District: located in Haidian District, it includes the Summer Palace, Jingyi Garden of the Fragrance Hill, Jingming Garden of Yuqun Mountain, etc. Also called the area of "three mountains and five gardens", it is the cream of the extant imperial garden.

(2) Lugou Bridge and Wanping City Conservation District: located in Fengtain District, it has the Marco Polo Bridge (Luogou Bridge) and Wenping City which are relics under the State and municipal conservation respectively. They are also the sites of the "Lugouqiao Incident" which shocked China and the world, thus having important historical and revolutionary significance.

(3) Moshikou Conservation District: located northwest of Shijingshan District, this ancient route west of the Beijing is situated between Jindingshan Road and Jingmen Highway. The traditional villages north of Moshikou Street are well preserved, boasting cultural relics like Cheng'en Temple, Tianyi tomb and Fahai Temple.

(4) Sanjiadian Conservation District: located at the northern bank of Yongding River of Mentoukou District, this village has many architecture and guild halls related to the coal industry, reflecting the features west of Beijing.

(5) Chuandixia Village Conservation District: a natural village in Zhaitang Town, Mentougou District, it was nestled at the foot of the mountain, and has many courtyard-type residences of the Ming and Qing Dynasties. The architectural art is highly skilled and the features completely preserved.

(6) Yulinbao Conservation District: located southwest of Kangzhuang town, Yanqing County, it was one of the important courier stations on the traffic line north of Beijing during the Yuan, Ming and Qing Dynasties. Its site is in the form of an inverted 'T'.

(7) Chadao City Conservation District: located at Badaling Town, it used to be a crucial military stronghold and courier station on the way from Beijing to northwest. Close to the Badaling Great Wall, it still retains city walls and gates.

(8) Gubeikou Old City Conservation District: located northeast of Gubeikou Town, Miyun District, it has been the place of contention since ancient times, boasting now the Theater of the Temple of the King of Medicine, Temple of the God of Wealth, Ancient Pass Site, North and South Streets and other relics. The features there are comparatively well preserved.

(9) Conservation Sites of Yaoqiaoyu and Xiaokou Castles: located east of Xinchengzi Village, Miyun District, Yaoqiaoyu Castle was build in 1599 under the rein of emperor Wanli of the Ming Dynasty. A the form of a square, it has a city wall in the middle of its southern side, which is well preserves. Xiaokou castle is located north of Xinchengzi village, 4 kilometers from anther castle. A frontier garrison in the Mind Dynasty, it is round in the north and square in the south, and is well preserv.

(10) Jiaozhuanghu Conservation District: this tunnel warfare site located at Jiaozhuanghu, Longwantun town, Shunyi District. In 1943, local Party led the masses engage the Japanese soldiers in tunnels, hence the famous names "Tunnel Warfare" and the "No. One People's Stronghold" in that period.

5.4.4 The number of first and second groups of conservation districts in Beijing reaches 40, out which there are 30 in the old city with a total land area of 1,278 hectares, accounting for 21% of the total area of the old city.

5.4.5 The total area of the conservation range and construction controlled areas of the first and second groups of the conservation districts and cultural relics under conservation in the old city covers an area of 2,617 hectares, accounting for 42% of the total area of the old city.

5.4.6 Once the list of the second group of conservation districts are approved by the municipal government, conservation plan for them must be worked

7.2 市区河湖水系保护规划

7.2.1 现有河湖水系的保护规划

与北京城市发展密切相关、在各个历史时期发挥过重要作用的河湖水域列为重点保护目标，划定保护范围并加以整治。

（1）护城河水系：重点保护河道为北护城河、南护城河、北土城沟和筒子河。

（2）古代水源河道：重点保护河道为莲花河和长河，以及莲花池和玉渊潭。

（3）古代漕运河道：重点保护河道为通惠河、坝河和北运河。

（4）古代防洪河道：重点保护河道为永定河和南旱河。

（5）风景园林水域：重点保护湖泊水域为六海、昆明湖、圆明园水系。

（6）重点保护的水工建筑物：后门桥、广济桥、卢沟桥、朝宗桥、白浮泉遗址、琉璃河大桥、广源闸、八里桥、麦钟桥、银锭桥、金门闸、庆丰闸、高梁桥、北海大桥等。

7.2.2 恢复河道的规划

规划将转河、菖蒲河、御河（什刹海——平安大街段）予以恢复。

（1）转河属于通惠河水系，恢复转河可将长河与北护城河连接起来。

（2）菖蒲河是故宫水系的一部分，与内城护城河水系、六海水系、外城护城河水系相连通。

（3）御河（什刹海——前三门大街段）始于元代，北起后门桥，南至前三门。规划将御河上段（什刹海——平安大街）予以恢复。

7.2.3 恢复湖泊的保护规划

（1）鱼藻池是金中都的太液池，应按原貌恢复。

（2）莲花池是金中都最早开发利用的水源地，应将其西南角水面按原状恢复。

7.2.4 控制前三门护城河规划用地内的新建项目

前三门护城河是贯穿北京旧城的一条重要历史河道，它的恢复对于保护北京旧城风貌、改善市中心生态环境具有积极作用，在远期应予以恢复，目前要控制新建项目。控制范围为西起南护城河，东至东护城河，前三门大街道路红线以南70m（包括河道及相应的绿化带）。

8. 城市中轴线的保护和发展

北京城市中轴线由旧城传统中轴线、北中轴线和南中轴线组成，全长约25km。

8.1 传统中轴线的保护规划

8.1.1 北京传统中轴线从永定门到钟鼓楼为7.8km，到北二环路为8.5km。其保护规划必须遵循以保护为主，保护与发展、继承和创造相结合的原则，重点研究钟鼓楼、景山——前门、永定门三个节点的保护与规划。

8.1.2 钟鼓楼节点：作为传统中轴线的端点，钟鼓楼在该地区拥有标志性建筑的地位，其周边以四合院民居为主。钟鼓楼周边建筑高度控制必须符合历史文化保护区保护规划的规定。

8.1.3 景山——前门节点：由景山、故宫、天安门、正阳门城楼和箭楼等组成，空间层次丰富，秩序严谨，起伏有致，必须严格加以保护。

8.1.4 永定门节点：应复建永定门城楼，对实现传统中轴线的完整性、有效衔接南中轴线意义重大，必须严格控制永定门城楼周边的建筑高度。

8.2 北中轴线的保护发展规划

8.2.1 北中轴线是从北二环到奥林匹克公园，应重点规划三个节点。

8.2.2 奥林匹克公园中心区节点：是北中轴线的端点，应重点规划，形成北京城市的新标志。端点以北地区为森林公园，作为北中轴线的背景。

8.2.3 北土城节点：可结合北土城遗址与北中轴80m宽道路中央绿化带，创造具有一定意义的城市公共空间，强化和丰富北中轴线。

8.2.4 北二环路北节点：在北二环路至安德路之间，中轴线两侧的用地宜规划为重要的城市公共空间。

8.3 南中轴线的保护发展规划

8.3.1 南中轴线是从永定门到南苑。南中轴线两侧在做好用地功能调整的同时，应注意丰富中轴线的空间结构，重点规划三个节点。

8.3.2 木樨园节点：结合木樨园商业中心区的建设，形成城市的公共空间。

8.3.3 大红门节点：在中轴路与南四环路交叉口处，塑造重要的城市景观。

8.3.4 南苑节点：作为南中轴线的端点，以大片森林公园相衬托。

8.4 北京城市中轴线的保护控制范围

8.4.1 以中轴路道路中心线为基准，距道路两侧各500m为控制边界，形成约1000m宽的范围作为北京城市中轴线的保护和控制区域，严格控制建筑的高度和形态。

8.4.2 位于中轴线保护和控制区域以外，对中轴线有重要影响的特殊区域，如天坛、先农坛、六海等，必须按文物及历史文化保护区的保护规定执行。

9. 皇城历史文化保护区的保护

9.1 规划范围的确定

9.1.1 本次规划将皇城整体设为历史文化保护区。

9.1.2 保护区范围四至为：东至东皇城根，南至现存长安街北侧红墙，西至西黄城根南北街、灵境胡同、府右街，北至平安大街，总用地约6.8km²。

9.2 皇城的历史文化价值

9.2.1 明清皇城以其杰出的规划布局、建筑艺术和建造技术，成为中国几千年封建王朝统治的象征，具有极高的历史文化价值。

9.2.2 皇城的惟一性：明清皇城是我国现存惟一保存较好的封建皇城，它拥有我国现存惟一的、规模最大、最完整的皇家宫殿建筑群，是北京旧城传统中轴线的精华组成部分。

9.2.3 皇城的完整性：皇城以紫禁城为核心，以明晰的城市中轴线为纽带，城内有序集合皇家宫殿园囿、御用坛庙、衙署库坊、民居四合院等设施，呈现出皇权至高无上的规划理念和完整的功能布局。

9.2.4 皇城的真实性：皇城中的紫禁城、筒子河、三海、太庙、社稷坛和部分御用坛庙、衙署库坊、民居四合院等传统建筑群至今保存较好，充分反映了古代皇家生活、工作、娱乐的历史信息和明、清、民国等历史演变的过程，体现了历史的真实性。

9.2.5 皇城的艺术性：皇城在规划理念、建筑布局、建造技术、色彩运用等方面具有很高的艺术性。

out as soon as possible.

6. Conservation of the General Layout of the Old City

The conservation of the old city of Beijing should be considered in a comprehensive way, embracing 10 aspects: historical river and lake systems, traditional axis, imperial city, the city walls in the old city in the form of inverted "T", roads and alleys, building height, urban landscape, corresponding views at the street ends, architectural color, and old and famous trees.

7. Conservation of the Historical Water System

7.1 Plan Objectives

Primarily protect the water systems closed linked with the evolution of Beijing, and partially renovate lake and river systems of important historical value, in order to form a complete urban water system.

7.2 Conservation Rules and Regulations of the urban water system

7.2.1 Conservation Rules and Regulations of the existing urban water system

Conservation priority should be given to those lakes and rivers that are close related to the development of Beijing, and have played an important role in various historical periods. Determine the range of conservation for renovation and dredging.

(1) City moat system: rivers under priority conservation are north city moat, south city moat, north Tucheng canal, and Tongzi moat.

(2) Ancient river courses of headwaters: river courses under priority conservation are Lianhua and Chang Rivers, and Lianhua Pond and Yuyuantan.

(3) Ancient river course for transporting grain to the capital: rivers under priority conservation are Tonghui and Ba Rivers, and North Grand Canal.

(4) Ancient flood-control rivers: rivers under priority conservation are Yongding and South Han Rivers.

(5) Landscape water system: waters under priority conservation are Liuhai, Kunming Lake, Yuanmingyuan, or the Old Summer Palace water system.

(6) Water works under priority conservation: Houmen, Guangji, Lugou and Chaozong Bridges, Baiquanfu Ruins, Liulihe Bridge, Guangyuan Floodgate, Bali, Maizhong, Yinding Bridges, Jinmen, Qingfeng Floodgates, Gaoliang and Beihai Bridges.

7.2.2 Restoration Plan of Rivers

Zhuan, Changpu and Yu (on the Shishahai-Ping'an Avenue section) Rivers will be renovated according to the plan.

(1) The renovation of Zhuan River, which belongs to the Tonghui River system, will link up the Chang River and the North City Moat.

(2) As part of the Forbidden City water system, Changpu River are connected with the inner City Moat, Liuhai and outer City Moat water systems.

(3) Yu River (on the Shishahai-Qiansanmen Street secion) was created in the Yuan Dynasty, originating from Houmen Bridge in the north, and reaching Qiansanmen in the south. The upper reaches of the this river (Shishahai-Ping'an Aevenue) will be renovated according to the plan.

7.2.3 Conservation Plan for the Restoration of Lakes.

(1) Yuzaochi was the Taiyechi of the Jin Dynasty, and must be rebuilt.

(2) Lianhuachi(Lotus Pond) is the earliest developed and utilized source of water, so its southwestern part should be recovered according to the original pattern.

7.2.4 Control of New Projects within the Land under the Plan in the Qiangsanmen City Moat Area

An important historical river flowing across the old city of Beijing, Qiansanmen City Moat, once renovated, will have a positive impact on recovering the features of the old city and improving the city center's ecological environment. Recovery is a long-term task, and the present goal is to restrain the new projects. The control area ranges from South City Moat in the west, East City Moat in the east, and 70 meters south of the red line on the Qiansanmen Street (including river courses and relevant greenbelts).

8. Conservation and Development of the City Axis

The axis in Beijing is composed of the axis in the old city, North Axis and South Axis, stretching about 25 kilometers.

8.1 Conservation Plan for the Traditional Axis

8.1.1 The old axis in Beijing is 7.8 kilometers from Yongdingmen to Bell and drum Tower, and 8.5 kilometers to North Second Ring Road. The conservation plan must proceed in line with the principle of giving priority to conservation. Conservation and development, inheritance and creation must also be combined. Research should be focused on the conservation and plan for the three places of the Bell and Drum Tower, Jingshan-Qiangmen, and Yongdingmen.

8.1.2 Node at the Bell and Drum Tower: Being the terminal of the traditional axis, the Tower, surrounded by courtyard houses, has the status of the landmark architecture in that area. The height control for buildings around the Tower must follow the regulations stipulated in the plan for conservation districts of historic sites.

8.1.3 Node at Jingshan-Qianmen: Consisting of Jingshan Park, Forbidden City, Tian'anmen, Zhengyangmen Gate Tower and Watch Tower, the area has a rich gradation of spaces, orderly and undulating layout. So it must be strictly protected.

8.1.4 Node at Yongdingmen: The restoration of Yongdingmen Gate Tower is of great significance for it can keep the completeness of the traditional axis and effectively connect itself with the South Axis line.

8.2 Conservation and Devleopment Plan for the Northern Axis

8.2.1 This axis stretches from the North Second Ring Road to the Olympic Park, out which three locations must be given priority in conservation.

8.2.2 Node at the Central Area of the Olympic Park: priority should be given in planning this area, the end of the North Axis, to make it a new landmark in Beijing. The Forest Park north of this area serves as backdrop of the North Axis.

8.2.3 Node at North Tucheng: The Tucheng ruins and central greenbelt along the 80-meter-long roads in the north axis can be utilized to create significant public urban spaces so as to enhance and enrich the North Axis.

8.2.4 Area North of the North Second Ring Road: Between North Second Ring Road and Ande Road. The land along the sides of the axis could be designed into important public urban spaces.

8.3 Conservation and Development of the South Axis

8.3.1 Goes from Yongdingmen to Nanyuan. While the functions of the land along the sides of the axis should be adjusted, the spatial structures of the axis must be enriched. Three places should be under priority conservation.

8.3.2 Node at Muxuyuan: Create urban public spaces in combination with the construction of Muxuyuan Business Center.

8.3.3 Node at Dahongmen: at the intersection of the axis and the south Fourth Ring Road. Importance urban landscape must be created.

8.3.4 Node at Nanyuan Area: as the end of the south axis, the place should be set off with large tracks of forest parks.

8.4 Conservation Territory for the Axis in Beijing

8.4.1 Based on the central line of the axis, the control boundary stretches to 500 meters on both sides, so that the architectural height and shape must be strictly restricted in this protected and controlled area of 1,000 meters.

8.4.2 Special areas outside of the protected and controlled area of the axis, but that have important impact on the axis such as the Altar of the God of Agriculture, Liuhai (Six Seas), and Temple of Heaven must be treated in line with the regulations for the conservation of conservation districts of historical sites.

9. Conservation of the Imperial City Conservation District

9.1 Determination of Conservation Territory

9.1.1 This plan has designated the whole imperial city as a conservation district.

9.1.2 The four boundaries of the protected area are: Dong Imperial City in the east, the red wall north of the Chang'an Avenue in the south, North and South Street at west Imperial Wall, Lingjing Hutong and Fuyou street in the west, and Ping'an Avenue in the north, occupying a total land area of 6.8 square kilometers.

9.2 Historical Value of the Imperial City

9.2.1 The Imperial City of the Ming and Qing Dynasties was symbol of feudal ruling class for thousands of years with its outstanding layout, architectural art and construction technique, thus having a high historical and cultural value.

9.3 皇城保护的措施

9.3.1 明确皇城保护区的性质:以皇家宫殿、坛庙建筑群、皇家园林为主体,以平房四合院民居为衬托的,具有浓厚的皇家传统文化特色的历史文化保护区。

9.3.2 建立皇城明确的区域意向,使人可明确感知到皇城区界的存在。

9.3.3 结合旧城外的土地开发,与皇城的保护和改造内外对应,降低保护区中的居住人口密度。

9.3.4 必须停止审批建设3层及3层以上的楼房和与传统皇城风貌不协调的建筑。

9.3.5 皇城内尚有部分文物保护单位利用不合理,应加以调整和改善。

9.3.6 皇城保护区内的道路改造应慎重研究,以保护为前提,逐步降低交通发生量。

9.3.7 必须将皇城内现有平顶的多层住宅改为坡顶。

9.3.8 制定皇城历史文化保护区保护管理条例。

10.明、清北京城"凸"字形城廓的保护

10.1 明清北京城的"凸"字形城廓是北京旧城的一个重要形态特征,必须采取措施加以保护。

10.2 在旧城改造中,沿东、西二环路尽可能留出30m绿化带,形成象征城墙旧址的绿化环。

10.3 保护北护城河与环绕外城的南护城河,规划沿河绿带。

10.4 保护现有的正阳门城楼与箭楼、德胜门箭楼、东便门角楼与城墙遗址、西便门城墙遗址,复建永定门城楼。

11.旧城棋盘式道路网和街巷胡同格局的保护

11.1 旧城主要交通对策

11.1.1 旧城区内的交通出行必须采取以公共交通为主的方式。

11.1.2 加快地铁建设,在主要干道上开设公交专用道,并布设小区公交支线网,方便市民出行。

11.1.3 实施严格的停车管理措施,控制车位供应规模,限制或调节驶入城区的汽车交通量。

11.1.4 采取切实可行的管理措施和调控手段(包括经济手段),限制私人小汽车在旧城区的过度使用。

11.1.5 控制旧城区建筑规模和开发强度,从根本上压缩机动车交通生成吸引量。

11.2 旧城路网调整原则

11.2.1 调整旧城路网规划和道路修建方式,协调好风貌保护与城市基础设施建设的关系,以此为前提确定路网的适当容量。

11.2.2 道路路幅宽度的确定应在满足文物和风貌保护的同时,协调处理好交通出行、市政设施、城市景观和生态环境等各项功能的需要。

11.3.3 同一等级道路,在旧城以外和旧城以内、在旧城的内城和外城、在历史文化保护区和非保护区,应采用不同的路幅宽度。

12.旧城建筑高度的控制

12.1 整个旧城的建筑高度控制规划应按照三个层次进行。

12.2 第一个层次为文物保护单位、历史文化保护区,是旧城保护的重点区域,这些区域必须按历史原貌保护的要求进行高度控制。

12.3 第二个层次为文物保护单位的建设控制地带及历史文化保护区的建设控制区,必须遵循文物及保护区保护规划的要求进行高度控制。

12.4 第三个层次为文物保护单位的建设控制地带、历史文化保护区的建设控制区之外的区域,建筑控高必须严格按《北京市区中心地区控制性详细规划》的要求执行,不得突破。

13.城市景观线的保护

13.1 《北京城市总体规划》规定的7条城市景观线必须加以严格保护,包括银锭观山、(钟)鼓楼至德胜门、(钟)鼓楼至北海白塔、景山至(钟)鼓楼、景山至北海(白塔)、景山经故宫和前门至永定门、正阳门城楼、箭楼至天坛祈年殿。

13.2 景观线保护范围内新建筑的高度,应按测试高度控制,严禁插建高层建筑。

14.城市街道对景的保护

14.1 对于历史形成的对景建筑及其环境要加以保护,控制其周围的建筑高度。对有可能形成新的对景的建筑,要通过城市设计,对其周围建筑的高度、体量和造型提出控制要求。

14.2 在旧城改造中必须处理好街道与重要对景建筑的关系,如北海大桥东望故宫西北角楼,陟山门街东望景山万春亭、西望北海白塔,前门大街北望箭楼,光明路西望天坛祈年殿,永定门内大街南望永定门城楼(复建),北中轴路南望钟鼓楼,地安门大街北望鼓楼,北京站街南望北京站等。

15.旧城建筑形态与色彩的继承与发扬

15.1 旧城内新建建筑的形态与色彩应与旧城整体风貌相协调。

15.2 对旧城内新建的低层、多层住宅,必须采用坡屋顶形式;已建的平屋顶住宅,必须逐步改为坡顶。

15.3 旧城内具有坡屋顶的建筑,其屋顶色彩应采用传统的青灰色调,禁止滥用琉璃瓦屋顶。

16.古树名木的保护

16.1 在危改区或新的建设区,严禁砍伐古树名木及大树。

16.2 历史文化保护区内的绿地建设包括街道、胡同和院落绿化。

16.3 旧城内的改造区应尽量增加公共集中绿地,绿地建设应采用适合北京特点的植物品种。

17.旧城危改与旧城保护

17.1 应树立旧城危改与名城保护相统一的思想。

17.2 历史文化保护区内的危房,必须严格按历史文化保护区保护规划实施,以"院落"为单位逐步更新,恢复原有街区的传统风貌。

17.3 历史文化保护区以外的危改地区,必须加强对文物及有价值的历史建筑的核查、保护,严格执行各级文物保护单位的保护范围和建设控制地带及《北京市区中心地区控制性详细规划》中的高度控制等有关规定。

17.4 建设单位必须处理好与保护有关的工作才能申报危旧房改造方案。危改项目的前期规划方案必须包括历史文化保护专项规划,内容包括街区的历史沿革、文物保护单位的保护、有价值的历史建筑及

9.2.2 The Uniqueness of the Imperial City: as the only well preserved imperial city of the Ming and Qing Dynasties, it is the sole largest and most complete imperial architectural complex, forming the peak in the traditional axis in the old city of Beijing.

9.2.3 The Completeness of the Imperial City: centered around the Forbidden City, and located on the clear-cut axis, it boasts a concentrated and orderly imperial palaces and gardens, temples and altars, government agencies and warehouses and courtyards for ordinary dwellers, reflecting the planning concept of the absolute imperial power and the complete functional layout.

9.2.4 Authenticity of the Imperial City: the Forbidden City, Tongzi Moat, Sanhai, Imperial Ancestral Temple, Altar to the God of the Land and Grain, many temples, government offices in feudal times, courtyards houses and many other traditional architecture have been well preserved, providing evidences of the historical information on the life, work and entertainment of the royal families, as well as the evolutionary process of the Ming and Qing Dynasites and the Republic of China (1912-1949), hence a real reflection of history.

9.2.5 Aesthetic Value of the Imperial City: it has attained a high artistic quality in planning concept, architectural layout, construction technique, and the use of colors.

9.3 Measures for Conserving the Imperial City

9.3.1 Clarify the Nature of the Imperial City Conservation District: conservation district with the imperial palaces, architectural complex of temples and royal gardens as the main body, combined with courtyard houses all having strong royal traditionally cultural feautures.

9.3.2 Create a clear concept of imperial city area to make sure that every one is aware of the existence of the boundary of such an area.

9.3.3 The land development outside of the old city should take into account of the conservation and renovation of the Imperial City, with a view of reducing the population within the protected area.

9.3.4 The examination and approval of buildings of or above three stories and those not in agreement with the traditional style of the Imperial City must be stopped.

9.3.5 The unreasonable use of some cultural relics sites within the Imperial City should be adjusted and improved.

9.3.6 The road renovation in the Imperial City must be conducted in a prudent way with conservation as the priority. Traffic should be gradually reduced.

9.3.7 The existing flat-roofed, multi-storied residences within the Imperial City must be transformed into buildings with pitched roofs.

9.3.8 Draw up a conservation ordinance for the Imperial City conservation district.

10. Conservation of the Inverted "T"-Shaped City Wall of the Ming and Qing Dynasties in Beijing

10.1 As the city walls in the form of inverted "T" is an importance feature of the old city of Beijing, measures must be adopted to protect them.

10.2 In the renovation of the old city, 30 meters of greenbelt along the east and west Second Ring Roads should be left to make room for a green ring symbolizing the old city wall.

10.3 Protecting the north City Moat and south City Moat surrounding the outer city, and plan the greenbelt along the rivers.

10.4 Protecting the existing Zhengyangmen Gate Tower and Watch Tower, the Deshengmen Watch Tower, the Dongbianmen Corner Tower and city wall ruins, the Xibianmen city wall ruins, and restore the Yongdingmeng Gate Tower.

11. Conservation of the Checkerboard Road System and Alley Layout in the Old City

11.1 Solutions to the Traffic of the Old City

11.1.1 Public transportation must be the primary way of circulation within the Old City.

11.1.2 Speed up the construction of subways, open special lanes for public transportation on the main roads, and set up public transportation feeder network in residential quarters for the convenience of the dwellers.

11.1.3 Strictly exercise the measures for parking management, restrict the parking places and control or regulated the traffic volume heading into the city.

11.1.4 Restrict the overuse of private cars in the old city by adopting feasible management and regulating measures (including economic means).

11.1.5 Control the construction scale and development intensity in the old city, so as to reduce the inbound traffic.

11.2 Adjusting Principles for the Road Network in the Old City

11.2.1 Adjust the plan for road network and road repair methods in the old city, strike a good balance between the conservation of features and the infrastructure construction in the city, upon which appropriate traffic load can be determined.

11.2.2 The width of roads must be in line with the requirements of protecting the cultural relies and traditional features, as well as meeting various functions such a traffic, urban facilities, city landscapes and ecological environment.

11.2.3 Roads of the same quality should be different in width depending on their places-inside or outside the old city, in the inner or the outer city within the old city, and in the conservation district or outside the district.

12. Control of Building Height in the Old City

12.1 The overall plan for the building height control in the old city must proceed in three levels.

12.2 The first level includes cultural relics sites and conservation districts of historic sites. Height control in these protected key places in the old city must meet the requirements of protecting the original features.

12.3 The second level refers to the construction controlled areas of the cultural relics sits and conservation districts. Height control in these areas must be implemented according to the requirements stipulated in the cultural relics and conservation district plans.

12.4 The third level includes the areas beyond the construction controlled areas of the cultural relics and conservation districts. Height control in theses areas must strictly abide by the requirements specified in the "Detailed Plan for the Central Controlled Area in Beijing".

13. Conservation of Urban Landscape Belts

13.1 "Master Plan for the Construction of Beijing" stipulates that 7 urban landscape belts must be strictly protected. They are: Yindingguan Mountain, Bell and Drum Tower to Deshengmen, Bell and Drum Tower to the Pagoda in Beihai Park, Jingshan Park to the Bell and Drum Tower, Jingshan Park to the Pagoda in Beihai Park, Jingshan Park to Yongdingmen through the Forbidden City, and Zhengyangmen Gate Tower and Watch Tower to the Hall of Payer for Good Harvests in the Temple of Heaven.

13.2 The height of new buildings within the landscape belts under conservation must be controlled by testing height, and no high-rises are allowed in these areas.

14. Conservation of Corresponding Views at Street Ends

14.1 Buildings and their surroundings at street ends formed in history must be conserved, and architectural height around them controlled. For buildings that are likely to form new views at street ends must be carefully designed, and the height, size and shape of surrounding buildings will have to meet the requirements of control.

14.2 The relationship between the streets and important end views in the course of renovation of the old city must be dealt with well. Some of the examples are: Beihai Bridge looking the northwest corner tower of the Forbidden City in the east, Zhidongmen Street echoing the Wancun Pavilion in Jiangshan Park in the east and the Pagoda in Beihai Park in the west, Qianmen street looking the Watch Tower in north, Guangminglou echoing the Hall of Prayer for Good Harvest in the Temple of Heaven in the west, Yongdingmennei Street looking the Yongdingmen Gate Tower (rebuilt) in the south, North Axis looking the Bell and Drum Tower in the south, Di'anmen Street looking the Drum Tower in the north, and Beijing Railway Station Street looking Beijing Railway Station in the south.

15. Inheritance and Development of Architectural forms and Colors in the Old City

15.1 The patterns and colors of the newly built architecture within the old city must be in harmony with the general features of the old city.

15.2 The low and multi-story buildings newly built in the old city must use sloping roofing; flat-roofed buildings already constructed must be gradually turned

北京历史文化名城北京皇城保护规划
Conservation plan for the Historic City of Beijing and Imperial City of Beijing

遗存的保护、古树名木和大树的保护、对传统风貌影响的评价、环境改善的措施等。

18. 传统地名的保护

18.1 传统地名是北京历史文化名城保护的重要内容之一，必须加以保护。

18.2 建立健全相关法规和技术规范，对传统胡同、街道的历史名称不得随意修改。

19. 传统文化、商业的保护和发扬

19.1 传统文化的保护和发扬

19.1.1 北京具有历史悠久的传统文化，与城市规划建设相关并较有代表性的有庙会、戏院、会馆等。

19.1.2 应尽量恢复各区有代表性的庙会，包括厂甸、白塔寺、护国寺等。

19.1.3 以昆曲和京剧为重点，进一步繁荣北京的传统戏曲事业，加强戏院和相关文化设施的建设。

19.1.4 应采取措施恢复和合理利用会馆。

19.2 传统商业的保护和发扬

19.2.1 传统商业是历史上长期存在的、符合当地民族生活习惯的、具有明显特色且不断继承发扬的商业、服务业。

19.2.2 传统商业的保护主要包括传统商业街区的保护与改造和老字号的恢复与保护两个方面。

19.2.3 传统商业街区的保护指重点保护大栅栏商业街、琉璃厂文化街、前门商业文化旅游区、什刹海地区的传统商业街（烟袋斜街、荷花市场等）、隆福寺商业街。

19.2.4 传统行业和老字号的保护以食品、餐饮、医药行业为多。如同仁堂、全聚德、王致和、稻香春等。

19.2.5 大力扶持老字号，继承发扬传统经营管理的特色。

20. 实施保障措施

20.1 加强对《北京历史文化名城保护规划》的宣传，在全社会形成"热爱名城、保护名城"的共识。

20.2 进一步落实"两个战略转移"的方针，积极推进城市建设重点逐步从市区向远郊区转移、市区建设从外延扩展向调整改造转移的步伐，疏解旧城区人口和功能，为保护历史文化名城创造良好条件。

20.3 在《北京历史文化名城保护规划》审批后，应尽快编制相关规划。主要包括：第二批历史文化保护区保护规划、旧城内道路系统的调整规划、旧城建筑高度控制的调整规划、第五批、第六批文物保护单位保护范围和建设控制地带的划定等。

20.4 制定历史文化名城保护的相关法规，包括《北京市历史文化名城保护条例》、《北京市历史文化保护区管理规定》等，作为法律依据在城市管理工作中严格执行。

20.5 研究制定历史文化保护区内的相关政策，主要包括：房屋产权改革和产权交易、人口外迁与疏解、房屋管理和修缮的相关政策等。

20.6 旧城内危房改造不宜采用房地产开发的方式。

20.7 除政府财政投入外，多渠道筹集资金，初步建立历史文化名城保护的资金保障机制。

20.8 各区要进行试点，探索文物保护单位和历史文化保护区保护、维修、整治、利用的有效途径。

20.9 加强对继承和发扬名城传统风貌和文化特色的研究。

2002年9月

into sloping roofed structure.

15.3 Architecture with sloping roofing in the old city must use traditional greenish gray color for their roofs. Overuse of Liu-Li tiles are forbidden.

16. Conservation of Well-known and Old Trees

16.1 Prohibit cutting down well-known, old and big trees in ramshackle building renovation and new construction areas.

16.2 The greening construction in the conservation districts include greening up streets, Hutongs and courtyards.

16.3 Public and concentrated green plots in the renovated areas of the old city should be enhanced, where plant species suitable to Beijing should be planted.

17. Ramshackle Building Renovation in the Old City and Conservation of the Old City

17.1 The renovation of crumbling houses in the old city should be integrated with the conservation of the historic city.

17.2 The renovation of dangerous houses in the conservation districts must be undertaken in strict compliance with the conservation plan for conservation districts of historic sites. Renovated in terms of the "courtyard", the traditional features of the original streets must be rebuilt.

17.3 When dealing with renovation areas outside the conservation districts, the verification and conservation of cultural relics and valuable historical buildings must be strengthened. The relevant regulations relating to the conservation range and construction controlled areas of cultural relics sites of various levels, and of the height control stipulated in the "Detailed Plan for the Central Controlled Area in Beijing" must be strictly implemented.

17.4 construction units can only submit plans for renovating old and risky houses for approval after having dealt with the conservation work in a satisfactory way. The initial plan of this kind must contain a special plan for the conservation of historical culture, including the historical change of streets, conservation of the cultural relics sites, conservation of valuable historical buildings and ruins, conservation of well-known, old and big trees, assessment of the influence of traditional features, and measures for improving environment.

18. Conservation of Traditional Names of Places

18.1 Traditional place names must be protected as one of the important contents in the conservation of the historic city of Beijing.

18.2 Work out relevant regulations and technical norms to guarantee that historical names of traditional Hutongs and streets can't be revised at random.

19. Conservation of Traditional Cultural and Commerce

19.1 Conservation and Development of Traditional Culture

19.1.1 Beijing boasts a traditional culture of long-standing. Temple Fairs, Theaters and guild halls are related with urban planning and construction and representative of the traditional cultural.

19.1.2 The representative temple fairs in various districts must be reinstated as soon as possible, including those in Changdian, Temple of White Pagoda, and Huguo Temple.

19.1.3 Further develop the traditional operas in Beijing by giving priority to Kunqu and Peking Opera. Construction of theaters and relevant cultural facilities must be enhanced.

19.1.4 Measures should be taken to renovate and make reasonable use of guild halls.

19.2 Conservation and Continuity of Traditional Commerce

19.2.1 Traditional commerce refers to business and service sectors that are of long-standing, suitable to the lifestyle of the local people, have obvious characteristics and can continue developing in the future.

19.2.2 Conservation of traditional commerce mainly involves the conservation and renovation of traditional commercial streets and rebuilding and conservation of stores of long standing.

19.2.3 Conservation of traditional commercial streets refers to the priority given to the conservation of Dashalan Commercial Street, Liulichang Cultural Street, Qianmen Commercial and Sightseeing Street, commercial streets in Shishahai area (Yandaixie Street and Lotus Market, etc), and Longfusi Commercial Street.

19.2.4 Conservation of traditional sectors and stores of long standing is mainly concentrated in food and beverage as well as medicine fields, such as Tongren, Quanjude, Wang zhihe and Daoxiangchun.

19.2.5 Vigorously promote stores of long standing, and carry forward the management characteristics of traditional operation.

20. Implementation of Conserving Measures

20.1 Promote highly the "Conservation Plan for the Historic City of Beijing", and try to reach a consensus of "love and protect the historic city" among citizens.

20.2 Further implement the policy of "two strategic shifts", quickening the pace of construction from inner city to outlying suburbs, and from expanding the urban boundary to adjustment and renovation, in order to mitigate the population in the old city and its functions, paying the way for protecting the historic city.

20.3 After the approval of the "Conservation Plan for the Historic City of Beijing", relevant plans should be drawn up as soon as possible, which include: the conservation plan for the second group of conservation districts of historic sites, adjustment plan for road network in the old city, regulating plan for architectural height control in the old city, and designation of the fifth and sixth groups of the conservation scope and construction controlled areas of cultural relics sites under conservation.

20.4 Work out relevant regulations for the conservation of the historic city, which will be used as legal basis to be implemented strictly in urban management. These include "Ordinance of Conservation for the Historic City of Beijing", "Management Regulations for the Conservation Districts of Historic Sites in Beijing".

20.5 Research and formulate relevant policies concerning the conservation districts of historic sites, including reform on property right and property trade, population migration and reduction, and relevant policies on housing management and renovation.

20.6 The approach used in real estate development should not be applied in the renovation of crumbling houses in the old city

20.7 Apart from investment from government, funds should be raised via many ways so as to establish a safeguard mechanism for the conservation of historic city.

20.8 Various districts should launch pilot projects, exploring effective approaches to protecting, maintaining, renovating and utilizing cultural relics sites and conservation districts.

20.9 Further encourage research on ways of carrying forward the traditional fabrics and cultural characteristics of historic cities.

September 2002

北京历史文化名城北京皇城保护规划
Conservation plan for the Historic City of Beijing and Imperial City of Beijing

故宫中轴线鸟瞰
Bird's eye view of the axis of the Forbidden City

《北京历史文化名城保护规划》说明

序言

北京是中华人民共和国的首都，也是世界著名的文化古都。经过千百年的发展历程，北京的历史遗产与文化内涵极其丰富，保护这些遗产是当代人的历史责任。

当前，北京危旧房改造速度加快，房地产开发力度很大，旧城改造与历史文化名城保护的矛盾日益尖锐，并受到首都社会各界和国内外的高度关注。北京市人大、政协和专家学者多次提出有关保护北京历史文化名城的提案建议。根据市领导的有关指示，我们编制了《北京历史文化名城保护规划》。

1. 北京历史文化名城的特征

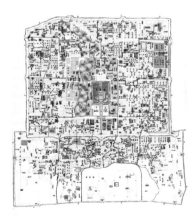

明清旧城地图
Old city maps of the Ming and Qing Dynasties

1.1 悠久的建城历史

北京建城的历史据考证应该从周武王分封蓟国（公元前1045年）的时候算起，距今已有3000多年。公元938年，辽代在蓟城的基础上，建南京城并成为陪都。金灭辽后于1153年迁都于此，改称中都。金中都是北京成为全国政治中心的重要标志，距今有800多年的历史。到了元朝，在金中都东北郊外创建了新城，称为大都。明代改称北京，1406年起新建宫殿、城池，距今已有近600年历史。清朝（公元1644年~1911年）继续建都于北京，基本延续了明北京城的格局，同时建造了北京西北郊的园林景区。

1.2 独特的城市地理形态

北京城位于华北大平原的北端，它的西部是西山，北部是军都山，东南一带则是一片逐渐低缓的平原，平原上水网密集，形成了典型的背山面水的"风水宝地。"

古代北京自蓟城到金中都时期仰赖于西湖水系（今莲花池一带），金代以后，北京城的发展依托于高粱河水系（今六海一带）。这些山脉、平原、湖泊和河流，造就了北京城市的演变，也形成了北京城市与山水相互映衬的独特的自然地理景观。

1.3 规模宏大的城市与宫殿

明清北京旧城占地62.5km^2，是中国历史上遗存下来的最大的帝王都城。旧城中心的紫禁城，是世界上保留最完整、规模最大的帝王宫殿。这是封建社会皇帝至高无上、惟我独尊思想在城市规划上的集中表现，也是古代劳动人民勤劳、智慧与血汗的结晶，在世界城市建设史上具有突出的地位。

1.4 完整的旧城风貌与格局

现在的北京旧城是以《周礼·考工记》的王城规划理想为出发点，结合地理形态进行的规划，其最大的特征就是城市风貌与格局的整体性和有机性。旧城规划水平之高，在世界城市中是公认的。

具体而言，主要有：

1.4.1 等级明确、层次分明的城廓。由内到外，北京旧城是由紫禁城、皇城、内城、外城组成，界面清晰，等级分明，管理有序，构成有别。内外城呈"凸"字形布局，形成北京标志性的城廓特征。

1.4.2 严谨有序的城市中轴线。以故宫和皇城为中心，南起永定门，北至钟鼓楼，绵延7.8km。在中轴线上或其两侧，布置了钟鼓楼、景山、紫禁城、太庙、社稷坛、天坛、先农坛等封建王朝最重要的建筑群，是严谨对称城市格局的集中体现。

1.4.3 有机活泼的城市水系。北京旧城的格局与水系密不可分。筒子河、护城河、六海等水系，形成了活泼的、贯穿城市中心的景观和生态廊道。

1.4.4 平直整齐的城市路网。元大都时北京旧城的棋盘式道路系统已基本形成，明清北京城的道路网延续了前朝系统且有其定制：分为大街、小街和胡同三类。

1.4.5 平缓开阔、起伏有致的城市天际线。在广大的平房四合院民居衬托下，形成以故宫为中心，以景山万春亭、钟鼓楼、正阳门、永定门城楼等建筑为控制点的，并由旧城墙和各城楼拱卫的，形成起伏有致的城市天际线。

1.4.6 独特的四合院式的城市居住形态。老北京城的民居均为灰墙灰瓦的单层四合院，它们大院套小院，层层叠叠，秩序统一而形态丰富，形成了皇宫的优美背景。

1.4.7 丰富多彩的城市街道对景。在平直整齐的路网系统里，旧城中各种城楼、牌楼和宫、坛、庙、亭、塔构成了丰富的街道对景，形成了优美的视觉走廊。

Introduction to the "Conservation Plan for the Historic City of Beijing"

Preface

Beijing is the capital of the People's Republic of China, and also one of the ancient cultural centers in the world. Throughout thousands of years of development, the historical and cultural heritages in Beijing are very rich, therefore, it is our historical obligation to protect these relics.

At present, owing to the fast pace of the housing renovation in Beijing, and the aggressive real estate development, the conflict between the urban renewal and the conservation of the historic city has become more intensified and drawn great attention of various circles in Beijing, and people from both home and abroad as well. Suggestions and Proposals concerning the conservation of the historic city of Beijing have been submitted many times by the Beijing People's Congress, Beijing Political Consultative Conference as well as scholars and specialists. We therefore have compiled the "Conservation Plan for the Historic City of Beijing" according to the relevant instructions of the leaders of Beijing municipal government.

I. Features of the Historic City of Beijing

1.1 Long History of the City

Documents show that the Beijing has existed for 3000 years as it was invested to the State of Ji as a fief by King Wu of the Zhou Kingdom in 1045 BC. In 938, Liao Dynasty set up the city of Nanjing on the modal of the city of Ji and made Nanjing a secondary capital. When Jin Dynasty eliminated Liao, it designated present Beijing as the capital in 1153 and changed its name into Zhongdu (middle capital). As an important political center of the country, the Zhongdu of the Jin Dynasty has had a history of 800 years. The Yuan Dynasty saw the construction of a new city on the northeastern suburbs of Zhongdu, which was called Dadu. It was later changed to Beijing in the Ming Dynasty. Palaces, city walls and moat were built 600 years ago in 1406. The Qing Dynasty (1644-1911) continued the construction of Beijing on the basis of the layout of the Ming period, and garden landscapes were added on the northwestern outskirts of Beijing.

1.2 Unique Urban Geographical Features

Located on the northern edge of the North China Plain, Beijing is bordered by the Western Mountains to the west, Jundu Mountain to the north, and a gently-sloping vast land to the southeast where there is network of lakes and rivers. According to the Chinese Fengshui theory, it is deemed a place of treasure thanks to its location with hills behind and water in front.

From the time of Ji to Zhongdu of the Jin Dynasty, Beijing relied on the Xihu river system (present Lianhuachi (lotus pond)). Since the Jin Dynasty, the Gaoliang river system (present Liuhai) has been playing a role in the development of Beijing. These mountains, plains, rivers and lakes not only facilitated the evolution of Beijing, but also helped the city and rivers match each other well in forming a unique geographical feature.

1.3 The Grandeur of the City and Its Palaces

钟鼓楼夜景
Night scene of the Bell and Drum Towers

Covering an area of 62.5 square kilometers, the old Beijing of the Ming and Qing Dynasties was the largest ever imperial capital in the Chinese history. The Forbidden City in the heart of the old city is the largest and most intact royal palace complex in the world. It is the urban manifestation of the absolute authority and overbearing attitude of monarchs of the feudal society. It is also the crystallization of the diligence, wisdom and sweat of the laboring people of the past, thus occupying a conspicuous place in the history of cities and architecture in the world.

1.4 Unbroken Layout and Urban Fabric of the Old City

The present old city of Beijing was designed as an imperial city following the concept of the Book of Diverse Crafts, blended with the geographical features of the place. The most obvious characteristic is its organic layout and the homogeneous fabric. The high level planning for the old city has been recognized all over the world. Specifically, the following should be noted:

1.4.1 Multilayered and hierarchical city walls. The old city of Beijing is composed of the Forbidden City, imperial city, inner city and outer city, with a clear-cut, hierarchical order. The inner and outer cities are arranged in the form of an inverted "T", forming a landmark city wall feature in Beijing.

1.4.2 The orderly urban axis. Using the Forbidden City and the imperial city as the center, the axis starts from Yongdingmen in the south to the Bell and Drum Tower in the north, stretching 7.8 kilometers. On or along the sides of the axis are important royal structures of feudal society such as the Bell and Drum Tower, Forbidden City, Imperial Ancestral Temple, Altar to the God of the Land and Grain, Temple of Heaven, Altar of the God of Agriculture, etc. They are the result of a strict symmetrical planning approach.

1.4.3 Organic and Lively River System. The layout of the old city of Beijing is inseparable with water system. Tongzi Moat, the City Moat, and the six lakes together form a lively landscape and ecological corridor going through the city center.

1.4.4 Orthographic and Neat Urban Road Network. The checkerboard network of roads were formed in the old city of Beijing during the Yuan Dynasty, which was continuously used in the Ming Dynasty with rules established: the roads were divided into avenues, streets and Hutongs.

1.4.5 Level, Vast and Undulating Urban Skyline. A meandering city skyline is created with the Forbidden City at the center, and the axis dotted with the Wanchun Pavilion of the Coal Hill Park, the Bell and Drum Tower, Zhengyanmen and Yongdingmen which are surrounded by old city walls and watch towers. In-between are rows of courtyard houses.

1.4.6 The Unique Habitat of Courtyard Housing. The residential houses in old Beijing are one-story quadrangles with gray walls and tiles. Overlapped and intertwined, they present orderly and rich shapes, setting off the royal palace in a graceful way.

1.4.7 Rich and Vibrant Visual Passages. In the even and straight streets, various temples, watcher towers, decorated archways, altars and pavilions constitute symmetrical sights, providing pleasant visual effects.

1.4.8 Primary and Secondary Colors. Large stretches of gray houses and green trees serve as foils to set off the royal architecture of red walls and yellow tiles, and the princes' residences and temples of green and blue glazed tiles, accentuating the primary colors of the city.

1.5 Abound Relics and Historical Ruins

A lot of invaluable historical relics are scattered in the vast surroundings of old Beijing. Among them are the Ming Tombs, Western Suburb Royal Garden of the Qing Dynasty, the Great Wall, Zhoukoudian (site of the fossils of the Peking Man), etc., all of which are important components of the historic city of Beijing.

1.6 Long-Standing Traditional Culture

Thousands of years of development of Beijing have helped to create a unique traditional culture, exerting a vital influence on the whole country. For example, the standard Chinese pronunciation based on the Beijing dialect, The Peking opera which is dubbed as the quintessence of Chinese culture, Kunqu, which is listed as the "world non-physical heritage". Besides, there are also distinctive Beijing customs, temple fairs and stores of long-standing, which together enrich the culture of the historic city of Beijing.

北海鸟瞰
Bird's eye view of Beihai Park

天坛
The Temple of Heaven

鸟瞰角楼、景山
Bird's eye view of the turret and Jingshan Park

1.4.8 主次分明的城市色彩。以大片青灰色房屋和绿树为基调，烘托出红墙黄瓦的皇家建筑群以及绿、蓝琉璃瓦的王府、坛庙，形成统一而重点突出的城市色彩。

1.5 丰富的文物与历史遗存

在北京旧城以外的辽阔区域里，分布着众多极为珍贵的文物和历史遗存。如明十三陵、清代的西郊皇家园林、长城、周口店猿人遗址等文化遗存，它们是北京历史文化名城的重要组成部分。

1.6 悠久的传统文化

北京几千年的城市发展史，形成了独特的传统文化，对全国有着重大的影响。如以"北京话"为标准音的普通话，被称为"国粹"的京剧。此外还有独具北京特色的市井习俗、庙会戏曲、老字商号等，丰富了北京历史文化名城保护的内涵。

2. 北京历史文化名城保护工作的回顾与存在的问题

2.1 北京历史文化名城保护工作的回顾

《中华人民共和国文物保护法》把历史文化名城定义为"保存文物特别丰富，具有重大历史价值和革命意义的城市"。北京作为国务院公布的第一批国家历史文化名城，具备了历史文化名城所要求的全部特征。回顾过去，北京在加强历史文化名城保护工作方面进行了不懈的努力和探索。

2.1.1 分批公布了文物保护单位的名单。

中华人民共和国成立以来，本市先后开展了三次全市范围的文物普查工作，陆续公布了六批市级文物保护单位。全市国家、市、区（县）级文物保护单位共1000余项。

2.1.2 划定了文物保护单位的保护范围和建设控制地带。

市政府已分四批颁布了202项市级以上文物保护单位的保护范围和建设控制地带。目前第五批、第六批市级文物的保护范围和建设控制地带正在编制中。

2.1.3 加大文物修缮力度，促进文物的合理利用。

文物修缮一直是文物保护工作的重点。20世纪90年代以来，在对天坛公园、颐和园、故宫筒子河等一批文物保护单位的环境进行整治的基础上，2000年5月，市政府决定三年拨款3.3亿元用于文物抢险修缮工作。"3.3亿工程"实施以来，已使历代帝王庙、后门桥、钟鼓楼等一批文物建筑得到抢修，北京文物保护工作取得了重大进展。文物不合理使用问题的解决取得了历史性突破。近两年，为保护文物搬迁居民近2000户，投入资金10多亿元，10多万m²的文物场所得到腾退。

2.1.4 大力开展地下考古挖掘工作。

中华人民共和国成立以来，北京市非常重视地下考古挖掘工作。20世纪80年代以来，先后对延庆山戎墓葬遗址、金中都水关遗址、东方广场旧石器时代晚期古人类遗址、老山西汉墓等进行考古发掘。本市已公布地下文物埋藏区36处。

2.1.5 编制北京旧城历史文化保护区保护规划。

1990年市政府公布了第一批北京历史文化保护区名单。1999年市政府批准了旧城历史文化保护区的保护和控制范围。2000年编制了《北京旧城25片历史文化保护区保护规划》，2002年2月，市政府批准了该项规划。25片历史文化保护区总占地1038hm²，占旧城总面积的17%。

2.1.6 将历史文化名城保护纳入城市总体规划中。

1953年上报中央的《改建与扩建北京市规划草案的要点》和1958年《北京城市建设总体规划初步方案（草案）》，确定了保护古代建筑物的原则。1983年中共中央、国务院原则批准的《北京城市建设总体规划方案》提出，保护古都风貌不仅要保护古建筑本身，也要保护周围环境。1993年经国务院批准的《北京城市总体规划》提出了从整体上保护历史文化名城的三个层次和十条措施。

2.1.7 制定市区中心地区控制性详细规划。

1995年至1997年，编制了《北京市市中心地区控制性详细规划》，对市区324km²范围内的用地提出了具体的控制要求。包括从保护历史文化名城的角度出发，对旧城格局、建筑高度、以及文物保护单位和25片历史文化保护区的保护及周围建设，提出了相应的控制规定。该规划于1999年经市政府批准执行。

2.1.8 综合治理城市水系。

从1998年起市政府投资10亿元治理市区城市水系，完成了故宫筒子河、六海、长河、京密引水渠昆玉段、玉渊潭至通惠河段的综合治理工程，恢复了广源闸、麦钟桥、紫玉湾御码头等文物古迹，恢复了莲花池和后门桥周围水道，使这些京城水系发源地的古迹重现了昔日的神韵。

2.1.9 制定保护文物与名城的法规规章。

依据《中华人民共和国文物保护法》和《中华人民共和国城市规划法》，北京市制定了保护历史文化名城的地方性法规和部门规章。主要有：《北京市城市规划条例》、《北京市文物保护管理条例》、《北京市文物保护单位保护范围及建设控制地带管理规定》、《北京市周口店北京猿人遗址保护管理办法》、《北京市文物建筑修缮工程管理办法》、《北京市人民政府关于严格控制颐和园、圆明园地区建设工程的规定》

规划说明

历代帝王庙
Temples of the past dynasties emperor

潭柘寺
Tanzhe Temple

修缮后的白塔寺
Renovated White Pagoda Temple

现在的筒子河
Present Tongzi River

2. Review and Problems of the Conservation of the Historic City of Beijing

2.1 Review of the Work Done for the Conservation of the Historic City of Beijing

"The Cultural Legacy Conservation Code of the People's Republic of China" defines the historic city as "a city with abundant cultural relics, thus having important historical and revolutionary significance." Beijing, as one of the historic cities in the first group designated by the State Counsel, possesses all the elements of a historic city. Looking back, we can see that relentless efforts and exploration have been made by Beijing in protecting this historic city.

2.1.1 Lists of Cultural Relics Sites Under Conservation are Made in Batches

Since 1949, three surveys on cultural relics were conducted citywide, which resulted in designating six groups of cultural relics sites at the city level. There are altogether 1,000 cultural relics under conservation at the State, city and district (county) levels.

2.1.2 Protective Boundary and Construction-Controlled Areas of the Cultural Relics Sites have been determined

The municipal government has proclaimed in four groups the conservation boundaries and construction-controlled areas of 202 cultural relics sites under conservation at and above the city levels. At the moment, the fifth and sixth groups of the extent of conservation and construction-controlled areas of cultural relics sites at the city level are under deliberation.

2.1.3 Enhance the Renovation of Cultural Relics and Make Sensible Use of Them

Renovation of cultural relics has always been the priority of the cultural relics conservation. Since the 1990s, renovations were done on the Temple of Heaven, Summer Palace, the Tongzi River of the Forbidden City as well as many other cultural relics. On the basis of this, the municipal government, in May 2000, decided to earmark 330 million yuan in three years to do rush repairs to cultural relics. Since then, a number of sites such as Houmen Bridge, Bell and Drum Tower and many temples have been repaired in time, making great strides in the cultural relics conservation in Beijing. The problem of inappropriately using cultural relics has also solved markedly. In the past two years, nearly 2,000 households have been relocated for the sake of protecting relics, involving over 1 billion yuan, and more than 100,000 square meters of cultural relics sites have been emptied of inhabitants.

2.1.4 Devote Major Efforts To Archeological Excavation

The Beijing municipal government has been attaching great importance to archeological work since 1949. Since the 1980s, excavation has been undertaken in Shanrong Grave in Yanqing, Shuiguan Ruins of the Dadu of the Yuan Dynasty, Ruins of the Ancient Men of late Old Stone Age at Oriental Plaza, The Western Han Dynasty Tomb in Laoshan, etc. The municipal government has made known 36 sites of underground relics.

2.1.5 Compile the Conservation Plan for the Preserved Regions of Historic Sites in Old City of Beijing

In 1990, Beijing municipal government publicized the first group of conservation districts of historic sites of Beijing. In 1999, the municipal government approved the extent of conservation and control for the conservation districts in the old city of Beijing. In 2000, the "Conservation Plan for the 25 Conservation Districts of Historic Sites in the Old City of Beijing" was drawn up, which was approved by the municipal government in February 2002. The 25 conservation districts cover an area of 1,38 hectares, accounting for 17% of the total area of the old city.

2.1.6 Integrate the Conservation of Historic City into the City's Master Plan

The "Key Points to the Draft Plan of Redeveloping and Expanding Beijing" and the "Preliminary Plan (Draft) of the Master Plan of Beijing" respectively submitted to the Party Central Committee in 1953 and 1958 laid down the principle of conservation of historic buildings. The "Master Plan of Beijing" approved in principle by the Party Central Committee and the State Counsel in 1983 stipulated that preservation of the old capital not only involve the conservation of the ancient buildings, but also their surroundings. The "Master Plan of Beijing" approved by the State Counsel in 1993 put forward the concept of preserving the historic city by way of 3 levels and 10 measures.

2.1.7 Development of the "Detailed Plan of Central Beijing"

From 1995 to 1997, the "Detailed Plan of Central Beijing" was formulated, imposing detailed control requirements for the land within the center area of 324 square kilometer in the city. Included are varied control regulations and guiding limits for the protecting of the layout of the old city and the conservation of the cultural relics sites and 25 conservation districts of historic sites as well as the construction of their surroundings, and building heights. This plan was carried out in 1999 after being approved by the municipal government.

2.1.8 Regulation of the City's Water System

Starting from 1988, the municipal government has invested 1 billion yuan to harness the urban river system, and among the projects completed are Dongzi River of the Forbidden City, the 6 lakes, Changhe River, Kunyu section in the diversion project from Beijing to Miyun reservoir, and the section from Yuyuantan to Tonghui River. Historical relics of Gangyuan Dam, Miazhong Bridge and Ziyuwan Wharf have been renovated, and the river courses around Lianhuachi and Houmen Bridge have also been rebuilt, thus recovering the former features of these ancient ruins of urban river system.

2.1.9 Formulate Law and Regulations Regarding the Conservation of Cultural Relics and Historic City

In line with the "The Cultural Legacy Conservation Code of the People's Republic of China" and "The Urban Planning Code of the People's Republic of China", Beijing also drafted the following local rules and regulations with regard to the conservation of historic cities: "Ordinance of Beijing Urban Planning", "Ordinance of Conservation of Historic Relics in Beijing", "Regulations for the Conservation Scope and Construction-Controlled Areas of the Cultural Relics Sites in Beijing", "Regulations for the Conservation of Peking Man Ruins at Zhoukoudian", "Regulations for Renovation Projects of Cultural Relics in Beijing", "Regulations of the Beijing Municipal Government for Strictly Controlling the Construction Projects around Summer Palace and Yuanmingyuan", etc. All these rules and regulation have played a positive role in protecting the historic city of Beijing.

2.2 Problems Concerning the Conservation of the Historic City of Beijing

Many problems have yet to be solved in protecting Beijing.

2.2.1 Awareness of Conservation Needs to be Further Raised

Owing to the lack of a systematic recognition of the historical value of the development of Beijing, the awareness of conservation is still inadequate, espe-

等法规规章,对历史文化名城的保护起到了积极的作用。

故宫
The Palace Museum (Forbidden City)

华润大厦
Huarun Tower

2.2 北京历史文化名城保护工作存在的问题

目前名城保护工作还存在许多有待解决的问题:

2.2.1 对历史文化名城保护的意识有待提高。

由于对北京城市发展的历史文化价值及特征缺乏系统的认识,对北京历史文化名城保护的意识不足,特别是对旧城的整体保护缺乏认识,表现在只重视文物及历史街区的局部保护,忽视旧城整体风貌的保护和控制。整体保护意识的缺乏、旧城规划控制高度不断突破,成为旧城风貌不断遭到"建设性破坏"的重要原因。

2.2.2 疏解旧城人口措施不力,功能过于集中。

旧城区内人口密度过大,是导致旧城居住环境难以改善的根本原因。长期以来由于没有找到有效的办法疏解人口,随着旧城改造的加快,高层居住区的增多,使一些地区的人口密度反而增加了。旧城功能过于集中,房地产超强度开发,使旧城内各种矛盾更难解决。

2.2.3 旧城改造与保护的矛盾日益尖锐。

随着北京"十五"计划的实施,在未来5年内要对164片近700万m²的危旧房进行改造,这对历史文化名城的整体保护是一个重大挑战。许多单位采用房地产开发的方式进行危改,为寻求经济平衡,追求容积率,成片地拆除平房四合院和有一定历史价值的传统建筑,建设密集的高层建筑,对古城风貌造成严重损害。

2.2.4 历史文化保护区的保护、整治和更新难度很大,缺少政策和资金支持。

历史文化保护区的保护、整治、修缮和更新,涉及到房屋权属、人口迁移、保护资金等许多问题,实施难度相当大。

2.2.5 历史文化名城保护资金短缺。

历史文化名城的保护是一个系统工程,它包括了文物、历史文化保护区、旧城整体及传统文化等层面的保护内容,需要大量资金支持。虽然政府在名城保护方面的投入力度逐年加大,但在市场经济条件下,吸引全社会参与名城保护的机制尚未建立,这将严重制约名城保护规划的实施。

2.2.6 文物保护仍有一些重要问题有待解决。

第一,文物的不合理使用、损毁严重等现象仍大面广。

第二,普查在册登记文物的保护措施不明确。在第三次文物普查中,普查登记在册项目有2523处,占全市文物资源总数的72%,但是对于文物普查登记项目没有保护规定,极易遭到破坏。

第三,在社会主义市场经济条件下,文物法规体系和执法体制有待进一步健全。

第四,地下文物的保护管理工作亟待加强。

3. 北京历史文化名城保护规划的指导思想、原则及思路

3.1 北京历史文化名城保护规划的指导思想

第一,坚持北京是全国政治中心、文化中心和世界著名古都的城市性质;第二,正确处理历史文化名城保护与城市现代化建设的关系,使传统风貌保护与现代化城市建设相互协调,在保护中求发展,在发展中促保护;第三,重点搞好旧城保护,最大限度地保护北京历史文化名城;第四,在保护与发展中要贯彻"以人为本"的思想,使历史文化名城在保护中得以可持续发展。

3.2 北京历史文化名城保护规划的原则

《北京历史文化名城保护规划》应遵循如下规划原则:

3.2.1 整体保护与分层次控制,重点保护与一般保护相结合的原则。

北京是一个整体的、全局的概念,不是个体的、局部的概念。整体保护意识是名城保护的基础,分层次控制是整体保护意识实现的手段,二者结合,才能将历史文化名城的保护落到实处。重点保护的建筑和地段控制要严,一般保护的可稍宽。

3.2.2 名城保护与现代化建设相结合的原则。

历史文化名城保护要兼顾风貌保护与城市发展,既要保护古都传统风貌,又要建设现代化的文明首都。必须坚持可持续发展原则,采取有力措施,降低旧城中的人口密度,改善城市的市政、交通条件,控制建筑容量,优化城市环境。

3.2.3 传统风貌保护与传统文化继承相结合的原则。

名城传统风貌的保护不仅仅是文物、保护区、历史水系等"硬件"的保护,还包括优秀的传统文化,如庙会、戏曲、老字号等"软件"的继承,二者互为补充,共同构筑了城市的历史和文化精髓。

3.3 北京历史文化名城保护总体规划的基本思路

本次历史文化名城保护规划属于《北京城市总体规划》的专项规划,规划工作的基本思路是:正确领会中央精神,充分挖掘历史内涵,深入调查现状情况,认真总结经验教训,广泛听取社会意见,提出切实可行的规划方案。

《北京历史文化名城保护规划》的主要内容可归纳为:"三个层次,一个重点和传统文化的继承。"

"三个层次"是:文物保护单位的保护、历史文化保护区的保护、历史文化名城的保护;

"一个重点"是:旧城区;

"传统文化的继承"是:传统文化、商业特色的继承和发扬,传统文化、商业和历史建筑、街区、城市的结合。

根据这一基本思路,北京历史文化名城保护规划可从四个层面进行研究:

第一,文物保护单位的保护;第二,历史文化保护区的保护;第三,旧城整体格局的保护;第四,传统文化、商业的继承和发扬。

这四个层面共同构成北京历史文化名城总体保护规划的具体内容。

cially in regard to the conservation of the old city in a comprehensive way. This is shown in the conservation only of cultural relics and historical streets, but not the old urban fabrics as a whole. Therefore, this lack of awareness of comprehensive conservation has resulted in the damage of the old city by unlimited breach of building heights there.

2.2.2 Inadequate Measures for Relocating Inhabitants in the Old City Where Functions Are Too

Concentrated The population density of the old city is the main difficulty of improving the living environment there. As effective relocation measures have not been available for a long time, and owing to the speedy urban redevelopment and increase of high-rises, the population density in some areas has even increased. The conflicts in old city are made more difficult to tackle by the concentrated functions of the old city and the rampant development of real estate industry.

2.2.3 The Contradiction between the Renovation and Conservation of the Old City is Further Aggravated

The implementation of the 10th "five-year-plan" in Beijing means that in the next 5 years, crumbling and old houses in 164 districts with an area of 7 million square meters will be renovated, thus posing a big challenge to the conservation of the historic city of Beijing. Many organizations are undertaking this project by way of real estate development. In order to seek financial balances and higher FARs, vast stretches of traditional courtyard houses of certain historical value have been demolished to make room for concentrated high-rises, which resulted in severe damage to the fabrics of the old city.

2.2.4 It Is Difficult to Preserve and Renovate the Conservation Districts of Historic Sites Due to Lack of Policy and Fund

The Conservation, renovation and updating of the conservation districts of historic sites involve property rights, population relocation, conservation funding and many other problem, so it is quite difficult to implement.

2.2.5 Short of Funding for the Conservation of Historic City

Conservation of historic city is a systematic program involving the conservation of cultural relics, conservation districts of historic sites, the whole old city and traditional culture, etc. Therefore large amount of fund is needed.

Although government is increasing financial support in this field each year, the mechanism of involving the whole society in the conservation of the historic city has not been established due to the development of the market economy. So the conservation plan will be greatly hindered in its implementation.

2.2.6 Some Important Problems Need to be Solved in Cultural Relics Conservation

First, It is still prevalent to abuse and damage cultural relics.

Second, Lack of conservation measures for cultural relics registered in surveys. There are 2,523 items of cultural relics in the third general survey, accounting for 72% of the cultural relics of whole city, but the registered items are vulnerable to damage as no conservation measures are available for them.

Third, Cultural relics law and enforcement system need to be further perfected under the socialist market economy.

3. Guidelines, Principles and Key Ideas for the Conservation Plan of the Historic City of Beijing

3.1 Guidelines of the Plan

First, maintain the nature of Beijing as the political and cultural center, as well as a world famous ancient capital. Secondly, correctly address the relationship between the conservation of the historic city and the construction of a modern metropolis, making the conservation of the traditional feature in tune with the modernization of the city. The goal is to seek development out of conservation, and vice versa. Thirdly, priority is given to the conservation of the old city, and strive for the conservation of the historic city of Beijing. Lastly, promote the man-based concept in conservation and development, so that the historic city can be further developed through conservation.

3.2 Principles of the Conservation Plan for the Historic City of Beijing

The "Conservation Plan for the Historic City of Beijing" should adhere the following planing principles:

3.2.1 Principles of combining overall conservation and control at various levels and combination of key and general conservations.

Beijing should be viewed in a holistic rather than localized way. While the awareness of holistic conservation is the basis, the control at various levels is the means of realizing the overall conservation concept. Only the two are combined can the conservation measures of historic city be carried through. Strict control must be exercised on buildings and areas under key conservation, while control on those under general conservation can be a little relaxed.

3.2.2 Principle of combining historic city conservation and construction of modernization

Conservation of historic city must take into account both the conservation past features and the urban development, not only preserving the traditional characteristics of the old capital, but also striving to build a modern and civilized capital. The principle of sustainable development must be followed and effective ways adopted to reduce the density of population in the old city, improve the urban and traffic conditions, put architectural height under control and beautify the urban environment.

3.2.3 Principle of combining the conservation of traditional features and carrying on the traditional culture

The conservation of traditional features not only involve the preservation of cultural relics, conservation districts, ancient river systems and other such "hard wares", but also include carrying forward the "soft wares" of excellent traditional culture such as temple fairs, local operas, stores of long-standing, etc. The two together constitute the history and culture of the city in a supplementary way.

3.3 Key Ideas on the Master Plan for the Conservation of the Historic City of Beijing

This conservation plan of historic city is a specialized item of the "Master Plan of Beijing". Its fundamental concepts are: work out a feasible plan on the basis of correctly understanding the directives of the Central Party Committee, exploring historical connotations, summing up experience and lessons and soliciting opinions of people from all walks of life.

The main content of the "Conservation Plan for the Historic City of Beijing" can be broken down into the following: three levels, one priority and inheritance of traditional culture.

The "three levels" refer to the conservation of cultural relics places, conservation of conservation districts, and conservation of historic city.

"One priority" refers to the old city.

"Traditional culture" refers to the carrying forward of the traditional culture and commercial features, and the combination of traditional culture, commercial and historical buildings, street blocs and the city.

In line with this concept, the Plan can be studied via four aspects:

First, conservation of the cultural relics places; secondly, conservation of conservation districts of historic sites; thirdly, conservation of the overall layout of the old city; and fourthly, inheritance and carrying forward of traditional culture and commerce.

These four aspects constitute the specific contents of the Master Plan.

4. Conservation of the Cultural Relics Places

4.1 Basic Conditions of Cultural Conservation Work

The Party and government has always been attaching great importance to the cultural relics conservation campaign in Beijing. Beijing boasts 5 world cultural heritages: The Palace Museum, the Great Wall, Zhoukoudian, (site of the fossils of Peking Man), Summer Palace, and the Temple of Heaven; 60 key historic relic sites on the conservation list of the State; 234 historic relic sties under the conservation of Beijing municipal government; 501 cultural relic sites protected by counties and districts; 237 cultural relic sites temporarily under the county's and district's conservation, and 2,521 registered cultural relic sites in a general survey. According to the third general survey on cultural relics, the resources of cultural relics in Beijing totals 3,553, which covers a floor space of about 2 million square meters, being the first in the country.

4.2 Levels and Contents of the Cultural Relics Conservation

4.2.1 Levels of Cultural Relics Conservation

Three levels are differentiated based on different localities of cultural relics

4. 文物保护单位的保护

4.1 文物保护工作的基本状况

长城
The Great Wall

颐和园
The Summer Palace

宣仁庙
Xuanren Temple

北京的文物保护事业一直受到党和政府的极大关注。目前，北京拥有世界文化遗产5处：故宫、长城、周口店北京猿人遗址、颐和园、天坛。国家级重点文物保护单位60处；市级文物保护单位234处；区、县级文物保护单位501处；区、县级文物暂保单位237处；普查登记在册文物2521处。据第三次文物普查统计，北京文物资源总量为3553处，文物建筑约200万 m^2，居全国首位。

4.2 文物保护的层次和内容

4.2.1 文物保护的层次

文物保护单位依所处区域不同可划分为三个层次：旧城范围、规划市区范围、市域范围。

旧城范围：四至以二环路界定，占地62.5km^2。有世界文化遗产2处：故宫、天坛；国家级重点文保单位34处；市级文保单位134处；区级文保单位115处；区级文物暂保单位39处；普查在册文物478处，总计800处。

规划市区范围：四至是东以定福庄，南以南苑，西以石景山，北以清河界定，占地1040km^2。除旧城范围以外，有世界文化遗产1处：颐和园；有国家级重点文保单位16处；市级文保单位49处；区级文保单位59处；区级文物暂保单位181处；普查在册文物310处，总计615处。

市域范围：辖十个远郊区县，总面积达16800km^2。除规划市区及旧城范围外，有世界文化遗产2处：长城、周口店北京猿人遗址；有国家级重点文保单位10处；市级文保单位51处；区、县级文保单位327处；区、县级文物暂保单位17处；普查在册文物1733处，总计2138处。在市级以上文物保护单位中，集中了八达岭、十三陵、居庸关、红螺寺、慕田峪长城、司马台长城、云居寺等文物景点。

全市范围内，各级文物保护单位总计3553处。

4.2.2 文物保护单位的分类

文物保护单位依内容不同可划分为十类：

古人类活动遗址：包括周口店北京猿人遗址、东胡林人遗址等。

万里长城在北京辖区内的遗存：包括八达岭、慕田峪、司马台长城等。

城市遗址：包括董家林商周遗址、金中都遗址等。

宫殿建筑群：以故宫为主。

皇家园林及风景名胜：包括颐和园、北海等。

重要的坛、庙、寺、塔：包括天坛、太庙、潭柘寺、妙应寺白塔等。

帝王陵寝：包括明十三陵、金陵等。

革命纪念物：包括天安门、北大红楼等。

古今名人故居：包括毛泽东、宋庆龄、鲁迅故居等。

四合院民居、戏楼、会馆等民间建筑。

4.3 文物保护规划

根据对文物保护单位"保护为主，抢救第一"的方针和"有效保护，合理利用，加强管理"的原则，今后一段时期文物保护规划的任务相当繁重。

4.3.1 新的文保单位名单的申报与公布

北京尚有十三陵、北海、云居寺、古观象台、卢沟桥在我国准备申报世界文化遗产的预备清单内。争取十三陵于2003年申报世界文化遗产成功。

继续公布市级文物保护单位名单。近年要公布第七批市级文物保护单位，使市级以上文物保护单位达到300处。各区县要分别公布一至两批区县级文物保护单位及区县级文物暂保单位。同时，完成第六批全国重点文物保护单位（北京部分）的论证及申报工作。

继续划定市级以上文物保护单位的保护范围和建设控制地带。近年完成第五批、第六批市级文物保护单位保护范围和建设控制地带的划定工作；逐步完善区县级文物保护单位保护范围和建设控制地带的划定工作。

4.3.2 文保单位的修缮与保护

对各级文物保护单位的保护，要单体建筑修缮与环境整治和改善相结合，景点保护与街、区成片保护相结合。

政府"3.3亿工程"的重点转向：整治"两线"景观、恢复"五区"风貌、重现京郊"六景"，营造北京历史文化名城的基本格局。

（1）整治"两线"景观

"两线"即中轴线、朝阜路沿线。它集中反映了明清北京城关于帝王之都的建设理念，恢复"两线"风貌是展示北京古都景观的核心。

（2）恢复"五区"风貌

"五区"指什刹海区、国子监区、琉璃厂区、皇城区、古城垣区。这些景区历史文化内涵丰富，在海内外知名度高，保护工作基础条件好，开展特色旅游潜力大。

（3）重现京郊"六景"

包括西郊风景名胜区、长城北京段景区、帝王陵寝区、京东运河文化带、宛平史迹区、京西寺庙区。

4.3.3 保护重要的近现代建筑

北京是一个有着光荣革命传统的历史文化名城，拥有一批历史、科学、艺术价值极高的近现代建筑，由于种种原因，未能列入文物保护单位。建议提出一批具有较高历史文化价值的近现代建筑名单，划定保护范围和建设控制地带，近期积极保护，远期列入文物保护单位。如

云居寺全景
The panorama of Yunju Temple

十三陵
The Ming Tombs

places: within the old city; within the planned urban area; and within the city suburbs.

Within the old city: the four boundaries are demarcated by the Second Ring Road with an area of 62.5 square kilometers. There are 2 world cultural heritages: the Forbidden City and Temple of Heaven; 34 key sites under the State conservation; 134 sites under the conservation of the municipal government; 115 sites under the conservation of districts; 39 sites temporarily protected by districts, and 478 registered sites in general survey, totaling 800.

Within the planned urban area: the four boundaries are defined by Dingfuzhuang to the east, Nanyuan to the south, Shijingshan to the west and Qinghe to the north, with an area of 1,040 square kilometers. There are 1 world cultural heritage: the Summer Palace; 16 key sites on the State conservation list; 49 sites under the conservation of the municipal government; 59 sites protected at the district-level; 181 sites under the temporary conservation of districts; 310 registered sites in general survey, totaling 615.

Within the city suburbs: excluding the planned area and old city, the area embraces 10 outlying counties with an area of 16,800 square kilometers There are 2 world cultural heritages: the Great Wall and Zhoukoudian, (site of the fossils of Peking Man); 10 key sites under the State conservation; 51 sites under the city-level conservation; 327 sites protected by counties and districts; 17 sites temporarily protected by districts and counties, and 1,733 registered sites in general investigation, totaling 2,138. Among the cultural relics under the city or State conservations are Badaling Great Wall, Juyong Pass, Hongluo Temple, Mutianyu Great Wall, Simatai Great Wall, Yunju Temple, etc.

There are 3,553 cultural relics places of various levels in the whole city.

4.2.2 Classification of Cultural Relics Places

Cultural relics units can be classified into 10 categories according to their different contents:

Ruins of Anthropological Values: including Paking Man Ruins at Zhoukoudian, Donghulin Man Ruins, etc.

Remnants of the Great Wall in Beijing Area: including Badaling, Mutainyu, Simatai, etc.

Urban Ruins: including Ruins of the Shang and Zhou Dynasties at Dongjialin, Jin capital Zhongdu, etc.

Palace complex: primarily the Forbidden City

Royal gardens and places of interest: including the Summer Palace, Beihai Park, etc.

Important temples and pagodas: including the Temple of Heaven, Imperial Ancestral Temple, Tanzhe Temple, the White Pagoda of Miaoying Temple, etc.

Imperial Tombs: including the Ming Tombs, Jin Tombs, etc.

Revolutionary monuments: including the Tian An Men, the Red Building of Peking University, etc.

Residences of ancient and modern celebrities: including residences of Mao Zedong, Song Qingling, Lu Xun, etc.

Civilian buildings such as courtyard housing, theaters, guilds, etc.

4.3 Conservation Plan of Cultural Relics

According to the policy of "give priority to conservation with salvage first"and the principle of "effective conservation, rational utilization and strict management", pursued by the cultural relics units, the task of protecting cultural relics in the future will be very arduous.

4.3.1 Submission and Publication of the New List of Cultural Relics Places under Conservation

The Ming Tombs, Beihai Park, Yunju Temple, Ancient Observatory and Marco Polo Bridge in Beijing are shortlisted for application for the world heritage, with the hope that the Ming Tombs will succeed in being approved as one of the world heritages in 2003.

Continue making public the list of cultural relics sites under the conservation of the municipal government with a view of having 300 such sites on the list. The seventh group of cultural relics sites under the conservation of the city will be announced in the near future. Various districts and counties should publicize one or two groups of historic sites protected by districts and counties and those protected by them on a temporary basis. In the meantime, complete in the near future the evaluation and application of the sixth group of key historical sites (those in Beijing area) under the conservation of the State.

Continue designating the conservation scope and construction control areas of the cultural relics sites under the conservation of the city and State. Complete the conservation scope and construction control area work for the fifth and sixth groups of historical sites under the conservation of the municipal government recently. Gradually perfect the conservation scope and construction control area work for the historical sites under the conservation of districts and counties.

4.3.2 Renovation and Conservation of Cultural Relics Places

The renovation and conservation of historical sites of various levels must integrate the renovation of individual buildings with the improvement of their surroundings, and integrate the preservation of scenic spots with the conservation of streets, districts and urban patches.

Shift of priority for the government's "330 million project": Create a basic layout of Beijing by renovating the scenic spots along the "two lines", restore the fabrics of the "five areas"and update the "six scenes"in the suburbs of Beijing.

(1) Renovate the scenic spots of the "two lines"

The "Two lines"refer to the Axis and Chao-Fu Road, which fully demonstrate the architectural concept of the imperial capital of the Ming and Qing Dynasties. The key to exhibiting the sights of old Beijing is to represent the features of the two lines.

(2) Restore the fabrics of the "five areas"

The "five areas"refer to Shishahai, Imperial College area, Liulichang (cultural street) area, imperial city area, and old city walls areas. These areas have rich historical and cultural relics, well-known to the world, have already done a spade work for conservation, and a good potential of developing tourism.

(3) Update the "six scenes"in the suburbs of Beijing

Include places of interest of the western suburbs, Great Wall at Beijing area, imperial burial area, canal cultural areas east of Beijing, historical monument area of Wanping, and temple area west of Beijing.

4.3.3 Conservation of Key Modern and Contemporary Buildings

Beijing is a historic city of glorious revolutionary tradition, with a large number of modern and contemporary architecture of high historical, scientific and artistic values. For various reasons, they have not been listed as cultural relics under conservation. It is suggested that a group of such architecture be designated and their conservation scope and construction control area defined, so that they can be protected and placed on the list of cultural heritage in the future. Examples are: The Great Hall of the People, Beijing Exhibition Hall, Beijing Agricultural Exhibition Hall and Beijing Railway Station.

4.3.4 Archeological Excavation

Publicize the burial locations of the fourth group of underground historical relics,

国子监
Imperial College

北京火车站
Beijing Railway Station

人民大会堂、北京展览馆、北京农展馆、北京火车站等。

4.3.4 考古勘探与发掘

公布第四批地下文物埋藏区，使全市地下文物埋藏区达到50处。

重视并加强城市考古工作，特别注意对辽南京、金中都、元大都城市遗址、遗迹的考古调查、勘探和发掘。

5. 历史文化保护区的保护

历史文化保护区是具有某一历史时期的传统风貌、民族地方特色的街区、建筑群、小镇、村寨等，是反映历史文化名城风貌比较集中、真实、完整的地区。

5.1 旧城25片历史文化保护区的保护规划

5.1.1 旧城第一批25片历史文化保护区的分布

市政府已于1999年正式批准了《北京旧城历史文化保护区保护和控制范围规划》，第一批25片历史文化保护区总占地面积为1038hm²，约占旧城总用地的17%。其中重点保护区占地面积649hm²，建设控制区占地面积389hm²。

北京旧城25片历史文化保护区中有14片分布在旧皇城区内：南、北长街、西华门大街、南、北池子、东华门大街、景山东、西、后、前街、地安门内大街、文津街、五四大街、陟山门街。另有7片分布在旧皇城外的内城：西四北头条至八条、东四三条至八条、南锣鼓巷地区、什刹海地区、国子监地区、阜城门内大街、东交民巷。还有4片分布在外城：大栅栏、鲜鱼口地区、东、西琉璃厂。

5.1.2 旧城25片历史文化保护区保护规划

《北京旧城25片历史文化保护区保护规划》的编制工作已于2001年初完成，2002年2月由市政府批准执行。

该项规划强调以"院落"为基本保护与更新单位，一个院落一个院落地逐步进行保护、修缮、改造和更新。对历史文化保护区内的建筑保护和更新分为六类进行规划管理：文物类建筑、保护类建筑、改善类建筑、保留类建筑、更新类建筑、沿街整饰类建筑。同时，对保护区内的用地性质变更、人口疏散、道路调整、市政设施改善、环境绿化保护等方面提出了具体的原则、对策和措施。

25片历史文化保护区保护规划对促进旧城的整体保护起到了十分重要的作用。

5.2 新增第二批历史文化保护区名单

划定第二批历史文化保护区是此次《北京历史文化名城保护规划》的重要内容之一。

此次工作在北京市域范围内开展。一方面在旧城内继续补充少量历史风貌较完整、历史遗存较集中和对旧城整体保护有重大影响的片区进行保护；另一方面在市域范围内确定一批文物古迹比较集中，能较完整地体现一定历史时期传统风貌和地方特色的街区或村镇，使其得到有效保护。

5.2.1 第二批历史文化保护区名单及分布

第二批历史文化保护区分为旧城内和旧城外两部分。

旧城内增加5片历史文化保护区，包括：皇城、北锣鼓巷、张自忠路北、张自忠路南、法源寺。

旧城外提出10片历史文化保护区，包括：海淀区西郊清代皇家园林、丰台区卢沟桥宛平城、石景山区模式口、门头沟区三家店村、川底下村、延庆县岔道城、榆林堡、密云区古北口老城、遥桥峪和小口城堡、顺义区焦庄户。

5.2.2 第二批历史文化保护区特色

(1) 旧城内新增的历史文化保护区特色

• 皇城历史文化保护区

本次规划将皇城作为历史文化保护区进行整体保护，其中包括内含紫禁城、太庙、社稷坛、北海、中南海以及第一批历史文化保护区的南北长街、南北池子、景山八片等14片历史文化保护区，面积约6.8km²。皇城是北京旧城整体保护的核心内容。

• 北锣鼓巷历史文化保护区

位于东城区，南至鼓楼东大街，北至车辇店、净土胡同，东至安定门内大街，西至赵府街，总面积约为46hm²。该地区西接钟鼓楼及什刹海历史街区，南邻南锣鼓巷历史街区，东连国子监历史街区，是钟鼓楼的重要背景和展示旧城风貌的重要边缘地带，是保护中轴线对称格局不可缺少的地段。

• 张自忠路北历史文化保护区

位于东城区，南至张自忠路，北至香饵胡同，东至东四北大街，西至交道口南大街，总面积约为42hm²。该街区有和敬公主府、段祺瑞执政府旧址、欧阳予倩故居等市级文物保护单位。

• 张自忠路南历史文化保护区

位于东城区，南至钱粮胡同，北至张自忠路，东至东四北大街，西至美术馆后街，总面积约为42hm²。该区域处于皇城与东四三条至八条保护区之间，现有胡同格局完整，风貌保存较好，有马辉堂花园等文物保护单位。

• 法源寺历史文化保护区

位于宣武区，南至南横西街，北至法源寺后街，东至菜市口南大街，西至教子胡同，总面积约20hm²。该街区内有法源寺、湖南会馆、绍兴会馆等文物保护单位，街区整体风貌保存完整。

旧城新增第二批历史文化保护区占地面积约249hm²，占旧城面积4%；加上第一批旧城25片历史文化保护区共占地约1287hm²，占旧城总面积的21%。

经过重新调整后，旧城内的北京历史文化保护区分别是：南长街、北长街、西华门大街、南池子、北池子、东华门大街、景山东街、景

in order to make the number of locations in Beijing reaching 50.

Attach importance to and enhance archeological work in the city, especially archeological and excavating work of the ruins and relics of Nanjing of the Liao Dynasty, Zhongdu of the Jin Dynasty and Dadu of the Yuan Dynasty.

5. Conservation of Conservation districts

Conservation districts refer to the streets, architectural complex, small towns and villages with traditional features and national or local characteristics of a certain historical period. They are a concentrated, authentic and complete area to reflect the features of a historic city.

5.1 Conservation Plan for the 25 Conservation districts in the Old City of Beijing

5.1.1 Distribution of the First Group of 25 Conservation districts in the Old City

The "Plan of Conservation and Control Scope for the Conservation districts in the Old City of Beijing" has been approved by the municipal government in 1999. The total area of the first group of 25 conservation districts is 1,083 hectares, accounting for about 17% of total land of the old city, out which key protected area is 649 hectares, and construction control area is 389 hectares.

Out of the 25 districts, 14 are located within the old imperial city: Nan and Bei Changjie Streets, Xihuamen Street, Nanchizi and Beichizi Streets, Donghuameng Street, Jingshandong Street, Jingshanxi Street, Jingshanhou Street and Jingshanqian Street, Di'anmen Street, Wenjin Street, Wusi Street, and Zhishanmen Street; 7 districts are scattered outside the royal city but within the Outer City Watll: the First to Eighth Lanes of Xisi Street, The Third to Eighth Lanes of Dongsi Street, Nanluoguxiang area, Shishahai area, the Imperial Collage area, Fuchengmennei Street, Dongjiaominxiang; another 4 districts are located in the Outer City: Dashilan, Dongliulichang and Xiliulichang, Xianyukou area.

5.1.2 Conservation Plan for the 25 Conservation districts in the Old City

The formulation of the "Conservation Plan for the 25 Conservation districts in the Old City of Beijing" was completed at the beginning of 2001, and approved by the municipal government on February 2002.

The "Plan for the 25 Conservation districts in the Old City of Beijing" stresses that conservation and renovation must be done with "courtyard unit" as a basis. The Plan divides the buildings in the conservation districts into 6 categories for management and conservation: buildings of historic relics, protected buildings, buildings needing improvement, buildings that should be retained, buildings needing renovation and buildings needing polishing. Meanwhile, detailed principles and measures were also stipulated in the Plan with regard to the nature of land use, population relocation, road adjustment, improvement of civic facilities and environmental conservation within the conservation districts.

5.2 The Added List of the Second Group of Conservation districts

Designating the second group of conservation districts is one of the important contents in the Plan.

The work was done in citywide area. On the one hand, in the old city, continue to identify a mall number of districts that have fairly complete historical features, concentrated historical sites and big impact on the conservation of the old city. On the other hand, try to designate citywide a number of streets or towns with cluster of cultural relics for conservation. These places should reflect in a comprehensive way the traditional features and local flavor of certain historical periods so that they can be protected effectively.

5.2.1 List and Distribution of the Second Group of Conservation districts

The districts of the second group are divided into two parts, one within the old city, the other outside.

The 5 added conservation districts of the second group within the old city are: imperial city, Beiluoguxiang, Area North of Zhangzizhong Road, area South of Zhangzizhong Road and Fuyuan Temple.

The 10 conservation districts outside of the old city are: Qing Dynasty Imperial Garden in the suburbs of Haidian District, Wanping City in Luogou Bridge in Fengtai District, Moshiko in Shijingshan District, Sanjiadian in Mentougou District, Chuandixia Village, Chadao city in Yanqing County, Yulinbao, Yaoqiaoyu and Xiaokou Castle, and Jiaohu Village in Shunyi County.

5.2.2 The Characteristics of the Second Group of Conservation districts

(1) The Characteristics of the newly added conservation districts in the old city

● The conservation district of the imperial city

According to the plan, the imperial city as a whole is one conservation district. It includes Forbidden City, the Imperial Ancestral Temple, the Altars to the Gods of Earth and Grain, Beihai Park, Zhongnanhai, and the first group of 14 conservation districts such as Nanchangjie and Beichangjie, Nanchizi and Beichizi Streets and Coal Hill Park with an area of 6.8 square kilometers. The Imperial City is the key content in the comprehensive conservation of old Beijing.

● Beiluoguxiang conservation district

Located in the Dongcheng District with an area of 46 hectares, it stretches to the East Gulou Street in the south, Cheniandian and Jingtu Hutong in the north, Andingmennei Street in the east, and Zhaofu Street in the west. An important backdrop of the Imperial City, it borders the historical streets of the Drum and Bell Tower and Shishahai to the west, Nanluoguxiang to south and the Imperial Collage to the east, thus an indispensable section in protecting the overall features of the old city and in forming a symmetrical layout along the axis.

● Conservation district north of Zhangzizhong Road

Located in Dongcheng District with an area of 42 hectares, it stretches to the Zhangzizhong Road in the south, Xiang'er Lane in the north, Dongsibeijie Street in the east, and Nanjiaodaokou Street in the west. The place boasts many cultural relics under the city- and district-level governments, such as the Mansion of Princess Hejing, Site of Duanqirui Government, and the former residence of Ouyan Yuqian.

● Conservation district south of Zhangzizhong Road

Located in Dongcheng District with an area of 42 hectares, it stretches to Qianliang Lane in the south, Zhangzizhong Road in the north, Dongsibeijie Street in the east and Meishuguanhou Street in the west. Sandwiched between the Impe-

鸟瞰钟鼓楼
Bird's eye view of the Bell and Drum Towers

法源寺
Fayuan Temple

礼士胡同
Lishi Hutong

山西街、景山前街、景山后街、陟山门街、地安门内大街、文津街、五四大街、阜城门内大街、西四北头条至八条、什刹海地区、南锣鼓巷地区、国子监地区、东四北三条至八条、东交民巷、大栅栏、东琉璃厂、西琉璃厂、鲜鱼口、皇城、北锣鼓巷地区、张自忠路北、张自忠路南、法源寺，总计30片。

5.2.3 旧城外新增历史文化保护区特色

(1) 西郊清代皇家园林历史文化保护区特色

西郊清代皇家园林位于海淀区，包括：颐和园、圆明园、香山静宜园、玉泉山静明园等，是我国现存皇家园林的精华。

(2) 卢沟桥宛平城历史文化保护区特色

卢沟桥、宛平城位于丰台区，是国家和市级文保单位，也是震惊中外的"卢沟桥事变"发生地，具有重要的历史和革命纪念意义。

(3) 模式口历史文化保护区特色

模式口位于石景山区西北部，金顶山路与门京公路之间，古为京西古道。现在模式口大街以北，传统村落的风貌保存较好，并有承恩寺、田义墓、冰川擦痕、法海寺等文物。

(4) 三家店历史文化保护区特色

位于门头沟区永定河北岸，东与石景山区相邻。村中现存多处文物，与煤业发展有关的建筑群、会馆等成为此地独特的景观，带有典型的地方特色。

(5) 川底下历史文化保护区特色

川底下村是门头沟区斋堂镇的一个自然村。村落四面环山位于阳坡，建筑依山而建。村中现保留着70个院落约300间明清时的四合院民居，其建筑艺术精湛，保存完好。

(6) 榆林堡历史文化保护区特色

榆林堡位于京郊延庆县康庄镇西南，元、明、清时期是京北交通线上的重要驿站之一。榆林堡占地约16hm²，平面呈"凸"字形，距今约600余年。

(7) 岔道城历史文化保护区特色

位于延庆县八达岭镇，城为不规则长方形，占地面积约8.2hm²。该城是北京通往西北的重要军事据点和驿站，其紧邻八达岭长城，至今城墙残破，城门尚在。

(8) 古北口老城历史文化保护区特色

古北口老城位于密云区古北口镇的东北部，地处卧虎、盘龙两山间的峡谷东侧，北依长城，西临潮河，东靠山脚，为自古以来兵家必争之地。现存药王庙戏楼、财神庙、古关址等文物和南北大街，风貌较完整。

(9) 遥桥峪城堡、小口城堡历史文化保护区特色

遥桥峪位于密云区新城子乡东部，占地约1.6hm²。遥桥峪村及其古城堡建于明万历二十六年（公元1599年），此堡呈方形，南面正中一座城门，至今保存完好。

小口城堡位于密云区新城子乡北部，距遥桥峪约4km，村落占地约1.7hm²，是明代戍边营城。城墙"北圆南方"北依小山而筑，呈弧形，内有松柏，南面城墙为方形，内为民居村落，现城墙基本保存完好。

(10) 焦庄户历史文化保护区特色

焦庄户地道战遗址位于顺义区龙湾屯镇焦庄户村。1943年，当地党组织和群众利用地道和日寇周旋作战，创造了抗战时期闻名的"地道战"，被誉为"人民第一堡垒"。

在第二批历史文化保护区名单批准后，应组织编制各保护区的保护规划。

北京第一批、第二批历史文化保护区合计后，总共有40片。其中，旧城内有30片历史文化保护区。

旧城第一批历史文化保护区和文物保护单位保护及控制范围的总面积为2383hm²，约占旧城总面积的38%。旧城第一批、第二批历史文化保护区和文物保护单位保护及控制范围的总面积为2617hm²，约占旧城总面积的42%。

6. 旧城整体格局的保护

除了文物保护单位和历史文化保护区的保护外，要从整体上考虑北京旧城的保护，具体体现在如下10个层面的保护规划。

6.1 历史河湖水系的保护

水是城市生存的命脉。北京在3000多年的发展中，为了保证防洪、灌溉、漕运的需要，历代王朝建造的众多的水利工程，奠定了北京市区河湖水系的基本格局，反映了北京城市发展的历史。

目前，在北京规划市区1040km²范围内，有通惠河、凉水河、清河、坝河四条主要排水河道及其支流，现有河道总长226km，总水面面积460hm²；现状湖泊23个，水面面积700hm²。

6.1.1 规划目标

重点保护与北京城市历史沿革密切相关的河湖水系，部分恢复具有重要历史价值的河湖水面，使市区河湖形成一个完整的系统。

6.1.2 市区河湖水系保护规划

(1) 现有河湖水系的保护规划

北京的河湖水系曾在漕运、防洪、输水、防御外敌入侵等方面发挥了重要作用，记录了北京城市发展和变迁的历史。与北京城市发展密切相关，在各个历史时期发挥过重要作用的河湖水域应列为重点保护目标，划定保护范围并加以整治。

圆明园
Yuanmingyuan

宛平城
Wanping City

模式口田义墓
Tianyi Tomb in Moshikou

三家店龙王庙
Temple of the Dragon King in Sanjiadian

川底下
Chuandixia

榆林堡南北街道
The north and south streets of Yulinbao

岔道城城墙
City wall of Chadao City

古北口药王戏楼
The theater of the medicine King in Gubeikou

rial City and the First and Eighth Lanes of Dongsi, it has a complete hutong layout, and cultural relics such as Mahuitang Garden.

• Conservation district of Fayuan Temple

Located in Xuanwu District with an area of 20 hectares, it stretches to the Nanhengxi Street in the south, Fayuansihou Street in the north, South Caishikou Street in the east, and Jiaozi Lane in the west. It has Fayuan Temple, Hunan and Shaoxing Guild Halls and other historic sites with a fairly preserved street layout.

The newly-added second group of conservation districts in the old city covers an area of 249 hectares, accounting for 4% of the total area of the old city. Added the first group of 25 conservation districts, the total area is 1,287 hectares, accounting for 21% of the total area of the old city.

After readjustment, the conservation districts in the old city are: Beichangjie Street, Nanchangjie Street, Xihuamen Street, Nanchizi Street, Beichizi Street, Donghuamen Street, Jingshandong Street, Jingshanxi Street, Jingshanhou Street, Zhishanmen Street, Di'anmennei Street, Wenjin Street, Wusi Street, Fuchengmennei Street, The North 1st to 8th Alleys of Xisi, Shichahai Area, Nanluoguxiang Area, Imperial Collage Area, the North 3rd to 8th Alleys of Dongsi, Dongjiaominxiang Street, Dashilan Street, Dongliulichang Street, Xiliulichang Street, Xianyukou, Imperial City, Beiluoguxiang Area, North of Zhangzizhong Street, South of Zhangzizhong Street, and Fayuan Temple, altogether 30 sites.

5.2.3 Characteristics of newly added conservation districts outside of the Old City

(1) Characteristics of the conservation district of the Qing Dynasty Imperial Garden

Located in Haidian District, it includes the Summer Palace, Yuanmingyan, Jingyi Garden of the Fragrant Hill, Jingming Garden of Yuqun Mountain, etc. It is the cream of the extant imperial garden.

(2) Characteristics of conservation district of Marco Polo Bridge and Wenping City

Located in Fengtai District, it has the Marco Polo Bridge (Luogou Bridge) and Wenping City which are relics under the State and municipal conservation respectively. They are also the sites of the "Lugouqiao Incident" which shocked China and the world, thus having important historical and revolutionary significance.

(3) Characteristics of conservation district of Moshikou

Located northwest of Shijingshan District, this ancient route west of the Beijing is situated between Jindingshan Road and Jingmen Highway. The traditional villages north of Moshikou Street are well preserved, boasting cultural relics like Cheng'en Temple, Tianyi tomb, marks left by glaciers and the Fahai Temple.

(4) Characteristics of conservation district of Sanjiadian

Located at the northern bank of Yongding River of Mentoukou District, this village has many architecture and guild halls related to the coal industry, reflecting the typical features of locatlity.

(5) Characteristics of conservation district of Chuandixia Village

A natural village in Zhaitang Town, Mentougou District, it was nestled at the foot of the mountain and faces south amid the hills. It has 300 courtyard-type residences of the Ming and Qing Dynasties in 70 couryards. The architectural art is highly skilled and the features completely preserved.

(6) Characteristics of conservation district of Yulinbao

Located southwest of Kangzhuang town, Yanqing County, it was one of the important courier stations on the traffic line north of Beijing during the Yuan, Ming and Qing Dynasties. Covering an area of 16 hectares, It is in the form of an inverted "T" in its plane and has a history of 600 years.

(7) Characteristics of conservation district of Chadao City

Located at Badaling Town, it is irregular in shape and covers an area of 8.2 hectares. It used to be a crucial military stronghold and courier station on the way from Beijing to northwest. Close to the Badaling Great Wall, it still retains the crumbling city walls and gates.

(8) Characteristics of conservation district of Gubeikou Old City

Located northeast of Gubeikou Town, Miyun District, it is perched east of the valley between Wohu and Panlong Mountains. With the Great Wall to the north, Chao River to the west and the foot of mountain to east, it has been the place of contention since ancient times, boasting now the Theater of the Temple of the King of Medicine, Temple of the God of Wealth, Ancient Pass Site, North and South Streets and other relics. The features there are comparatively well preserved.

(9) Characteristics of conservation district of Yaoqiaoyu and Xiaokou Castles

Located east of Xinchengzi Village, Minyun District, Yaoqiaoyu Village and its Castle were build in 1599 under the rein of emperor Wanli of the Ming Dynasty, covering an area of 1.6 hectares. In the form of a square, it has a city wall in the middle of its southern side, which is well preserved

Xiaokou Castle is located north of Xinxiangzi, 4 kilometers from anther castle, and covers an area of 1.7 hectares. A frontier garrison in the Ming Dynasty, it is round in the north and square in the south. Situated at the foot of a hill in north, its northern part is in the shape of an arch, containing pines and cypresses. Its southern wall is square, sheltering residential houses. The walls are well preserved.

(10) Characteristics of conservation district of Jiaozhuanghu

This tunnel warfare site is located at Jiaozhuanghu, Longwantunzhen, Shunyi District. In 1943, local Party led the masses engage the Japanese soldiers in tunnels, hence the famous names "Tunnel Warfare" and the "No. One People's Stronghold" of that period.

After the list of second group of conservation districts is approved, conservation plans should be worked out for each conservation district.

The number of first and second groups of conservation districts in Beijing reaches 40, out which 30 are in the old city.

The total area of the first group of conservation districts and the extent of conservation and control areas of cultural relics sites in the old city is 2,383 hectares, accounting for 38% of the total area of the old city. The total area of the first and second groups of conservation districts and the scope of conservation and construction control areas of the cultural relics under conservation is 2,617 hectares, accounting for 42% of the total area of the old city.

6. Conservation of the General Layout of the Old City

Apart from the conservation of cultural relics sites and conservation districts, the conservation of the old city of Beijing should be considered in a comprehensive way, involving the following 10 aspects:

6.1 Conservation of the Historical Water System

Water is the lifeblood of city. In the 3,000 years of development of Beijing, various dynasties constructed many irrigation works for the purpose of flood-control, irrigation and transporting grains to the capital, setting up a basic river system layout in Beijing which reflects the urban evolvement of the city.

Presently, in the 1,040 square meter planned areas of Beijing are Tonghui,

早春什刹海
Shichahai in early spring

御码头
The Imperial Harbor

长河
Changhe River

莲花池
Lianhuachi

- 护城河水系

重点保护河道为北护城河、南护城河、北土城沟和筒子河。

- 古代水源河道

重点保护河道为莲花河和长河，以及莲花池、玉渊潭。

- 古代漕运河道

重点保护河道为通惠河、坝河和北运河。

- 古代防洪河道

重点保护河道为永定河和南旱河。

- 风景园林水域

重点保护湖泊水域为六海、昆明湖、圆明园水系。

- 重点保护的水工建筑物

重点保护的水工建筑物为后门桥、广济桥、卢沟桥、朝宗桥、白浮泉遗址、琉璃河大桥、广源闸、八里桥、麦钟桥、银锭桥、金门闸、庆丰闸、北海大桥等。

（2）恢复河道的规划

北京城市河湖水系在中华人民共和国成立前后，由于种种原因，先后将御河、菖蒲河、东西护城河、前三门护城河等改为暗沟或填平，造成了北京城市水系的严重破坏。恢复北京城的河湖水系，对再现北京城的历史风貌、改善城市生态环境具有重要的意义。

恢复河湖水系应本着因地制宜、分期实施的原则，将水环境设计与传统风貌相统一。

规划将转河、菖蒲河、御河（什刹海——平安大街段）予以恢复。

- 转河

转河属于通惠河水系，恢复转河可将长河与北护城河连接起来。

- 菖蒲河

菖蒲河于20世纪70～80年代改为暗沟，是故宫水系的一部分，与内护城河水系、六海水系、外城护城河水系相连通。恢复菖蒲河进一步体现了历史的景观。

- 御河（什刹海——平安大街段）

御河始于元代，北起后门桥，南至前三门暗沟，20世纪50年代改为暗沟。

规划将御河上段（什刹海——平安大街）予以恢复，远期考虑结合皇城根遗址公园将御河下段（北河沿——前三门大街）纳入恢复范围。

（3）恢复湖泊的保护规划

- 鱼藻池

鱼藻池是金中都的太液池，按原貌恢复鱼藻池，是北京最早成为都城的历史见证。

- 莲花池

莲花池是金中都最早开发利用的水源地，已经过整治建成了公园，但西南角水面仍未按原状恢复。应将其复原，新辟水面面积约4.9hm^2。

（4）控制前三门护城河规划用地内的新建项目

前三门护城河是明代开挖的河道，也是内城排水的总出路，它贯穿旧城东西，从前门通过，无论从城市景观，还是生态来看，其作用极其重要。20世纪60～70年代因地铁一期工程改为暗沟。20世纪70年代中后期，现状沿河一带已建有许多高层建筑，给恢复前三门护城河带来相当难度。因拆除建筑数量太多，实施难度太大，本次规划暂不考虑恢复河道，但要控制新建项目。控制范围为西起南护城河，东至东护城河，前三门大街道路红线以南70m（包括河道及相应的绿化带）。

6.2 城市中轴线的保护和发展

北京传统中轴线最初形成于元代，明清北京城"凸"字形平面布局形成之后，从永定门始，向北穿过紫禁城和景山中峰，最后止于鼓楼与钟楼，传统中轴线成形。因市区扩大，规划把传统中轴线向南北延伸，形成了北中轴线和南中轴线，规划总长度约25km。

6.2.1 传统中轴线的保护规划

北京传统中轴线从永定门到钟鼓楼为7.8km，到北二环路为8.5km。其保护规划应遵循以保护为主，保护与发展、继承和创造相结合的原则，重点研究景山——前门、永定门、钟鼓楼三大节点的保护与规划。

钟鼓楼节点：钟鼓楼在古代具有报时功能，古代的鼓楼是君主发表重要文告或行为的象征，钟楼是一种统一秩序的标志。作为传统轴线的终点，钟鼓楼一直拥有标志性建筑的地位，其周边环境以四合院民居为主，陪衬钟鼓楼的宏伟。保护钟鼓楼尤其要保护其环境，不宜建造大尺度的现代建筑，破坏钟鼓楼传统的空间感受。周边建筑高度控制应按历史文化保护区保护的要求进行。

景山——前门节点：景山——前门是北京最具历史价值的空间，它由景山、故宫、天安门、正阳门城楼和箭楼等组成，空间层次丰富，秩序严谨，起伏有致，应严格加以保护。

在条件成熟时，恢复前三门护城河水系将使天安门广场的景观和形态更加完整、壮美。

永定门节点：永定门由于历史原因被拆除，导致传统中轴线的南部起点的消亡。恢复永定门城楼，对实现传统中轴线的完整性与庄严性、有效衔接南中轴线意义重大。应严格控制永定门城楼周边环境的建筑高度。

6.2.2 北中轴线的保护发展规划

北中轴线从北二环路到奥林匹克公园长约6.7km，其保护发展规划应抓住三大节点。

奥林匹克公园中心区节点：是北中轴线的端点，应重点规划，形成北京城市的新标志。端点以北地区为森林公园，作为北中轴线的背景。

北土城节点：奥林匹克公园的序幕与衔接。可结合北土城遗址与北中轴80m宽道路中央绿化带，创造具有一定意义的城市空间，强化和丰富北中轴线。

Liangshi, Qing and Ba Rivers, all of which and their tributaries are drainage river courses, with a total length of 226 kilometers. Their total water areas are 460 hectares. The number of rivers is 23 with water areas of 700 hectares.

6.1.1 Objectives of the Plan

Primarily protect the river systems closely linked with the evolution of Beijing, and partially renovate lake and river systems of important historical value, in order to form a complete urban water system

6.1.2 Conservation rules and regulations of the urban river system

(1)Conservation rules of current river system

The rivers and lakes of Beijing had played an important role in flood-control, transporting grains to the capital, water supply and resisting foreign invasion, thus witnissing the history and evolution of the city. Conservation priority should be given to those lakes and rivers that are closely related to the development of Beijing, and have played an important role in various historical periods. The range of conservation should also defined for purpose of renovation and dredging.

● City moat system

Rivers under priority conservation are north city moat, south city moat, north Tuchenggou, and Dongzi River.

● Ancient river courses of headwaters

River courses under priority conservation are Lianhua and Chang Rivers, and Lotus Pond and Yuyuantan.

● Ancient river course for canals

Rivers under priority conservation are Tonghui and Ba Rivers, and North Grand Canal.

● Ancient flood-control rivers

Rivers under priority conservation are Yongding and Nanhan Rivers.

● Landscape water system

Waters under priority conservation are Liuhai, Kunming Lake, and Yuanmingyuan, or the Old Summer Palace river system.

● Water works under priority conservation

They are Houmen, Guangji, Lugou and Chaozong Bridges, Beiquanfu Ruins, Liulichang Bridge, Guangyuan Floodgate, Bali, Maizhong, Yindian Bridges, Jinmen, Qingfeng Floodgates, and Beihai Bridges.

(2) Restoration Plan for the Rivers

After the founding of new China, out of many reasons, many rivers in Beijing such as Yu, Changpu Rivers, East and West City Moats, and Moat of Qiansanmen were transformed into underground drainage ditch or filled up, severely damaging the river system in Beijing. So salvaging the river system in Beijing is of great significance to recover the historical features of Beijing and improve ecological environment.

The principle of suiting the measures to local conditions and implementing by stages should be adhered to in salvaging the river system, so the design of water environment and that of traditional features must be unified.

Zhuan, Changpu and Yu (on the Shishahai-Ping'an Avenue section) Rivers will be renovated according to the plan.

● Zhuan River

The renovation of Zhuan River, which belongs to the Tonghui River system, will link up the Chang River and the North City Moat.

● Changpu River

Transformed into an underground drainage ditch during the 1970s and 80s, and being a part of the Forbidden City water system, Changpu River is connected with the inner City Moat, Liuhai and outer City Moat river systems. Sights of historical interest will be recovered by salvaging this river.

● Yu River

Yu River (on the Shishahai-Ping'an Streets secion) was created in the Yuan Dynasty, originating from Houmen Bridge in the north, and reaching underground drain at Qiansanmen in the south. It was turned into an underground drainage ditch in the 1950s.

The upper reaches of this river (Shishahai-Ping'an Aevenue) will be renovated according to the plan. In long run, the lower reaches of the river (Beiheyan-Qiansanmen Street) connected with the imperial wall garden relics will possibly also be recovered.

(3) Conservation Plan for the Restoration of Lakes.

● Yuzaochi

Yuzaochi was the Taiyechi of the Jin Dynasty. One of the earliest historical monuments in Beijing, it must be rebuilt.

● Lianhuachi

The earliest water body developed in the Jin Dynasty, Lianhuachi (Lotus Pond) has been rebuilt into a garden except the water of southeastern corner. This part must be recovered with a water area of 4.9 hectares.

(4) Control of New Projects within the Land under the Plan in the Qiangsanmen City Moat Area

Constructed in the Ming Dynasty, the Moat used to be the final exit of city drainage system. It is an important river in terms of urban landscape and ecology flowing across the old city of Beijing from east to west, and passing Qiangmen. Owing to the first phase of subway construct, it was turned into an underground drain in the 1960s to 70s. Recovering it was rendered difficult due to fact that during the 1970s, many high-rises have been put up along its banks. As it would be extremely difficult to tear down so many buildings, salvaging the river course is not considered in the Plan, but new projects there will be restricted. The control area ranges from South City Moat in the west, East City Moat in the east, and 70 meters south of the red line on the Qiansanmen Street (including river courses and relevant greenbelts).

6.2 Conservation and Development of the City Axis

The old axis in Beijing is generated in the Yuan Dynasty, which, after the inverted "T"plane layout was formed in Beijing during the Ming and Qing Dynasties, starts from Yonngdingmen and ends in the Drum and Bell Towers after passing northward though the Forbidden City and the Central Peak of the Coal Hill. As the city expands, the Plan extends the old axis both in the north and south directions, forming the North Axis and South Axis, stretching about 25 kilometers.

6.2.1 Conservation Plan for the Traditional Axis

The old axis in Beijing is 7.8 kilometers from Yongdingmen to Bell and drum Tower, and 8.5 kilometers to the North Second Ring Road. The conservation plan must proceed in line with the principle of giving priority to conservation. Conservation and development, inheritance and creation must also be combined. Research should be focused on the conservation and plan for the three nodes of the Bell and Drum Tower, Jingshan-Qiangmen, and Yongdingmen.

Node at the Bell and Drum Tower: The Drum and Bell Towers had the function of giving correct time with the ancient Drum Tower serving the symbol of issuing monarch's edicts and the Bell Tower the token of unified order. Being the terminal of the traditional axis, the Towers, surrounded by courtyard houses, are grand and has the status of the landmark architecture in that area. The conservation of these two buildings must take into account the conservation of environment, allowing no large modern buildings to be constructed in case the traditional space is undermined. The height control for buildings around the Tower must follow the regulations stipulated in the plan for conservation districts.

Node at Jingshan-Qianmen: Being the space of the greatest historical value and Consisting of Jingshan Park, Forbidden City, Tain'anmen, Zhengyangmen Gate Tower and Watch Tower, the area has a rich gradation of spaces, orderly and undulating layout. So it must be strictly protected.

When conditions are ripe for recovering the river system of the Qiangsanmen City Moat, The sight of Tian'anmen would look grander and beautiful.

钟鼓楼
Bell and Drum Towers

故宫中轴线
The axis of the Forbidden City

北二环路北节点：在北二环路至安德路之间，中轴线两侧用地宜规划为重要的城市公共空间。

6.2.3 南中轴线的保护发展规划

南中轴线鸟瞰
Bird's eye view of the southern axis

南中轴线从永定门到南苑长约9.8km。南中轴线两侧调整改造任务较大，在做好用地功能调整的同时，注意丰富中轴线的空间结构，规划标志性节点。

天安门
Tian'anmen

现状南中轴线是一个道路虚轴，增加轴线节点是空间形态规划的目标，应重点研究三个节点。

木樨园节点：结合木樨园商业中心区，在中轴路与南三环路交叉口的四个角用地上，以群体建筑的有机围合形成重要的城市公共空间，建筑高度控制在45m。

大红门节点：在中轴路与南四环路交叉口的东北、西北两个角用地上，规划一组建筑，塑造重要的城市景观。

南苑节点：规划在建筑体量、用地规模上与北中轴奥运公园相当的城市公共空间，这将成为南中轴线的端点。端点以南地区，以大片森林公园相衬托。

6.2.4 城市中轴线的保护控制范围

以中轴路道路中心线为基准，距道路两侧各500m为控制边界，形成约1000m宽的范围作为北京城市中轴线的保护和控制区域，严格控制建筑的高度和形态。

位于中轴线保护和控制区域以外，对中轴线有重要影响的特殊区域，如天坛、先农坛、六海等，应按文物及历史文化保护区的保护规定执行。

6.3 皇城历史文化保护区的保护

明清皇城在历史文化名城保护中具有重中之重的核心地位，本次规划将皇城整体设为历史文化保护区。

6.3.1 规划范围的确定

由于皇城城墙的拆除，目前很难完全恢复和确定其准确位置，因此，皇城重点保护区范围是在参考明清北京皇城城廓平面的基础上，结合现状道路加以调整确定的。

皇城重点保护范围四至为：东至东皇城根，南至现存长安街北侧红墙，西至西黄城根南北街、灵境胡同、府右街，北至平安大街，总用地约6.8km²。

6.3.2 皇城的历史文化价值

明清皇城是在元大都皇城基础上发展起来的，自明永乐十八年（公元1420年）新建宫殿竣工，至今已有500多年的历史。虽然历经500余年的沧桑岁月，但皇城以其杰出的规划布局、建筑艺术和建造技术，成为中国几千年封建王朝统治的象征，至今保存基本完好，具有极高的历史文化价值。

（1）惟一性

皇城是我国现存惟一的、规模最大、最完整的皇家宫殿建筑群。是我国封建社会晚期城市规划艺术和建筑艺术的集大成者，是北京旧城中的精华。它们占据了城市的主导地位，象征皇权的至高无上。

皇城内分布有8处国家重点文保单位，20处市级文保单位，9处区级文保单位，34处普查在册文物，14片历史文化保护区。在一个区域内集中如此众多的文物古迹、历史遗存，是罕见的。

（2）真实性

皇城中的宫城、筒子河、三海、太庙、社稷坛、民居四合院等建筑群，布局严谨，保存完好，真实地再现了古代皇家生活、工作、娱乐的场景。

（3）完整性

皇城遵循了《周礼·考工记》上所记的"前朝后市，左祖右社"的王城规划制度，具有完整的规划布局。其东侧、南侧、西侧至今仍存有部分皇城城墙或城墙遗址，北侧有城市道路为界限。故宫北面有景山，西面有三海，南面沿中轴线对称布置了太庙和社稷坛，四周是平房四合院及庙寺，形成了完整统一的皇城形象。

（4）艺术性

皇城在规划理念、建筑布局、建造技术、皇城色彩等方面具有很强的艺术性，历史文化价值极高。

6.3.3 皇城保护存在的主要问题

（1）皇城四至边界有待进一步标明

应采取措施逐渐界定皇城的四至边界，特别是北部边界和部分西部边界。

（2）皇城保护区内人口密度很大

皇城内现状重点保护区内总人口77572人，其中，西城区43346人，东城区34226人。居住区中人口密度达500人/hm²以上。

（3）皇城内尚有部分文物利用不合理

皇城内有些文物被一些单位占用，使用不合理，文物资源浪费严重和破损。

（4）皇城内建筑高度控制有待加强

现状皇城内已有许多多层建筑，严重影响了皇城整体景观。

（5）皇城保护区内的道路改造要慎重研究

皇城作为历史文化保护区，其道路改造应以保护为前提。

6.3.4 皇城保护的措施

（1）明确皇城保护区的性质

以皇家宫殿、坛庙建筑群、皇家园林为特征，以平房四合院民居为衬托，具有浓厚的皇家传统文化特色的历史文化保护区。

（2）建立皇城区域意向

历史上皇城由皇城墙所围合，要通过实地勘测界定出原皇城墙的位置，从而建构起皇城完整的区域意象，使人可明确感知到皇城的存在。

（3）降低人口密度

结合旧城外的土地开发，与皇城保护区对号入座，疏导人口，降低保护区中的居住人口密度。

Node at Yongdingmen: the Yongdingmen Gate was demolished due to historical reasons, resulting in the disappearance of the southern end of the old axis. The rebuilding of Yongdingmen Gate Tower is of great significance for it can keep complete and solemn the traditional axis and effectively connect itself with the South Axis line. Building height nearby the Gate must be strictly restricted.

6.2.2 Conservation and Development Plan for the North Axis

This axis stretches from the North Second Ring Road to the Olympic Park, out of which three nodes must be given priority in conservation.

Node at the Central Olympic Park: priority should be given in planning to this area, the end of the North Axis, to make it a new landmark in Beijing. The Forest Park north of this area serves as a backdrop of the North Axis.

Node at the North Tucheng: As a prelude and a connection to the Olympic Park, The Tucheng ruins and the central greenbelt along the 80-meter-wide roads in the north axis can be utilized to create significant public urban spaces so as to enhance and enrich the North Axis.

Node to the North of the North Second Ring Road: Between the North Second Ring Road and Ande Road. The land along the sides of the axis could be designed into important public urban spaces.

6.2.3 Conservation and Development of the South Axis

It goes from Yongdingmen to Nanyuan with a length of 9.8 kilometers. The task of adjusting the two sides of the south axis is arduous. While the functions of the land along the sides of the axis should be adjusted, the spatial structures of the axis must be enriched to generate landmark places

The current south axis is only a nominal traffic road, so the plan objective is to add new places to create a rich space. Three nodes should be under priority conservation.

Node at Muxuyuan: Create urban public spaces in combination with the CBD of Muxuyuan on the land at the four corners of the intersection of the axis and the South Third Ring Road. The spaces are enclosed by architectural complex whose height will not exceed 45 meters.

Node at Dahongmen: on the land northeast and northwest of the intersection of the axis and the south Fourth Ring Road, a group of buildings and important urban landscape must be created.

Node at Nanyuan: urban spaces similar to the Olympic Park on the north axis in terms of size and land used will be planned. As the end of the south axis, the place should be set off with large tracks of forest parks to its south.

6.2.4 Controlled Zone for the Conservation of the Axis in Beijing

Based on the central line of the axis, the control boundary stretches to 500 meters on both sides, so that the architectural height and shape must be strictly restricted in this protected and controlled area of 1,000 meters.

Special areas outside the protected and controlled area of the axis, but that have important impact on the axis such as the Altar of the God of Agriculture, Liuhai (Six Seas), and Temple of Heaven must be treated in line with the regulations for the conservation of conservation districts of historical sites.

6.3 Conservation of the Conservation District of the Imperial City

The imperial city of the Ming and Qing Dynasties is at the core of the conservation of historic city. The Plan designates the imperial city as a conservation district as a whole.

6.3.1 Identification of Conservation Scope

As the walls of the imperial city have long been pulled down, their exact positions are hard to be identified and recovered completely. So the protected area is defined on the basis of studying the wall planes of the imperial city of the Ming and Qing periods, coupled with the current road conditions.

The four boundaries of the key protected area of the imperial city are: east Imperial Wall in the east, the red wall north of the Chang'an Avenue in the south, North and South Street at west Imperial Wall, Lingjing Lane and Fuyou street in the west, and Ping'an Avenue in the north, occupying a total land area of 6.8 square kilometers.

6.3.2 Historical and Cultural Values of the Imperial City

The Imperial City of the Ming and Qing Dynasties was constructed on the basis of the Dadu of the Yuan Dynasty and completed in 1,420 under the reign of emperor of Yong Le of the Ming Dynasty. Although it has a history of 500 years, it is still the symbol of thousands of years of the feudal ruling class with its outstanding layout, architectural art and construction technique, thus having a high historical and cultural value. It is well preserved.

(1) Uniqueness

Being the only well preserved imperial city of the Ming and Qing Dynasties, the imperial city is the sole largest and most complete imperial architectural complex, forming the cream of the old city of Beijing. As the highest development of urban planning and architectural art of the late feudal society of China, it occupies the dominant position in the city, symbolizing the absolute authority of the royal monarch.

In this area, there are 8 cultural relics sites under the key conservation of the State, 20 under the conservation of the city, 9 under the conservation of district, 34 registered sites in the general survey and 14 conservation districts. It is very rare to have so many relics in one area.

(2) Authenticity

Architectural complex of the Imperial City such as the City Wall, Dongzi River, Sanhai, Imperial Ancestral Temple, Altar to the God of the Land and Grain, and courtyard houses have been well preserved, providing solid evidences of the historical information on the life, work and entertainment of the royal families.

(3) Completeness

The Imperial City has a complete planning layout by adhering the planning principle of "court in the front, market in the back, ancestral temple on the left and fair in the right" stipulated in Zhou Rites-Records of Craftsman. Even today, there are still remnants of imperial walls and ruins on the eastern, western and southern sections, while the north is delimited by urban roads. The Forbidden City is surrounded by courtyard housing and temples with the Coal Hill Park to its north, three seas to its west and the Imperial Ancestral Temple and Altar to the Gods of Earth and Grains along the axis to the south, thus an image of unified and complete imperial city is created.

(4) Artistic Quality

The Imperial City has attained a high artistic quality in planning concept, architectural layout, construction technique, and color application. So it boasts high historical and cultural value.

6.3.3 Main problems regarding the conservation of the Imperial City

(1) The four boundaries of the Imperial City need further clarifying

Measures should be taken to gradually define the four boundaries of the Imperial City, especially its northern and part of western boundaries.

(2) The population density of the Imperial City is too high

The total population in the key protected area in the Imperial City is 77,572, out of which 43,346 are in the Xichang District, and 34,22 in the Dongcheng District, with the population density in the residential area equals 500 per hectare.

(3) Some of cultural relics in the Imperial City are being abused

The unreasonable use of some cultural relics sites occupied by certain units within Imperial City is very severe, resulting in their damage and wastes.

(4) Building height control in the Imperial City must be intensified

Many multi-storied buildings in the imperial city have a negative impact on the overall landscape of the area.

(5) Renovation of the road system in the Imperial City must be studied in a careful way

The road renovation in the Imperial City, deemed one of the conservation districts, must be conducted in a prudent way with conservation as the priority.

6.3.4 Measures of Conservation of the Imperial City

(1) Clarify the nature of the conservation of the Imperial City

Conservation district with the imperial palaces, architectural complex of temples and royal gardens as the main body, combined with courtyard houses all having strong royal traditionally cultural features.

(2) Create a clear image of the Imperial City area

Historically, the Imperial City was enclosed by the imperial walls, so the positions of the original walls must be identified by investigation to make use that every one is aware of the existence of the boundary of such an area.

(3) Reduce population density

(4) 停止审批建设三层以上的楼房

必须停止审批三层及三层以上的房屋和与传统皇城风貌不协调的建筑，杜绝"建设性"破坏。逐步推行"平改坡"，使现有平顶的多层建筑逐步改造成坡屋顶建筑。

(5) 以"院落"为单位整治、更新保护区环境

皇城内有许多危旧平房，在改造中应以"院落"为单位进行整治、修缮和更新，不能用大片推平的方式进行改造。

(6) 合理利用文物

皇城内尚有部分文物保护单位利用不合理，应加以调整和改善。

(7) 道路改造要慎重

皇城保护区内的道路改造应慎重研究，以保护为前提，逐步降低交通发生量。

(8) 制定保护管理条例

应尽快制定皇城历史文化保护区保护管理条例，健全管理体系，加大对皇城内建设的控制力度。

6.4 明、清北京城"凸"字形城廓的保护

明清北京城的城墙及大部分城楼已经拆除，代之以二环路和道路立交，但"凸"字形城廓仍是北京旧城的一个重要形态特征。应采取如下措施进行保护：

6.4.1 在旧城改造中，沿东、西二环路内侧留出30m绿化带，形成象征城墙址的绿化环。

6.4.2 保护北护城河与环绕外城的南护城河，规划沿河绿带。

6.4.3 保护沿城墙位置分布的正阳门城楼与箭楼、德胜门箭楼、东便门角楼与城墙遗址、西便门城墙遗址，复建永定门城楼。

6.5 旧城棋盘式道路网和街巷胡同格局的保护

北京棋盘式道路网始于元，定型于明清。当今北京旧城道路网继承了明清北京旧城路网格局和胡同体系，形成了北京特色的道路网。但规划道路红线拓宽与文物保护单位、历史文化保护区的保护发生矛盾，过宽的道路带来城市空间的突变，促使建筑高度的提高，不利于旧城风貌的整体保护。

护城河
City moat

德胜门现状
Present Deshengmen Gate

为了减轻旧城区的交通压力，总体设想将穿越旧城核心区的交通吸引到内环路上，再通过内环路与二环路之间的干道疏散到二环路上，从而减少历史文化保护区内道路的交通压力。

6.5.1 旧城主要交通对策

(1) 旧城区内的交通出行必须采取以地铁和地面公共交通为主的方式。

(2) 加快地铁建设，在主要干道上开设公交专用道，并布设小区、胡同公交支线网，方便市民出行。

(3) 实施严格的停车管理措施，控制车位供应规模，限制或调节驶入城区的汽车交通量。

(4) 采取切实可行的管理措施和调控手段（包括经济手段），限制私人小汽车在旧城区的过度使用。

(5) 控制旧城区建筑规模和开发强度，从根本上压缩机动车交通生成吸引量。

6.5.2 旧城路网调整原则

(1) 调整旧城路网规划和道路修建方式，协调好风貌保护与城市基础设施建设的关系，以此为前提确定路网的适当容量。

(2) 为了减轻旧城历史文化保护集中地区的交通压力，总体设想是降低区域内部道路等级，将穿越旧城核心区的交通吸引到内环路上，再通过内环路与二环路之间的干道疏散到二环路上，从而减少历史文化保护区内道路的交通压力。

(3) 内环路（及北延长线）围合区域内，穿过历史文化保护区的道路，原有路段规划红线宽度相应缩窄。

(4) 与历史文化保护区密切相关、尚未按原道路规划建成的主干路，该路段的道路红线宽度可按风貌保护的要求进行缩窄。

(5) 旧城内其他道路根据两侧用地性质，只调整规划道路的路幅宽度，红线宽度不变。

道路路幅宽度的确定应在满足文物和风貌保护的同时，协调处理好交通出行、市政设施、城市景观和生态环境等各项功能的需要。

(6) 同等级道路，在旧城以外和旧城以内、在旧城的内城和外城、在历史文化保护区和非保护区，应采用不同的路幅宽度。

(7) 路幅宽度设置，应针对不同情况和条件，采取相应措施保护文物。

6.5.3 旧城路网调整要点

(1) 按照路网调整原则，将处于旧城核心区统称朝阜干道的西安门内大街、文津街、景山前街、五四大街、东四大街，以及统称中轴路的地安门内大街、地安门外大街、景山后街、景山西街、景山东街、南北长街、南北池子等主干路降级为次干路。

(2) 朝阜干道中的文津街、景山前街、五四大街等处于历史文化保护区内的主要街道维持现状道路宽度，不再拓宽。阜成门内大街以北是历史文化保护区，并有多处国家级、市级文物保护单位，规划调整为保持其现状北道牙不动，将道路向南拓展到规划道路的南红线。

(3) 中轴路中的地安门大街、景山后街、景山东街、景山西街、南北长街、南北池子现状道路宽度基本保持不变。

(4) 取消处于东四、西四和张自忠北保护区内的三条东西向次干路，将处于皇城、什刹海、国子监、南北锣鼓巷、阜内和东交民巷保护区内的六条次干道降级为支路。

(5) 历史文化保护区内的支路依胡同宽度和走向进行调整，大于3m的胡同可按具体情况布设市政管线。

(6) 其他道路路幅宽度应按规划道路等级和所处区位的不同进行相应调整。

The land development outside the old city should take into account the conservation and renovation of the Imperial City, with a view of relocating and reducing the population within the protected area.

(4) Stop the construction of buildings of three-stories and above

The examination and approval of buildings of or above three stories and those not in agreement with the traditional style of the Imperial City must be stopped. Damage done owing to "construction" must be avoided. The existing flat-roofed, multi-storied residences must be transformed into buildings with pitched roofs.

(5) Updating and renovating the environment of the conserved areas by using "courtyard" as
basic module

The man risky and old bungalows in the Imperial City must be renovated and updated by using courtyard as basic module. The approach of leveling them to the ground must not be used.

(6) Make use of cultural relics in a reasonable way

For those cultural relics in the Imperial City that are not being used in a sensible way, adjustment and improvement must be done in this regard.

(7) Road renovation must be prudent

The road renovation in the Imperial City must be conducted in a prudent way with conservation as the priority. Traffic volume should be gradually reduced.

(8) Draw up a conservation ordnance

Conservation ordnance for the Imperial City conservation district must be drawn up as soon as possible and a management system perfected, so that control over construction within the area can be greatly enhanced.

6.4 Conservation of the Inverted "T" Shaped City Wall of the Ming and Qing Dynasties in Beijing

Although the walls and part of the watch towers built in the Ming and Qing Dynasties in Beijing have been replaced by the Second Ring Road and overpasses, the layout of inverted "T" still remains an importance feature of the old city of Beijing. So following measures must be adopted to protect them:

6.4.1 In the renovation of the old city, 30 meters of greenbelt along the inner sides of the east and west Second Ring Roads should be left to make room for a green ring symbolizing the old city wall.

6.4.2 Conserve the north City Moat and south City Moat surrounding the outer city, and plan the greenbelt along the rivers.

6.4.3 Preserve the structures along the city walls such as the Zhengyangmen Gate Tower and Watch Tower, the Deshengmen Watch Tower, the Dongbianmen Corner Tower and city wall ruins, the Xibianmen city wall ruins, and rebuild the Yongdingmeng Gate Tower.

6.5 Conservation of the Checkerboard Road System and the Hutong Fabric

The checkerboard road system originated in the Yuan Dynasty and finalized in the Ming Dynasty. The current road network allay system in old city of Beijing is passed down from the Ming and Qing Dynasties, generating a local feature. However, the inroad of the planned traffic network runs counter to the conservation of cultural relics sites and conservation districts. Wider road has caused prompt change to the urban space, resulting in appearance of more high-rises. This is detrimental to the overall conservation of the old city.

In order to alleviated the traffic pressure on the old city and conservation districts, it is suggested that traffic passing through the center of the old city be diverted to the inner ring road, which will, in turn, further be diverted to the Second Ring Road by the traffic network linking the inner and Second Ring Roads.

6.5.1 Solutions to the Circulation within the Old City

(1) Circulation within the Old City should be mainly supported by bus and subway systems.

(2) Speed up the construction of subways, open special lanes for public transportation on the main roads, and set up public transportation feeder network in residential quarters for the convenience of the dwellers.

(3) Implementation of stricter parking management, restrict the number of the parking stalls, limit or regulate the traffic heading into the city.

(4) Restrict the overuse of private cars in the old city by adopting feasible management and regulating measures (including economic means).

(5) Control the scale of construction and development within the old city, so as to reduce the inbound traffic.

6.5.2 Principles for the Renovation of Road System within the Old City

(1) Adjust the plan for road network and pattern of maintenance in the old city, strike a good balance between the conservation of features and the infrastructure construction in the city, upon which appropriate traffic load can be determined.

(2) In order to alleviated the traffic pressure on the old city and conservation districts, it is suggested that the grades of road in the protected areas must be reduced, and that traffic passing through the center of the old city be diverted to the inner ring road, which will, in turn, further be diverted to the Second Ring Road by the traffic network linking the inner and Second Ring Roads.

(3) For roads passing through the conservation districts within the enclosed area demarcated by the Inner Ring Road (including extended north axis), the original planned red line should be narrowed.

(4) The main roads that are closely related with conservation districts and have not been constructed in accordance with the original road plan, the red line width should be narrowed in line with the requirements of conservation.

(5) Other planned roads in the old city can only adjust the width, and ignore the width of red line. But this should be done by taking into account the nature of the land on the sides of the roads.

The width of roads must be in line with the requirements of protecting the cultural relies and traditional features, as well as meeting various functions such as traffic, urban facilities, city landscapes and ecological environment.

(6) Roads of the same level should be different in width depending on their places-inside or outside the old city, in the inner or the outer city within the old city, and in the conservation district or outside the district.

(7) Determination of the road width should be done in line with specific conditions and appropriate measures must be taken to protect the cultural relics.

6.5.3 Key Roads in Old City that Need To Be Adjusted

(1) In accordance with the road adjustment principle, the following main roads (generally called Chao-Fu Main Road) in the center part of the old city will be relegated to secondary roads: Xi'anmennei Street, Wenjin Street, Jingshanqian Street, Wusi Street and Dongsi Street. The same will be applied to the roads on the axis: Di'anmennei Street, Di'anmenwai Street, Jingshanhou Street, Jingshanxi Street, Jingshandong, Nan and Bei Changjie Streets, Nanchizhi and Beichizi Steets.

(2) Wenjin, Jiangshanqiang, Wusi, and other streets on the Chao-Fu Main Road and within the conservation districts will not be widened. North of Fuchengmennei Street are conservation districts where cultural relics sites under State and city conservation abound. The adjustment plan stipulates that the northern curb of the street will remain unchanged, but it will extend southward till the southern red line of planned road.

(3) Di'anmen Street, Jingshanhou Street, Jingshandong Street, Jingshanxi Street, Nan and Bei Changjie Streets, Nanchizhi and Beichizi Steets on the axis will remain unchanged in width.

(4) Three secondary main roads going east to west within the conservation districts of Dongxi, Xisi and north part of Zhangzizhong Road will be removed. The six secondary main roads in the conservation districts of Imperial City, Shichahai, Imperial Collage, Nan and Bei Luoguxiang, Fuchengmengnei and Dongjiaominxiang will be relegaed to side roads.

(5) The sides roads within conservation districts are subject to adjustment in line with the width and direction of allays. Public piping and wiring may be installed in allays exceeding three meter wide.

(6) The width of other roads are subject to adjustment in accordance with the different grades and locations specified in the Plan.

(7) The remaining land in the red lines after adjustment of road width must be used for greening. Any new and renovated buildings must be undertaken beyond the red lines.

6.6 Control of Building Height in the Old City

6.6.1 Present conditions of building height in the old city

Investigation reveals that the problems of architectural height in the old city are as follows:

(1) One-story and courtyard housing within the conservation districts are well

(7) 调整路幅宽度后道路红线内的剩余用地应用于城市绿化，新建、改建建筑仍需在道路红线以外建设。

6.6 旧城建筑高度的控制

6.6.1 旧城建筑高度的现状情况

综合旧城的现状情况，旧城的建筑高度和主要问题有以下几个方面：

(1) 以平房四合院为主的区域，穿插多、高层建筑。

处于历史文化保护区范围内的平房四合院用地保存比较完整，保护区以外平房四合院用地的完整性已经受到了很大的破坏。这些区域中零散穿插着许多建筑，高度一般都超过12m，大多与环境不协调。

(2) 高层建筑的数量增多、布局分散、破坏力强。

目前，旧城中的超高层建筑逐渐增多，分布杂乱，造成的破坏和影响很大。

(3) 旧城城廓面临被吞噬的危险。

保持明清北京的"凸"字形城廓平面，是名城保护的一项重要内容。

目前，二环路内侧的高层建筑群，使旧城的城廓难以分辨，旧城没有了边界，旧城也将面临被吞噬的危险，因此严格控制城廓内侧的建筑高度十分必要。

(4) 景观视廊破坏现象严重。

原有北京旧城，除重要建筑外，整个城市的建筑背景高度基本上在3～6m，因此重要建筑(如城门楼、景山、钟鼓楼、天坛祈年殿等)之间有着十分通畅的对视关系。目前这些视廊由于高、多层建筑的插建，受到了很大的破坏。

文昌胡同
Wenchang Hutong

育群胡同
Yuqun Hutong

(5) 多层建筑外型单调。

这类建筑大多外形简单、粗糙，建筑的屋顶形式、色彩很少考虑北京旧城建筑的特色，应在屋顶形式与色彩上加以改善。

五四大街
Wusi Street

6.6.2 北京旧城建筑高度控制规划

(1) 旧城建筑高度控制规划的三个层次。

整个旧城建筑高度控制规划可按照以下三个层次进行：

第一：文物保护单位保护范围、历史文化保护区的重点保护区。

景山东街
Jingshandong Street

文物保护单位保护范围、历史文化保护区的重点保护区是旧城保护的核心保护区域，应按历史原貌保护的要求进行高度控制。这些区域建筑的有机更新应在保持历史原貌的基础上进行，保持历史上原有的风貌特征和建筑高度。

第二：文物保护单位建设控制地带、历史文化保护区的建设控制区。

在保护范围与保护区外都有一个建设控制地带或区域，这些建设控制地带应遵循文物及保护区保护规划的要求进行高度控制。

第三：文物保护单位建设控制地带、历史文化保护区的建设控制区之外的区域。

旧城中，在文物及保护区保护和控制范围之外的区域，建筑控高应严格按《北京市区中心地区控制性详细规划》的要求执行，不得突破。

6.7 城市景观线的保护

北京市区以燕山山脉和太行山山脉为依托，随着时代的发展，旧城中产生了一些相当重要的城市景观线，对其两侧的建筑高度、形态应加以控制。

北京城市总体规划中提到的城市景观线(视线走廊)主要有7条，分别是：银锭观山、(钟)鼓楼至德胜门、(钟)鼓楼至北海白塔、景山至(钟)鼓楼、景山至北海(白塔)、景山经故宫和前门至永定门、正阳门城楼、箭楼至天坛祈年殿。

景观线保护范围内的新建筑的高度，应按测试高度控制，严禁插建高层建筑。

6.8 城市街道对景的保护

在一般丁字路口或曲折道路转角处形成的城市街道的对景建筑，是城市景观设计的重要组成部分，应注意处理好街道与建筑的关系。

对于历史形成的对景建筑及其环境要加以保护，控制其周围的建筑高度。对有可能形成新的对景的建筑，要通过城市设计，对其周围建筑的高度、体量和造型提出控制要求。

在旧城改造中必须处理好街道与重要对景建筑的关系，如北海大桥东望故宫西北角楼、陟山门街东望景山万春亭、西望北海白塔、前门大街北望箭楼、光明路西望天坛祈年殿、永定门内大街南望永定门城楼(复建)、北中轴路南望钟鼓楼、地安门大街北望鼓楼、北京站街南望北京站等。

6.9 旧城建筑形态与色彩的继承与发扬

旧城内新建建筑的形态与色彩应与旧城整体风貌相协调。北京旧城传统的主体建筑色彩是青灰色的民居烘托红墙、黄瓦的宫殿建筑群。目前对北京全市的建筑色彩做出统一规定的条件还不成熟。对旧城内新建的低层、多层住宅，应采用坡屋顶形式；已建的平屋顶住宅，应逐步采用"平改坡"的方式进行改造。旧城内具有坡屋顶的建筑，其屋顶色彩应采用传统的青灰色调，禁止滥用琉璃瓦屋顶。

6.10 古树名木的保护

在危改区或新的建设区，要在清点登记的基础上，加强古树名木及大树的保护，严禁砍伐。

历史文化保护区内的公共绿地建设包括街道、胡同绿化，并尽量

preserved and remain intact, but the completeness of the land for these kind of housing outside the districts have been damaged to a great extent. Many high-rises of 12 meter high are scattered in these areas disrupting the harmony with surroundings.

(2) The number of high-rise buildings is increasing, with their locations being more scattered and therefore more destructive to the cityscape

At present, more and more high-rises have appeared in the old city in a disorderly way, causing sever damage to the area.

(3) The wall contour of the old city is on the brink of being engulfed.

Protecting the city wall plane in the shape of in inverted "T" is one of the important contents of the conservation of historic city.

At present, the high-rise complex in the inner side of the Second Ring Road has blurred the wall contour of the old city. Without the boundary, the old city would face the risk of being annihilated. So it is imperative to control the architectural height within the city wall contour.

(4) The effect of visual passages with scenic ends have been severely damaged

In ancient Beijing, Apart from importance buildings, the height of all architecture ranges only from 3 to 6 meters, so there was an essential visual relationship among significant buildings (such as city Gates, Coal Hill, the Drum and Bell Towers, the Hall of Prayer for Good Harvest in the Temple of Heaven, etc.). Now this relationship has been damaged by many high-rises.

View the mountain from Yinding Bridge(one of the eight beautiful sceneries of Yanjing)

Drum Tower viewed from south

(5) Boring forms of multi-story buildings

This kind of architecture is usually lack of variety in form which is also very rough. The shapes and colors of roofing rarely match the architectural features of the old Beijing. Therefore, improvement should be made in this regard.

6.6.2 Plan for architectural height control in the old city of Beijing

(1) Three levels for the architectural height control plan in the old city.

The overall architectural height control plan for the old city can proceed in the following three levels:

First, Conservation range of cultural relics sites and key areas of the conservation districts

Height control in these protected key places in the old city must meet the requirements of protecting the original features. The architectural updating in

Bird's eye view of No. 36,Fuxue Hutong

these areas must be undertaken on the basis of keeping the original looks, including historical features and architectural height.

Second, Construction control areas of the cultural relics sites and construction control districts of conservation districts.

There is a construction control area in and outside any protected district, and the height control in these areas must be implemented according to the requirements stipulated in the cultural relics and conservation district plans.

Third, Areas beyond the construction control areas of the cultural relics and conservation districts.

Height control in areas beyond the conservation and control scope of cultural relics and conservation districts must strictly abide by the requirements specified in the "Detailed Plan for the Central Controlled Area in Beijing". No breach of this Plan is allowed.

6.7 Conservation of Urban Landscape

Beijing has been relying on the Yan and Taihang Mountain ranges. With the passing of times, a number of quite important urban landscape came into being in the old city. So control over the height and shapes of buildings on both sides of these landscapes must be imposes.

The "Master Plan of Beijing" mentions 7 key urban landscape lines (visual corridors). They are: View the mountain from Yinding bridge , Bell and Drum Tower to Deshengmen, Bell and Drum Tower to the Pagoda in Beihai Park, Jingshan Park to the Bell and Drum Tower, Jingshan Park to the Pagoda in Beihai Park, Jingshan Park to Yongdingmen through the Forbidden City, and Zhengyangmen Gate Tower and Watch Tower to the Hall of Payer for Good Harvests in the Temple of Heaven.

The height of new buildings within the landscape areas under conservation must be controlled by testing height, and no high-rises are allowed in these areas.

6.8 Conservation of Visually Corresponding Nodes in City Streets

The scenic focal points at T-shaped road junctures and street corners are consequential components of urban landscape design, so their relations with architecture and streets must be dealt with well.

Symmetrical architecture and their surroundings formed in history must be protected, and architectural height around them controlled. For buildings that are likely to form new symmetrical sights must be carefully designed, and the height, size and shape of surrounding buildings will have to meet the requirements of control.

The relationship between the streets and substantial symmetrical sights in the course of renovation of the old city must be dealt with well. Some of the examples are: Beihai Bridge looking the northwest corner tower of the Forbidden City in the east, Zhishanmen Street echoing the Wanchun Pavilion in the Coal Hill Park in the east and the Pagoda in Beihai Park in the west, Qianmen street looking the Watch Tower in north, Guangming Road echoing the Hall of Prayer for Good Harvest in the Temple of Heaven in the west, Yongdingmennei Street looking the Yongdingmen Gate Tower (rebuilt) in the south, North Axis looking the Bell and Drum Tower in the south, Di'anmen Street looking the Drum Tower in the north, and Beijing Railway Station Street looking Beijing Railway Station in the south.

6.9 Inheritance and Development of Architectural shapes and Colors in the Old City

The patterns and colors of the newly built architecture within the old city must be in harmony with the general features of the old city. The traditional and main architectural color of old Beijing is greenish gray which serves as foils to set off the palace complex of red walls and yellow tiles. The conditions are still not ripe for specifying a unified color for all the buildings in Beijing. The low and multi-story buildings newly built in the old city must use sloping roofing; flat-roofed buildings already constructed must be gradually turned into sloping roofed structure. Architecture with sloping roofing in the old city must use traditional greenish gray color for their roofs. Overuse of glazed tiles are forbidden.

6.10 Conservation of Well-known and Old Trees

Prohibit cutting down well-known, old and big trees in ramshackle building renovation and new construction areas on the basis of checking and registry.

The greening construction in the conservation districts include greening up

湖南会馆
Hunan Hall

大菊胡同
Daju Hutong

利用原四合院中的私人花园开辟为公共绿地。

发扬四合院宅院绿化的传统，充分考虑院落植树栽培方式。

旧城内的改造区应尽量增加公共集中绿地，绿地建设应采用适合北京特点的植物品种。

7. 旧城危改与旧城保护

市文物局、规委、建委和房地局共同制定了"关于在城市危改中加强文物保护工作"的文件，确定了在不同类型危改片中做好文物及风貌保护的原则。

必须树立危改与名城保护相统一的思想，历史文化保护区内的危房，应严格按历史文化保护区保护规划实施，以"院落"为单位逐步更新危房，恢复原有街区的传统风貌。

历史文化保护区以外的危改片，要加强对文物及有价值的历史建筑的核查、保护，强调古树名木、胡同等历史遗存的保护，严格执行各级文物保护单位的保护范围和建设控制地带的有关规定，严格执行《北京市区中心地区控制性详细规划》中的建筑高度控制等有关规定。

必须尽最大可能保护传统胡同的格局、肌理和尺度，建立健全相关技术规范，特别是传统胡同、街道的历史名称不得随意修改。

建设单位必须处理好与保护有关的工作才能申报危旧房改造方案。危改项目的前期规划方案应包括历史文化保护专项规划，内容包括街区的历史沿革、文物保护单位的保护、有价值的历史建筑的保护、对传统风貌影响的评价、环境改善的措施等。

市规划、文物、园林、房地等相关部门要建立相应的工作程序和制度，切实把好审批关，做好危旧房改造中的文物保护工作。

8. 传统地名的保护

传统地名是北京历史文化名城保护的重要组成部分，要保持并保护这些地名的历史延续性。新命名的地名，要注意与传统文化氛围相协调。

建立健全相关技术规范，对传统胡同、街道的历史名称不得随意修改。按照"尊重历史、照顾习惯、体现规划、好找好记"的原则，保护好历史地名。

9. 传统文化、商业的保护和发扬

9.1 传统文化的保护和发扬

9.1.1 历史与现状

北京文化的发展历史悠久，源远流长，涵盖多个方面及领域。现仅选取与建设相关的、较有代表性的庙会、戏院加以论述。

（1）庙会

北京的庙会是一种集吃喝玩乐于一体的民间性娱乐活动，具有悠久的历史传统。

近几年来，随着旅游事业的发展，北京地区恢复了春节庙会活动，使中断的民间活动得以复苏。规模较大的属地坛庙会、龙潭庙会、白云观庙会、大钟寺庙会等，庙会上有风味小吃，民间花会，技艺表演等，内容丰富多彩。

（2）戏院

北京的戏院历史久远，特色鲜明，种类丰富。满足了各类戏曲艺术的使用要求。如京剧、评剧、单弦、快板书、北京曲剧、皮影、相声、评书、北方昆曲等，过去都有适合自己的演出场所。

9.1.2 保护规划

（1）尽量恢复各区有代表性的庙会，包括厂甸、白塔寺、都城隍庙、护国寺、五显财神庙庙会。

（2）以昆曲和京剧为重点，进一步繁荣北京的传统戏曲事业，加强戏院建设。

（3）应采取措施恢复和合理利用会馆。

9.2 传统商业的保护和发扬

9.2.1 传统商业的基本概况

（1）传统商业的概念

是历史上长期存在的、符合当地民族生活习惯的、具有明显特色且不断继承发扬的商业服务业。

（2）传统商业的内容

一是传统商业街和老字号以及由坐商、摊贩、游动商贩等组成的旧北京商业供应服务网络体系；二是北京传统商业文化，它是长期商业实践活动中总结、积累下来的丰厚的经营管理和经商理念，"严格的管理手段、独特的经营理念、良好的商业信誉"构成了传统商业文化的特征。

9.2.2 北京传统商业的现状及问题

（1）北京传统商业街区的现状及问题

北京著名传统商业街区，随着岁月流逝，原有的基础设施已远不能适应当代的需求。目前存在的问题：一是交通状况差，道路狭窄，机动车通行困难；二是人口密集，购物拥挤不堪；三是现有建筑严重老化；四是市政基础设施差；五是商业特色不突出，商品缺乏竞争力；六是整体形象差。因此，改造传统商业街区势在必行。

（2）北京老字号的现状及问题

中华人民共和国成立以来，一些知名的老字号的发展几经曲折，主要表现在：

中华人民共和国成立初期比较重视老字号，并发挥了老字号在私改中的作用。后来调整商业网点，并店联营中，造成一批老字号开始消失。实行工商分别归口后，使许多老字号的"前店后场"系统破坏，从而失去了老字号质优价廉的最突出特色的根基。

9.2.3 北京传统商业的保护

保护传统商业主要包括两个方面，一是传统商业街区的改造与保护；二是老字号的恢复与保护。

（1）传统商业街区的保护

北京应重点保护如下几条传统商业街区：

- 大栅栏商业街
- 琉璃厂文化街

streets, Hutongs and courtyards. Private gardens in former courtyard housing should also be turned into green park.

Carry forward the tradition of cultivating plants in courtyards, try to study the approach of courtyard cultivation.

Public and concentrated green plots in the renovated areas of the old city should be enhanced, where plant species suitable to Beijing should be planted.

7. Ramshackle Building Renovation in the Old City and Conservation of the Old City

Supported by the municipal leaders, the Municipal Administration Cultural Heritage, Urban Planning Commission, Construction Commission and Bureau of Housing and Land issued a document titled "On the Conservation of Cultural Relics Work in Urban Redevelopment", which formulates the principles of protecting cultural relics and original features in different kinds of urban redevelopment districts.

The renovation of crumbling houses in the old city should be integrated with the conservation of the historic city. The renovation of dangerous houses in the conservation districts must be undertaken in strict compliance with the conservation plan for conservation districts. Renovated in terms of the "courtyard", the traditional features of the original streets must be rebuilt.

When dealing with renovation areas outside the conservation districts, the verification and conservation of cultural relics and valuable historical buildings must be strengthened. Conservation of famous and old trees as well as allays and other historical ruins must be highlighted. The relevant regulations relating to the conservation range and construction controlled areas of cultural relics sites of various levels, and of the height control stipulated in the "Detailed Plan for the Central Controlled Area in Beijing" must be strictly implemented.

construction units can only submit plans for renovating old and risky houses for approval after having dealt with the conservation work in a satisfactory way. The initial plan of this kind must contain a special plan for the conservation of historical culture, including the historical change of streets, conservation of the cultural relics sites, conservation of valuable historical buildings and ruins, assessment of the influence of traditional features, and measures for improving environment.

Municipal departments in charge of urban planning, cultural heritage, gardening and housing and land should set up relevant work procedure and system to deal with examination and approval work, so that conservation of cultural heritage in urban redevelopment can be guaranteed.

8. Conservation of Traditional Names of Places

Traditional place names must be protected as one of the important contents in the conservation of the historic city of Beijing. The historical continuity of these names must be kept and protected while new place names should be in harmony with traditional culture.

Work out and perfect sound rules and regulations and technical norms to guarantee that historical names of traditional alleys and streets cannot be revised at random. Historical place names should be protected in the principle of "respect history, show consideration to customs, adhere to the Plan and easy to identify".

9. Conservation and Inheritance of Traditional Culture and Commerce

9.1 Conservation and Inheritance of Traditional Culture

9.1.1 Past and present

Beijing boasts a traditional culture of long-standing, which covers all aspects of fields. The following only touches upon temple fairs and theaters that are related with architecture.

(1) Temple fairs

Temple fairs in Beijing has a long history and is a folk entertaining form integrating eating, drinking, and entertainment.

In recent years, with the development of tourism, temple fairs, which is one of the folk activities, were reinstated in Beijing. Large temple fairs include those at Temple of Earth, Loutan, Baiyunguan and Big Bell Temple, where local foods, flower display, arobatics performances and other varieties of activities are featured.

(2) Theaters

Theaters in Beijing have a long history, distinctive features and many kinds, satisfying the needs of all kinds of operatic arts. The forms of operas include Peking opera, pingju, story-tell and singing to musical accompaniment, rhythmic story-telling, Beijing quju, shadow play, cross-talk, story-telling, and kunqu. There were suitable performance venues for all of them in the past.

9.1.2 Conservation Plan

(1) The representative temple fairs in various districts must be reinstated as soon as possible, including those at Chagdian, Temple of White Pagoda, Town God's Temple, Huguo Temple, Wuxian Temple ot the God of Wealth.

(2) Further develop the traditional operas in Beijing by giving priority to Kunqu and Peking Opera. Construction of theaters and relevant cultural facilities must be enhanced.

(3) Measures should be taken to renovate and make reasonable use of guild halls.

9.2 Conservation and Development of Traditional Commerce

9.2.1 Basic situation of traditional commerce

(1) Concept of traditional commerce

Refers to business and service sectors that are of long-standing, suitable to the lifestyle of the local people, have obvious characteristics and can continue developing in the future.

(2) Components of traditional commerce

Firstly referring to the service network of old Beijing composed of traditional business streets, stores of long-standing, fixed vendors, street peddlers and itinerary vendors. Secondly referring to the traditional business culture of Beijing, which are the rich management and operational concepts accumulated during the long time of business practice. Traditional business culture is characterized by "strict managerial methods, unique operational concept and good business reputation".

9.2.2 Present situation and problems of Beijing's traditional business

(1) The famous traditional business streets in Beijing is no long able to meet the needs of the times in terms of their infrastructure. The current problems are: first, the traffic is bad with few roads wide enough to accommodate vehicles; secondly, there are so many inhabitants that the shops are always crowded; thirdly, the current buildings are too old; fourthly, infrastructure is in poor conditions; fifthly, business has no brand names and commodities lack competitiveness; lastly, the overall image is far from satisfying. So updating the traditional business must be done without delay.

(2) Current conditions of old stores in Beijing

Since the founding of new China, the development of some well-known old stores have undergone twists and turns:

At the beginning of the new China, great attention was attached to stores of long-standing, which were allowed to play a role in the reform of private enterprises. Later, a number of old stores disappeared due to adjustment of commercial networks and the merge of stores. After the industrial and commercial sectors were placed under different authorities, the system of "shop in the front and workshop at the back" practiced by many old stores was undermined. As a result, the most conspicuous feature of high quality and cheap price of the old stores lost its foundation.

9.2.3 Conservation of traditional business in Beijing

Conservation of traditional commerce mainly involves the conservation and renovation of traditional commercial streets and rebuilding and conservation of stores of long standing.

(1) Conservation of traditional commercial streets

Priority should be given to the conservation of the following traditional business areas in Beijing:

- Dashalan Commercial Street
- Liulichang Cultural Street
- Qianmen Commercial and Sightseeing Street
- traditional commercial streets in Shishahai tourist area (Yandaixie Street and Lotus Market, etc)
- Longfuxi Commercial Street.

(2) Conservation of traditional trades and stores

- 前门商业文化旅游区
- 什刹海文化旅游风景区中的传统商业街（烟袋斜街、荷花市场等）
- 隆福寺商业街

(2) 传统行业和老字号的保护

- 老字号的行业分布

本市老字号分布于多种行业，以食品、餐饮、医药行业为多。

- 目前老字号经营状况分类

继承发扬传统商业文化，恢复了前店后场，发展了连锁商店，取得了显著成效的，如同仁堂、全聚德等。

继承发扬传统，恢复前店后场，特色商品出现在大商场、超市上或发展连锁商店，取得相当效益的，如王致和腐乳、稻香春等。

继承传统并有所发扬，如茶叶行业，仍然沿用自采、自窨、自拼、自配方法，保持了名店特色。

有的老字号牌匾依旧，但等同于普通商店，因而只能勉强维持；甚至虽是老字号店名，靠出租柜台盈利，已是名存实亡。

(3) 传统商业保护的若干建议

- 重新认识老字号，大力扶持老字号。
- 继承发扬传统经营管理特色。
- 收集出版有关传统商业方面的宣传介绍材料，以供吸纳借鉴，古为今用。
- 举办传统商业展览，并在此基础上建设传统商业博物馆。

10. 实施保障措施

10.1 统一思想，广为宣传北京名城保护的重要性。

认真贯彻中央对北京城市规划建设和管理工作的一系列重要指示，立足于把北京建设成为第一流的现代化国际城市和历史文化名城的高度，充分认识实施《北京历史文化名城保护规划》的重要性，广为宣传，在全社会形成"热爱名城、保护名城"的共识。

10.2 进一步落实《北京城市总体规划》关于"两个战略转移"的方针。

前门大街
Qianmen Street

进一步落实《北京城市总体规划》的要求，积极推进城市建设重点逐步从市区向远郊区县转移、市区建设从外延扩展向调整改造转移的步伐，疏解旧城区人口和功能，落实与城四区危改相应地段的经济适用房项目，为保护历史文化名城创造良好条件。

10.3 在《北京历史文化名城保护规划》审批后，应尽快深化落实。

"十五"期间，组织编制落实北京历史文化名城保护规划的具体规划，主要包括：第二批历史文化保护区保护规划，旧城内道路规划的调整与定线，旧城建筑高度控制的调整规划，第五批、第六批文物保护单位保护范围和建设控制地带的划定等。

10.4 加快制定北京历史文化名城保护的相关法规。

由市文物、规划、建设等部门共同开展立法的前期调研工作，结合北京实际，借鉴外省市经验，广泛听取社会各界意见，制定《北京市历史文化名城保护条例》，报市人大审议。制定《北京市历史文化保护区管理规定》，作为法律依据在城市管理工作中严格执行。制定故宫、颐和园、长城、天坛等世界文化遗产的专项保护法规，明确管理责任，保证保护资金，规范建设和管理行为。

10.5 旧城内危改不宜采用开发的方式进行。

旧城内危改区建设的好坏对北京旧城的整体风貌影响很大，旧城危改区住宅建设项目不宜采用房地产开发方式。危改中要注意保护原有主要胡同、大树、古树和具有一定历史价值的建筑。

10.6 深入研究旧城房屋产权改革的相关法规政策并开展试点。

积极推进旧城历史文化保护区内房屋产权的改革，并研究制定相关实施政策，各区都要开展历史文化保护区规划实施试点，积累经验。

10.7 多渠道筹集资金，初步建立资金保障机制。

根据"国家保护为主，动员全社会参与"的主导思想，积极开展试点工作，探索文物保护单位和历史文化保护区保护、维修、整治、利用的有效途径，建立资金保障机制，除政府财政拨款外，还要通过开发旅游经济、开辟使用单位自筹资金、社会募捐等多种渠道，建立保护基金。

10.8 加强对北京传统风貌和文化特色的继承和发扬的研究。

历史文化名城的内涵既包括物质形态的城市格局、文物古迹、有特色的建筑等传统城市风貌，也包括精神形态的戏剧、文学、绘画、音乐等文化艺术活动，还包括具有地方特色的庙会等传统习俗，以及老字号等名店、名品和特色商务活动。要研究保持和发扬传统文化、艺术、民俗以及商务活动的有效措施。

总之，只要我们本着对历史负责、对人民负责的态度，重视和切实保护好历史文化遗产，挖掘历史文化名城的深刻内涵，在实现城市现代化过程中，采取具体有效措施加强对名城整体的保护，就一定能把北京建设成为一流的历史文化名城和现代化国际城市。

2002年9月

- Distribution of old stores by trade

The old stores are mainly concentrated in food and beverage as well as Chinese medicine fields.

- Classification of old stores according to their current operating status

Those stores that have inherited and carried forward traditional business culture, reinstated the practice of "store in the front and workshop at the back", developed chain stores and have scored great achievements include Tongrentang Medicine Shop and Quanjude Roast Duck Restaurant.

Those stores that have inherited and carried forward tradition, reinstated the practice of "store in the front and workshop at the back", had their unique products sold at supermarkets, department stores and chain stores, and have made considerable profits include Wang zhihe Fomented Beancurd and Daoxiangchun.

And those stores that have inherited tradition and carried it forward to a certain extent, that are still using the approaches of self-plucking, adding aroma to tea and mixing by themselves, so have retained the distinctive feature of a well-known store.

Some old stores have relegated itself to an ordinary shop, although they still use the same name. They can only make the ends meet by leasing out their counters, thus have ceased to exist but in name.

(3) Suggestions for conserving traditional business
- To re-recognize and lend support stores of long-standing.
- Carry forward the management characteristics of traditional operation.
- Collect and publish publicity materials concerning traditional business with a view of learning and drawing on the past experience.
- Hold exhibitions on traditional business and, on the basis of this, construct a museum of this kind.

10. Implementation of Conservation Measures

10.1 Seek Common Understanding and highly promote the Importance of Protecting the Historic City of Beijing

Seriously implement the series of important directives of the Party Central Committee regarding the urban construction and management of Beijing and look at the question of implementing the "Conservation Plan for the Historic City of Beijing" from the high plane of building Beijing into a first-class international metropolis and a historic city. Try to reach a consensus of "love and protect the historic city" among citizens by effective promotion.

10.2 Further implement the policy of "two strategic shifts" contained in the "Master Plan for Beijing"

Further carry out the requirements set in the "Master Plan of Beijing", quickening the pace of construction from inner city to outlying suburbs, and from expanding the urban boundary to adjustment and renovation, in order to mitigate the population in the old city and its functions, successfully undertake the project of building affordable housing in places where urban redevelopment is going on, thus paying the way for protecting the historic city.

10.3 After the approval of the "Conservation Plan for the Historic City of Beijing", measures must be taken to implement it quickly

During the tenth "five-year-plan", detailed plans should be worked out for implementing the conservation plan for the historic city of Beijing. These include: the conservation plan for the second group of conservation districts, adjustment plan for road network in the old city, regulating plan for architectural height control in the old city, and designation of the fifth and sixth groups of the conservation scope and construction controlled areas of cultural relics sites under conservation.

10.4 Work out relevant regulations for the conservation of the historic city

Initial research for legislative work will be done by municipal departments of cultural heritage, urban planning and construct. They will be charged to draw up the "Ordinance of Protecting the Historic City of Beijing" by taking into account the specific conditions of Beijing, drawing on the experience of other provinces and cities and by soliciting the opinions of people from all walks of life. This ordnance will be submitted to the municipal people's congress for approval. Other documents that will be formulated include "Management Regulations for the Conservation districts in Beijing", which will be used as legal basis to be implemented strictly in urban management; special conservation regulations concerning world heritages such as the Forbidden City, Summer Palace, Great Wall, Temple of Heaven, which will clarify accountability, guarantee preservation fund, standardize construction and management behavior.

10.5 Urban renewal in the Old City should not be accomplished by means of real estate development

The result of urban redevelopment in the old city will have a great impact on the general looks of the old city of Beijing, so The approach used in real estate development should not be applied in the renovation of crumbling houses in that area. Major Hutongs, big and famous trees and architecture of historic values must be protected in the process of redevelopment.

10.6 Do research on relevant regulations and policies on the reform of housing property in the old city and launch pilot projects

Actively push the reform on property right in the conservation districts of the old city and conduct research on and formulate relevant policies. Each district should launch pilot project on implementing the plan for protecting the conservation districts, and try to accumulate experience.

10.7 Raise fund via multiple channels, and set up a financial support system on a preliminary basis

Guided by the concept of "relying on the State and helped by the participation by all members of society", pilot projects should be launched in an active way to explore effective approaches to protecting, maintaining, renovating and utilizing cultural relics sites and conservation districts. A safeguard system for funding should also be established. Apart from investment from government, funds should be raised via many channels such as by developing tourist industry, money raised by units and donations by society, so as to establish a safeguard mechanism for the conservation fund.

10.8 Further encourage research on ways of inheriting and carrying forward the traditional features and cultural characteristics of Beijing

The contents of historic city include not only physical things such as urban layout, cultural heritages, architecture of unique styles and traditional urban features, but also spiritual things like plays, literature, painting, music, cultural and artistic activities. They also embrace local customs such as temple fairs, as well as old stores, unique brand name products and other business activities. Research must be done to work out effect measures to inherit and carry forward these time-honored culture, art, folk customs and business activities.

In a word, so long as we assume a responsible attitude towards history and people, attach importance to cultural heritage and take measures to protect them, explore the deep meanings inherent in the historic city, and work out effect ways to preserve the historic city in a comprehensive way in the process of urban modernization, we can definitely build Beijing into a first-class historic city and a modern international metropolis.

September 2002

北京市域文物保护单位分布图
The Distribution of the Preserved Cultural Relics in Beijing Proper

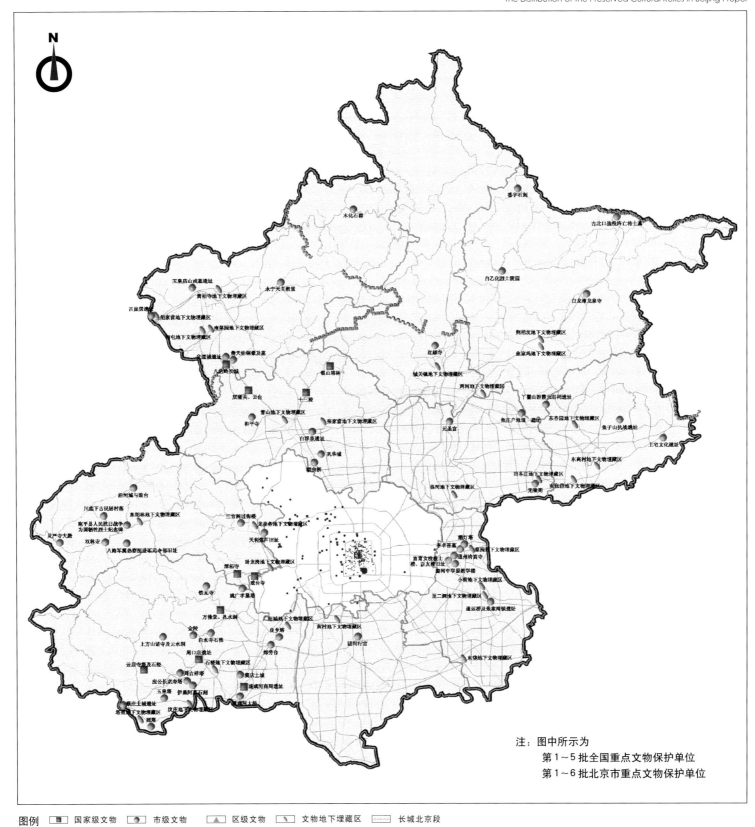

注：图中所示为
第1~5批全国重点文物保护单位
第1~6批北京市重点文物保护单位

规划图纸

北京市区文物保护单位分布图
The Distribution of the Preserved Cultural Relics in Urban Area of Beijing

北京历史文化名城保护规划·规划图纸

图例　■ 国家级文物保护单位　● 市级文物保护单位　▲ 区级文物保护单位　▽ 地下文物埋藏区

北京名城

北京历史文化名城北京皇城保护规划
Conservation plan for the Historic City of Beijing and Imperial City of Beijing

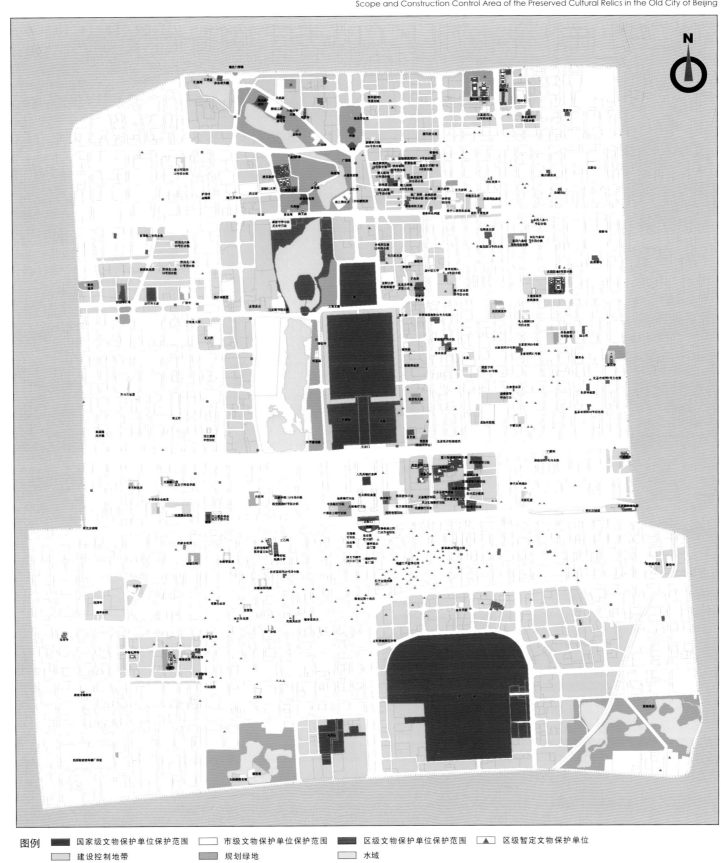

北京旧城文物保护单位保护范围及建设控制地带图
Scope and Construction Control Area of the Preserved Cultural Relics in the Old City of Beijing

规 划 图 纸

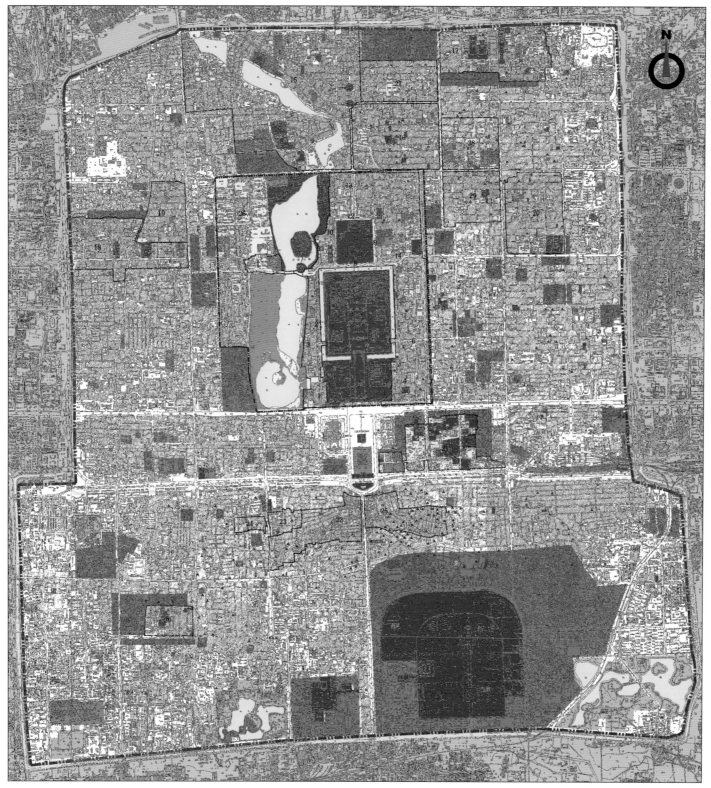

北京旧城文物保护单位及历史文化保护区规划图
Planning of the Preserved Cultural Relics and Conservation Districts in the Old City of Beijing

图例 ■ 国家级文物保护单位　■ 市级文物保护单位　■ 区级文物保护单位　▲ 区级暂定文物保护单位
□ 历史文化保护区　■ 文物建控地带　■ 绿地　□ 水域

历史文化保护区：1.南长街 2.北长街 3.西华门大街 4.南池子 5.北池子 6.东华门大街 7.文津街 8.景山前街 9.景山东街 10.景山西街 11.陟山门街 12.景山后街 13.地安门内大街 14.五四大街 15.什刹海地区 16.南锣鼓巷 17.国子监地区 18.阜成门内大街 19.西四北一条至八条 20.东四北三条至八条 21.东交民巷 22.大栅栏 23.东琉璃厂 24.西琉璃厂 25.鲜鱼口 26.皇城 27.北锣鼓巷 28.张自忠路北 29.张自忠路南 30.法源寺

北京历史文化名城保护规划·规划图纸

北京历史文化名城北京皇城保护规划
Conservation plan for the Historic City of Beijing and Imperial City of Beijing

北京旧城历史文化保护区分布图（第一批、第二批）
The Distribution of the Conservation Districts in the Old City of Beijing (the First Group and the Second Group)

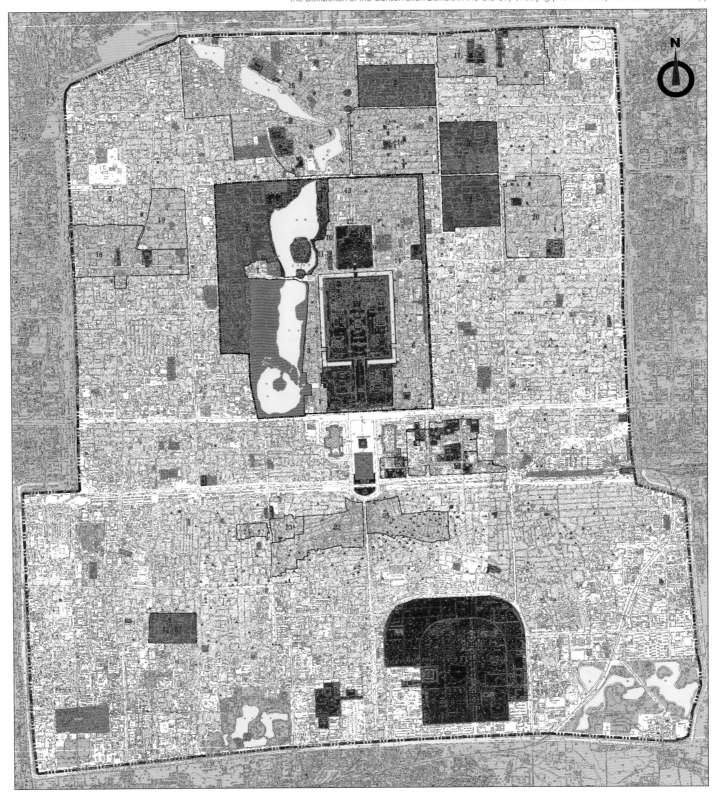

图例： ■ 国家级文物保护单位　■ 市级文物保护单位　■ 区级文物保护单位　▲ 区级暂定文物保护单位
□ 第一批历史文化保护区保护范围　■ 第二批历史文化保护区保护范围　□ 绿地　□ 水域

第一批历史文化保护区：1.南长街 2.北长街 3.西华门大街 4.南池子 5.北池子 6.东华门大街 7.文津街 8.景山前街 9.景山东街 10.景山西街 11.陟山门街 12.景山后街 13.地安门内大街 14.五四大街 15.什刹海地区 16.南锣鼓巷 17.国子监地区 18.阜成门内大街 19.西四北一条至八条 20.东四北三条至八条 21.东交民巷 22.大栅栏 23.东琉璃厂 24.西琉璃厂 25.鲜鱼口

第二批历史文化保护区：① 皇城　② 北锣鼓巷　③ 张自忠路北　④ 张自忠路南　⑤ 法源寺

规划图纸

皇城历史文化保护区现状分析图
Analysis of Conservation District of the Imperial City

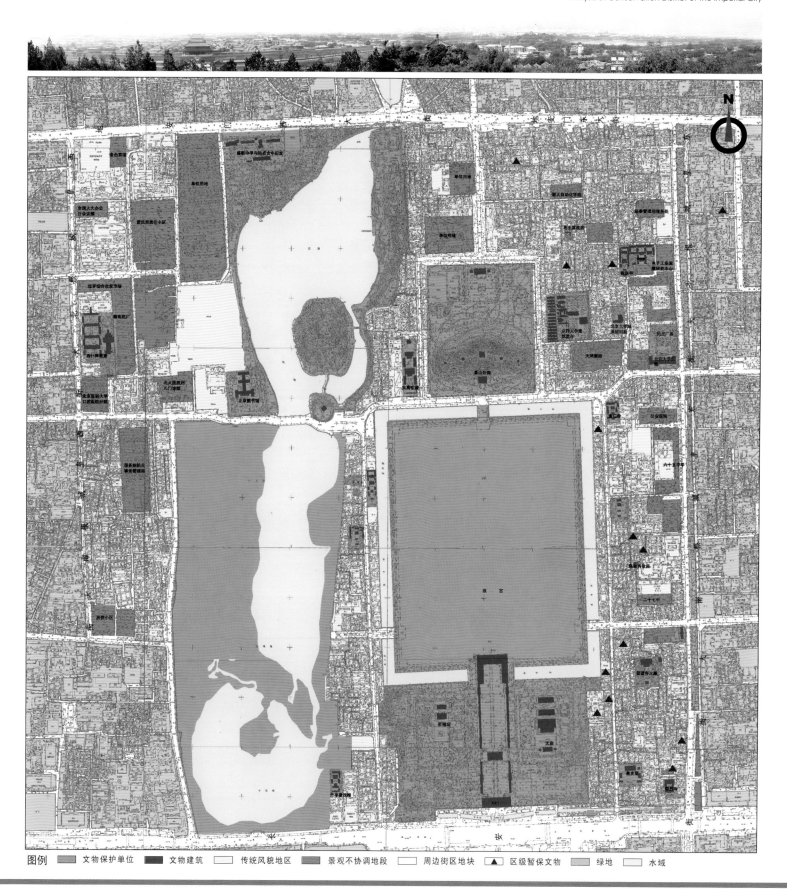

北京历史文化名城保护规划·规划图纸

图例　文物保护单位　文物建筑　传统风貌地区　景观不协调地段　周边街区地块　▲ 区级暂保文物　绿地　水域

北锣鼓巷历史文化保护区现状分析图
Analysis of Beiluoguxiang Conservation District

规划图纸

张自忠路北历史文化保护区现状分析图
Analysis of Conservation District North of Zhangzizhong Street

北京历史文化名城保护规划·规划图纸

北京历史文化名城北京皇城保护规划
Conservation plan for the Historic City of Beijing and Imperial City of Beijing

张自忠路南历史文化保护区现状分析图
Analysis of Conservation District South of Zhangzizhong Street

规划图纸

法源寺历史文化保护街区现状分析图
Analysis of Conservation District of Fayuan Temple

图例： 文物保护单位 ■ 文物建筑 ■ 传统风貌地区 □ 景观不协调建筑 ■

北京历史文化名城保护规划·规划图纸

北京市域第二批历史文化保护区分布图（旧城外）
The Distribution of the Second Group of Conservation Districts in Beijing Proper (out of the Old City)

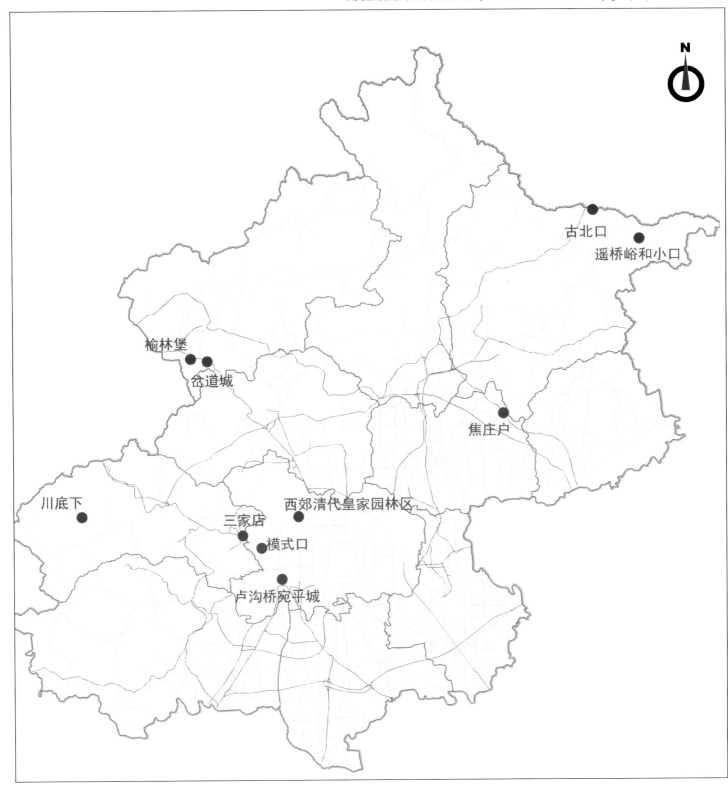

图例　●　第二批历史文化保护区位置示意

规划图纸

西郊清代皇家园林历史文化保护区位置示意及风貌特色图
The Location and Characteristics of the Conservation District of Qing Dynasty Imperial Garden in the Northwestern Outskirt

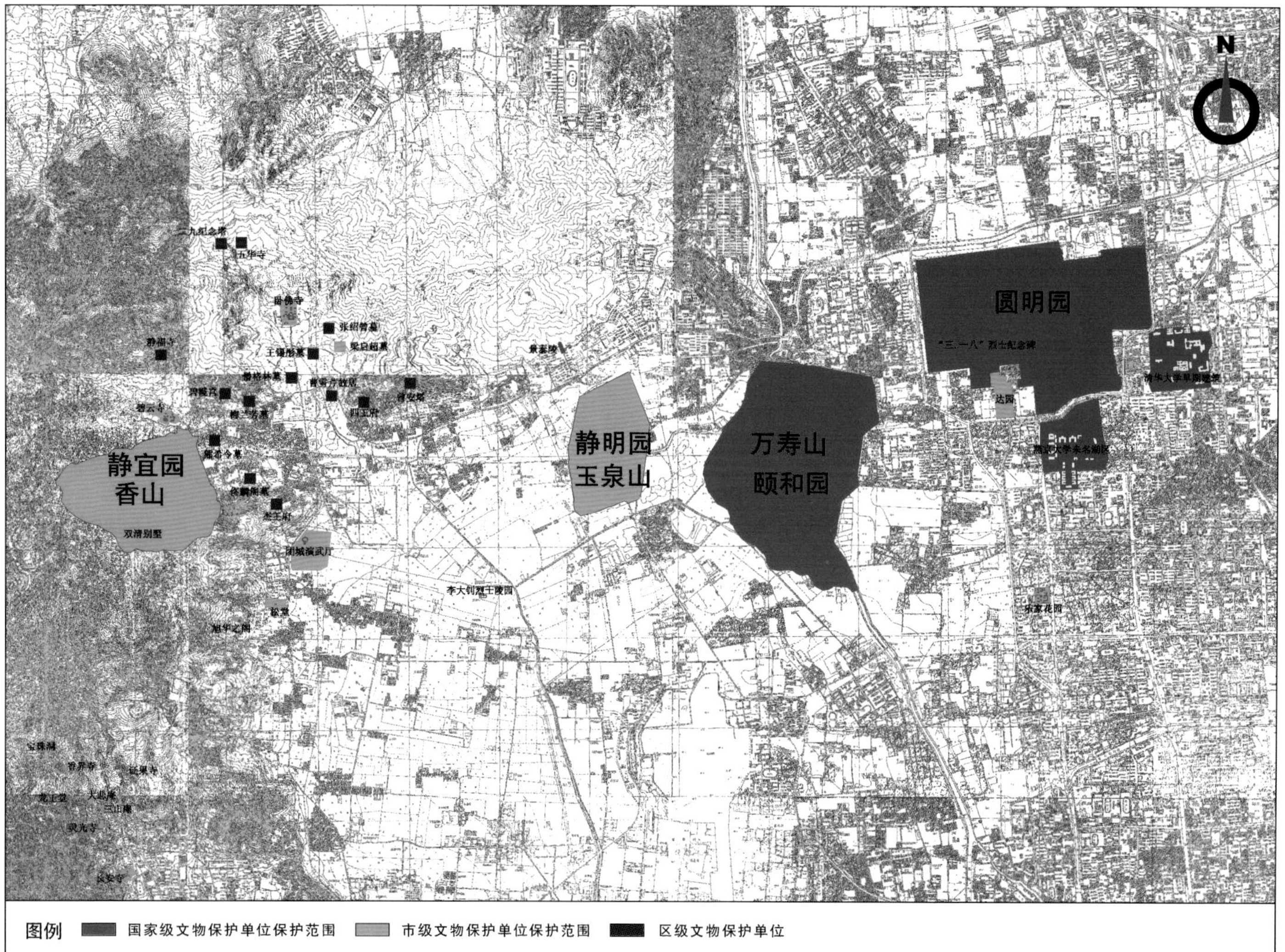

圆明园

静明园 玉泉山

万寿山 颐和园

静宜园 香山

图例：■ 国家级文物保护单位保护范围　■ 市级文物保护单位保护范围　■ 区级文物保护单位

昆明湖与玉泉山　香山　颐和园　圆明园

北京历史文化名城保护规划·规划图纸

北京历史文化名城北京皇城保护规划
Conservation plan for the Historic City of Beijing and Imperial City of Beijing

北 京 名 城

卢沟桥宛平城历史文化保护区位置示意及风貌特色图
The Location and Characteristics of the Conservation District of Lugou Bridge and Wanping City

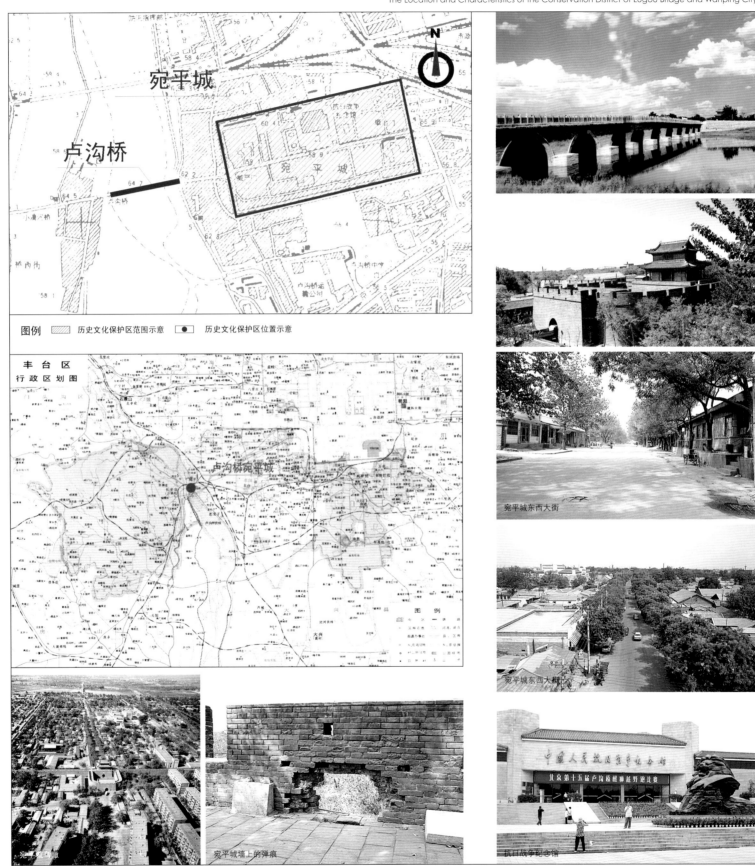

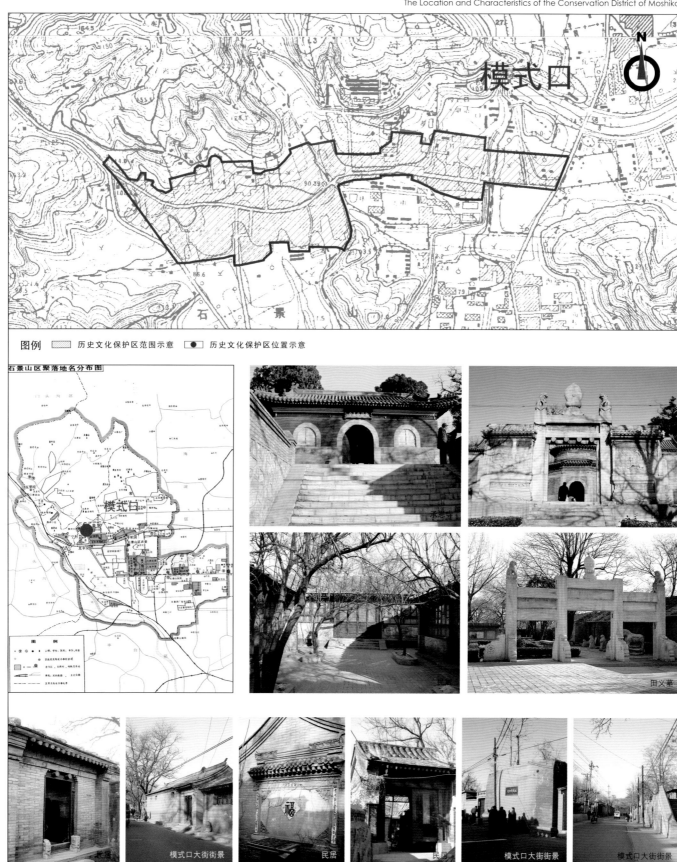

模式口历史文化保护区位置示意及风貌特色图
The Location and Characteristics of the Conservation District of Moshikou

三家店历史文化保护区位置示意及风貌特色图
The Location and Characteristics of the Conservation District of Sanjiadian

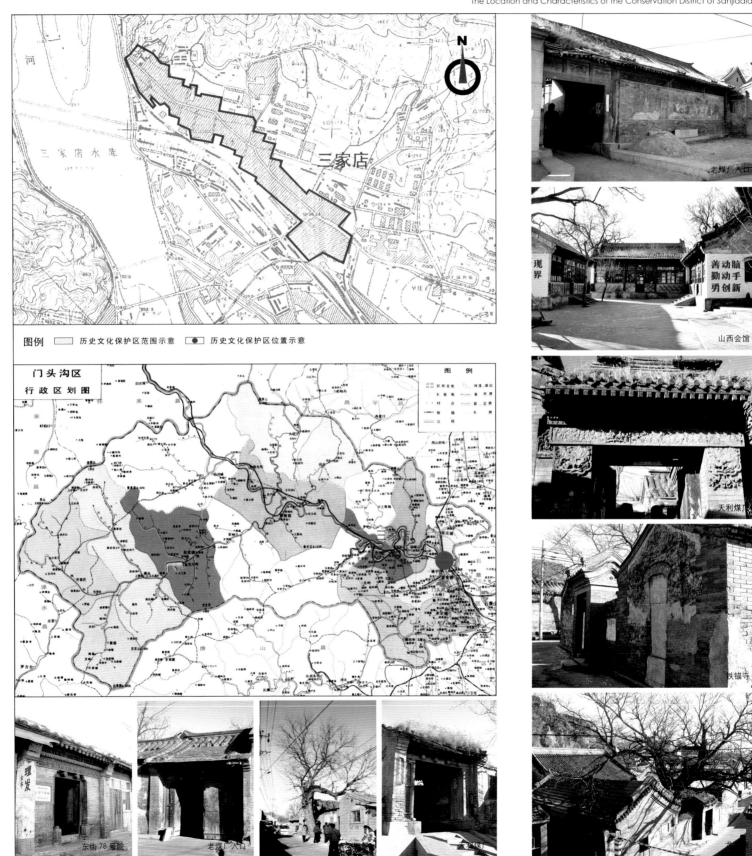

川底下历史文化保护区位置示意及风貌特色图
The Location and Characteristics of the Conservation District of Chuandixia Village

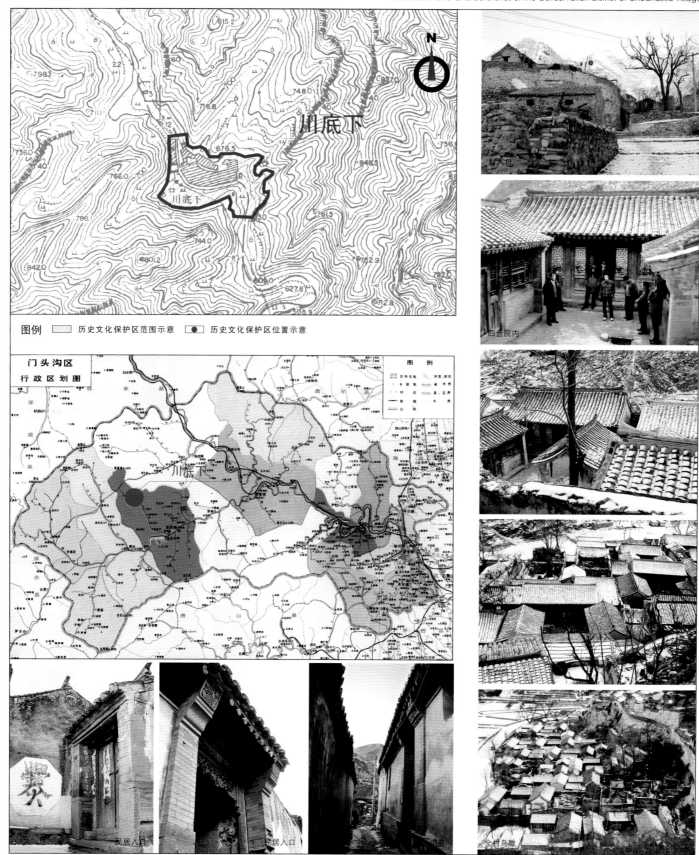

北京历史文化名城保护规划·规划图纸

岔道城历史文化保护区位置示意及风貌特色图
The Location and Characteristics of the Conservation District of Chadao City

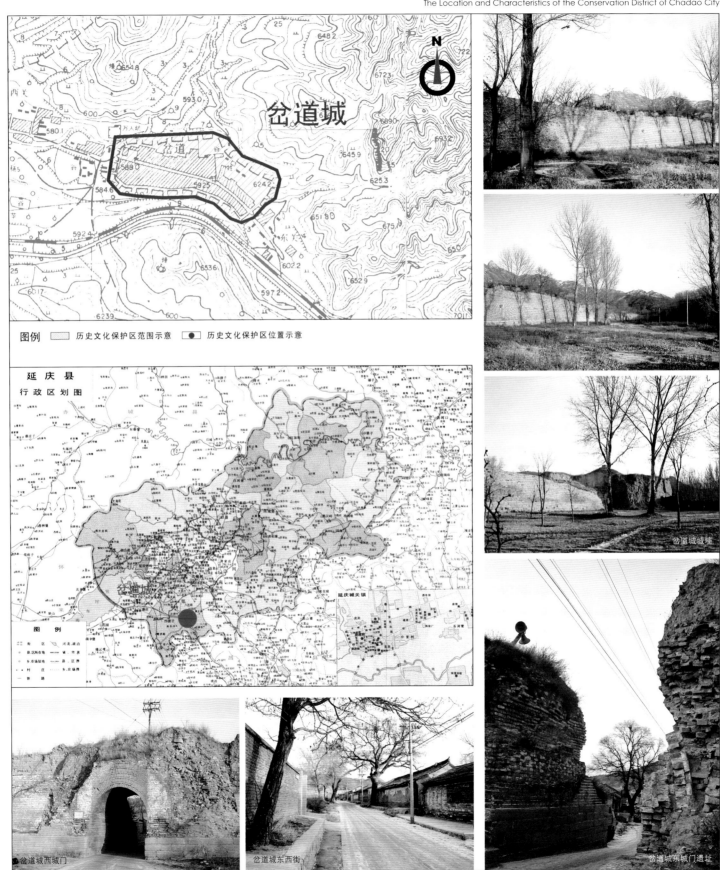

规划图纸

榆林堡历史文化保护区位置示意及风貌特色图
The Location and Characteristics of the Conservation District of Yulinbao

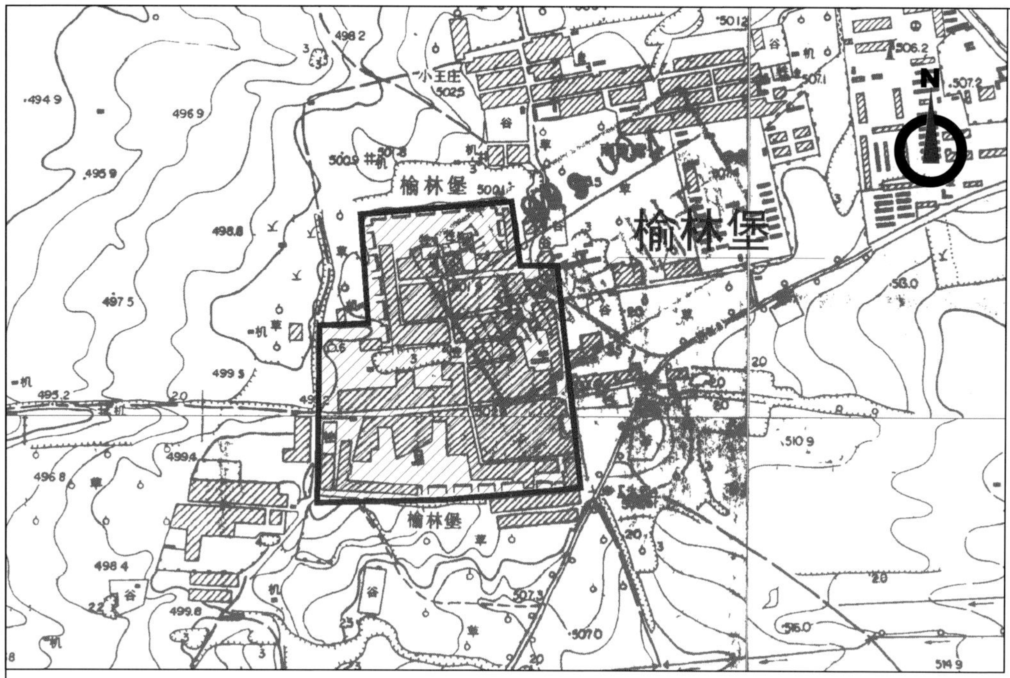

图例 ▨ 历史文化保护区范围示意   ● 历史文化保护区位置示意

延庆县行政区划图

历史文化名城保护规划·规划图纸

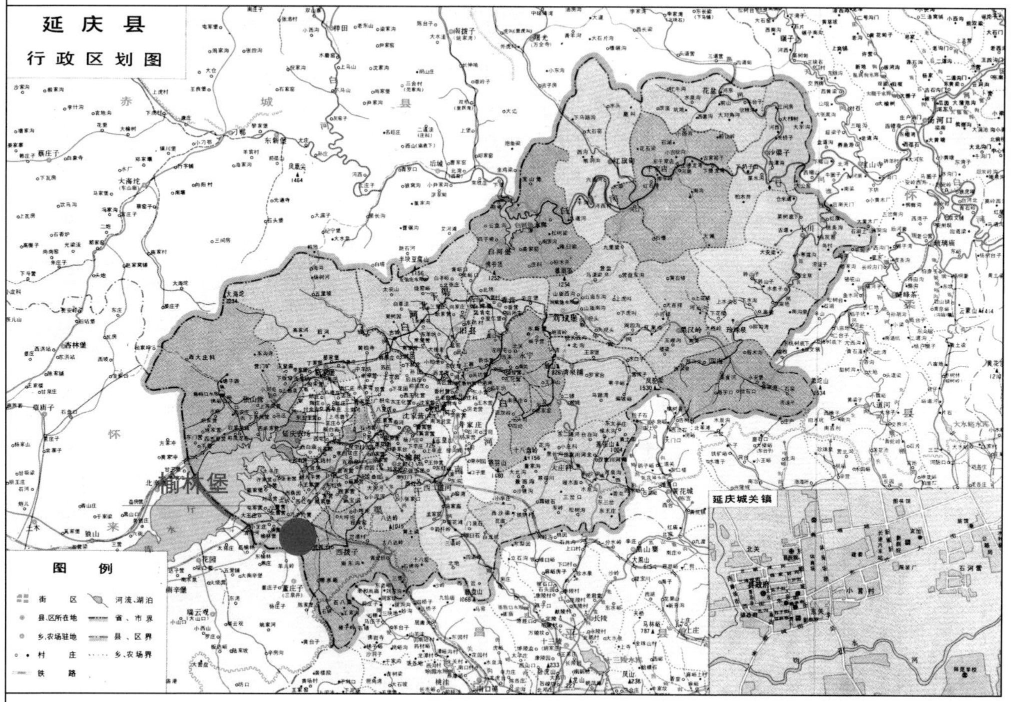

焦庄户历史文化保护区位置示意及风貌特色图
The Location and Characteristics of the Conservation District of Jiaozhuanghu

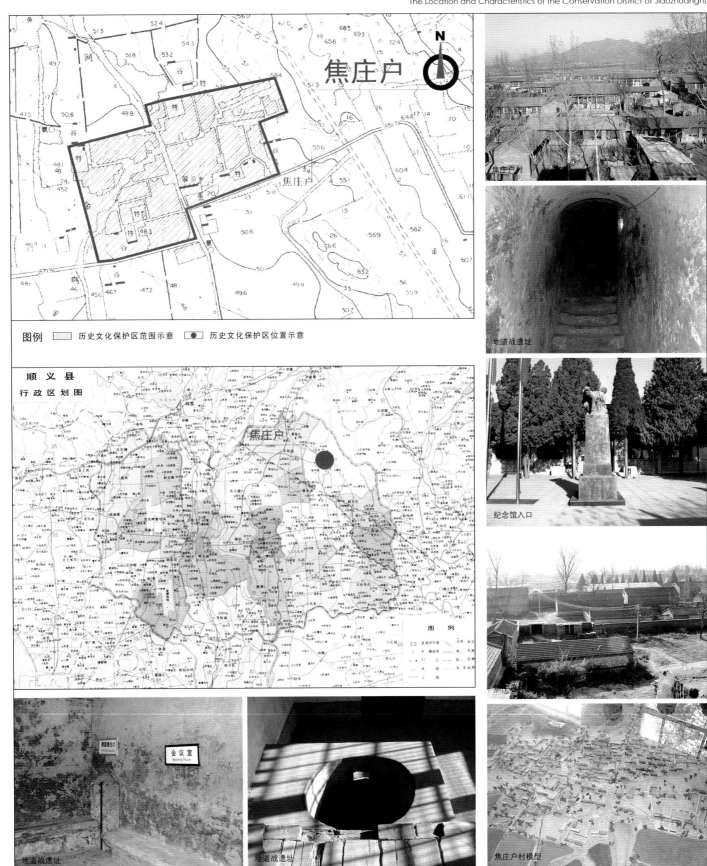

古北口历史文化保护区位置示意及风貌特色图
The Location and Characteristics of the Conservation District of Gubeikou Old City

北京历史文化名城保护规划·规划图纸

北京历史文化名城北京皇城保护规划
Conservation plan for the Historic City of Beijing and Imperial City of Beijing

遥桥峪和小口历史文化保护区位置示意及风貌特色图
The Location and Characteristics of the Conservation District of Yaoqiaoyu and Xiaokou Castle

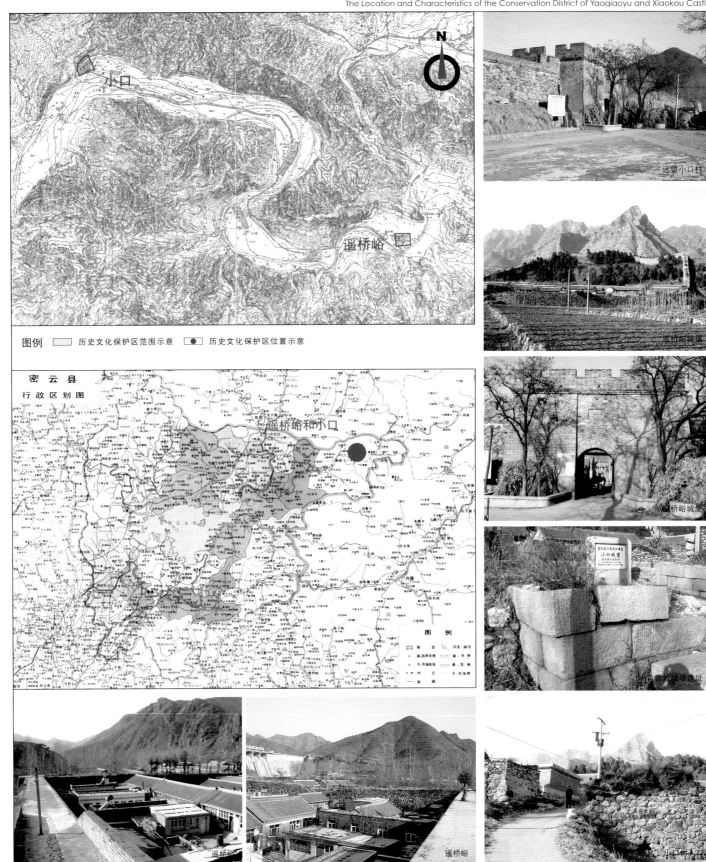

规 划 图 纸

北京市区河湖水系现状示意图
Lakes and Rivers System in Urban Area of Beijing

北京历史文化名城保护规划·规划图

图例　现状暗沟　现状湖泊　护城河道　现状河道　水源河道　漕运河道　防洪河道　风景园林水域

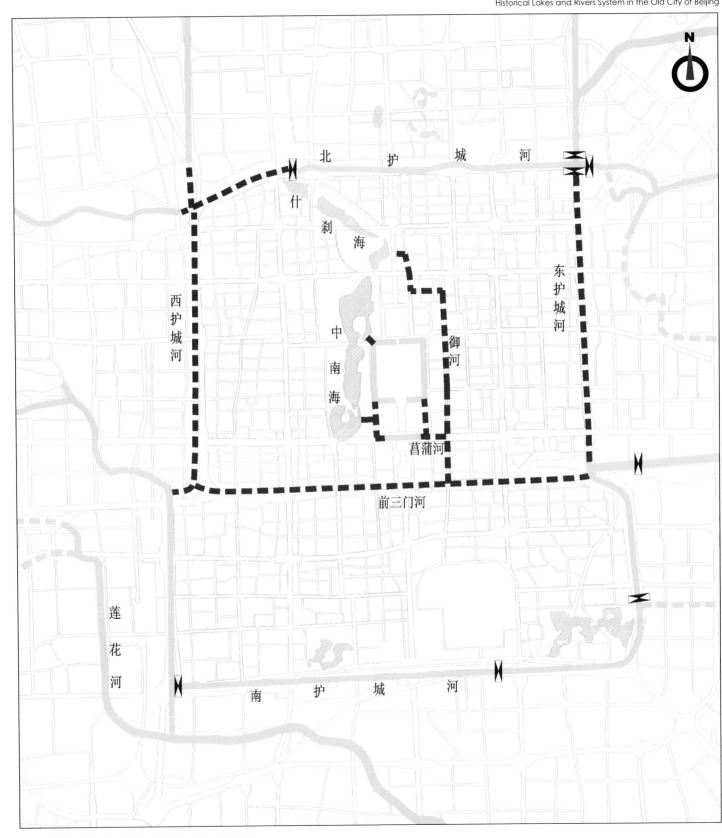

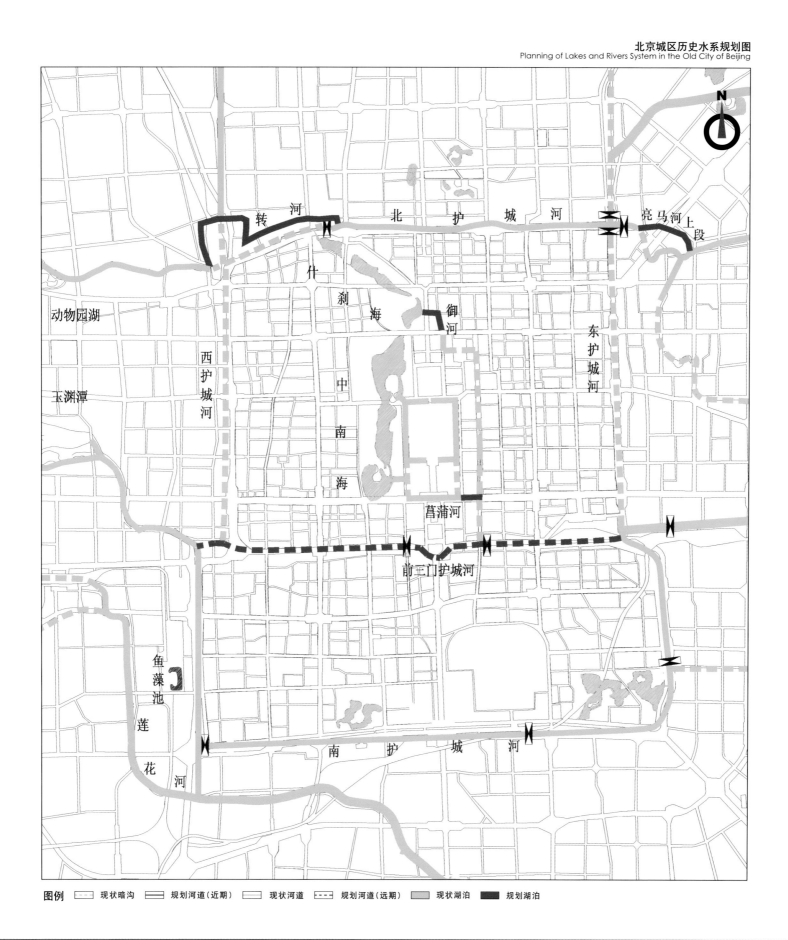

北京名城

北京历史文化名城北京皇城保护规划
Conservation plan for the Historic City of Beijing and Imperial City of Beijing

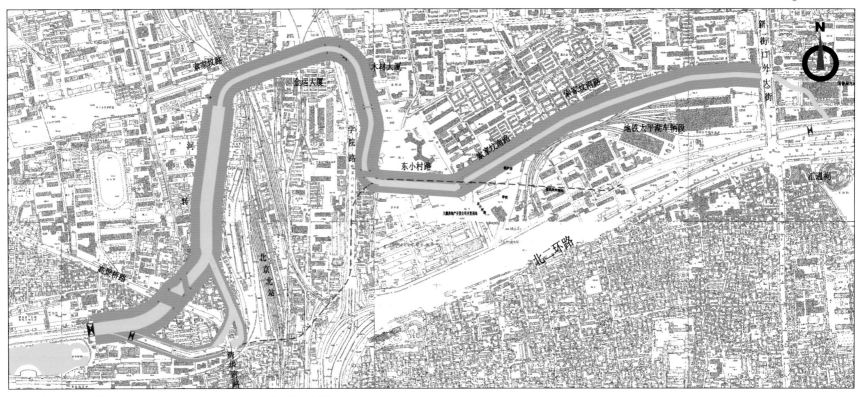

转河恢复规划平面示意图
Restoration Planning of Zhuan River

图例　现状河道　规划河道　规划绿地　现状雨水管道

菖蒲河恢复规划平面示意图
Restoration Planning of Changpu River

图例　规划河道　规划绿地　(高级)住宅用地　文化旅游设施用地　文物古迹用地　市政公用设施用地

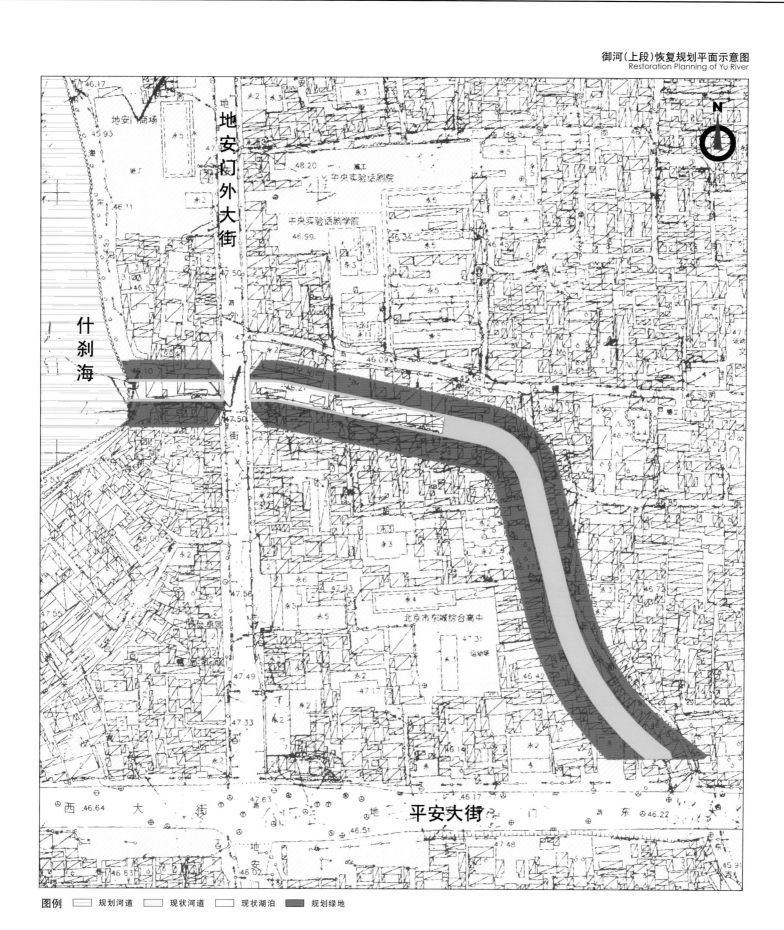

御河（上段）恢复规划平面示意图
Restoration Planning of Yu River

北京历史文化名城北京皇城保护规划
Conservation plan for the Historic City of Beijing and Imperial City of Beijing

前三门护城河规划控制范围图(1)
Planning Scope of Qiansanmen Moat (1)

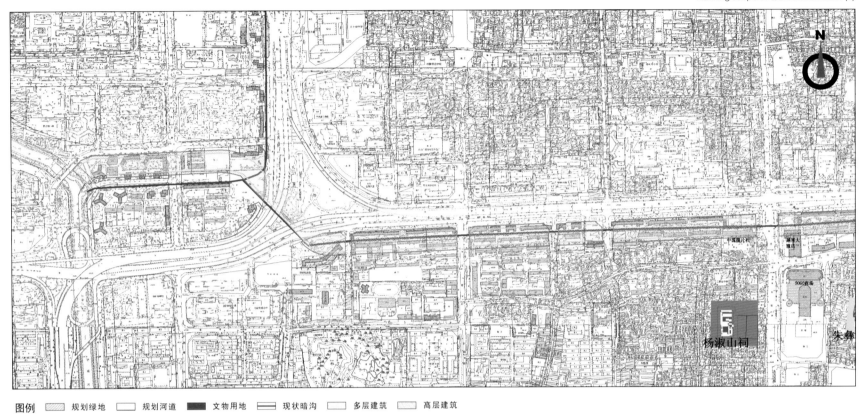

图例　规划绿地　规划河道　文物用地　现状暗沟　多层建筑　高层建筑

前三门护城河规划控制范围图(2)
Planning Scope of Qiansanmen Moat (2)

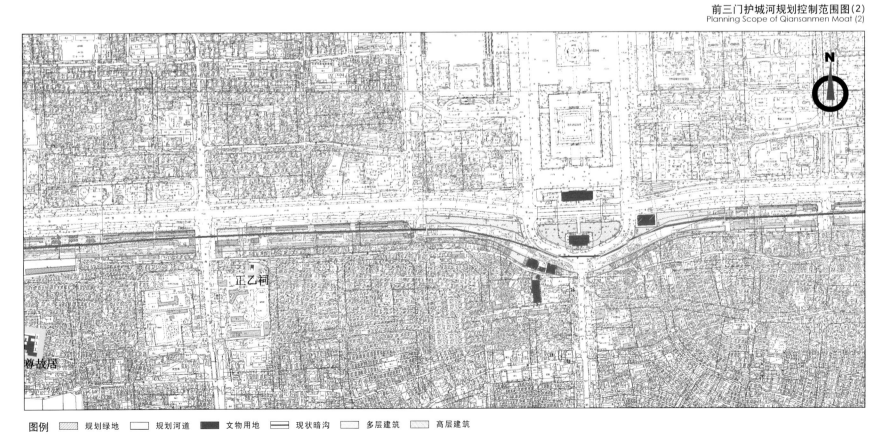

图例　规划绿地　规划河道　文物用地　现状暗沟　多层建筑　高层建筑

规划图纸

前三门护城河规划控制范围图(3)
Planning Scope of Qiansanmen Moat (3)

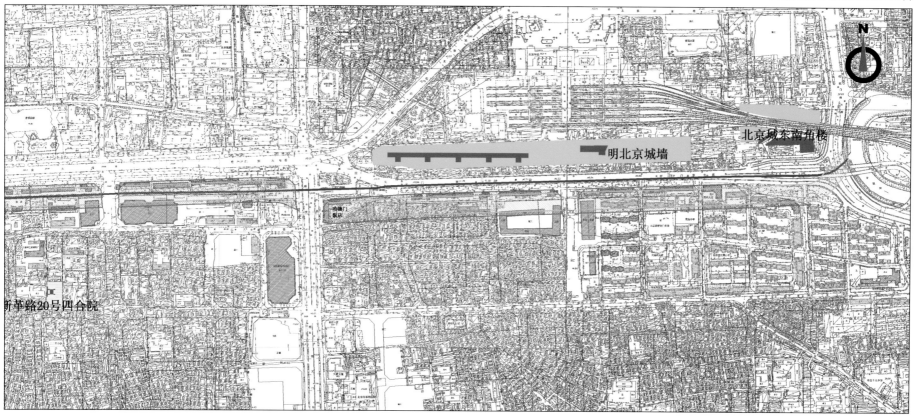

图例 规划绿地 规划河道 文物用地 现状暗沟 多层建筑 高层建筑

北京历史文化名城保护规划·规划图纸

鱼藻池恢复规划平面示意图
Restoration Planning of Yuzao Pond

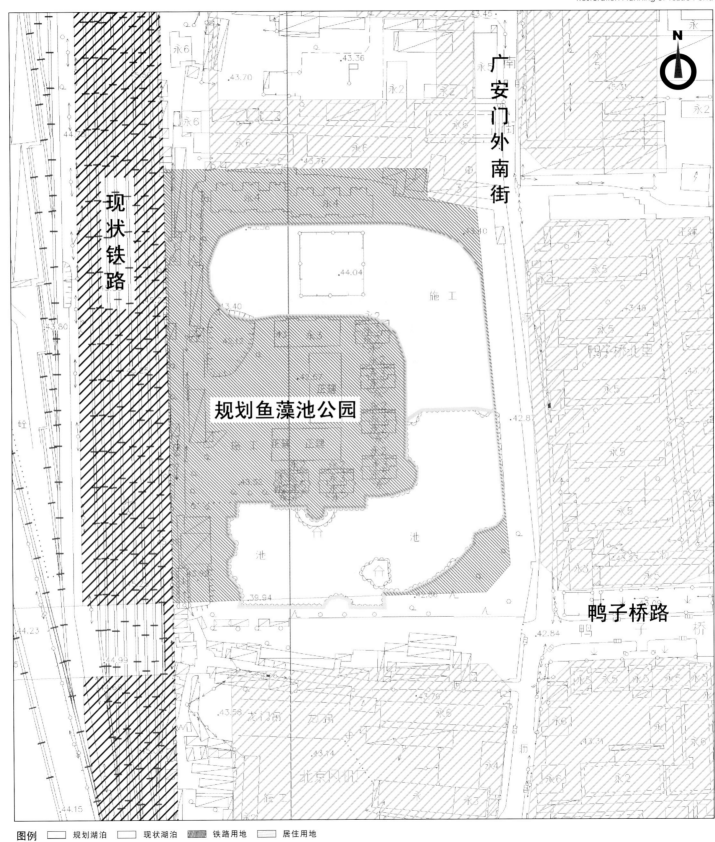

规 划 图 纸

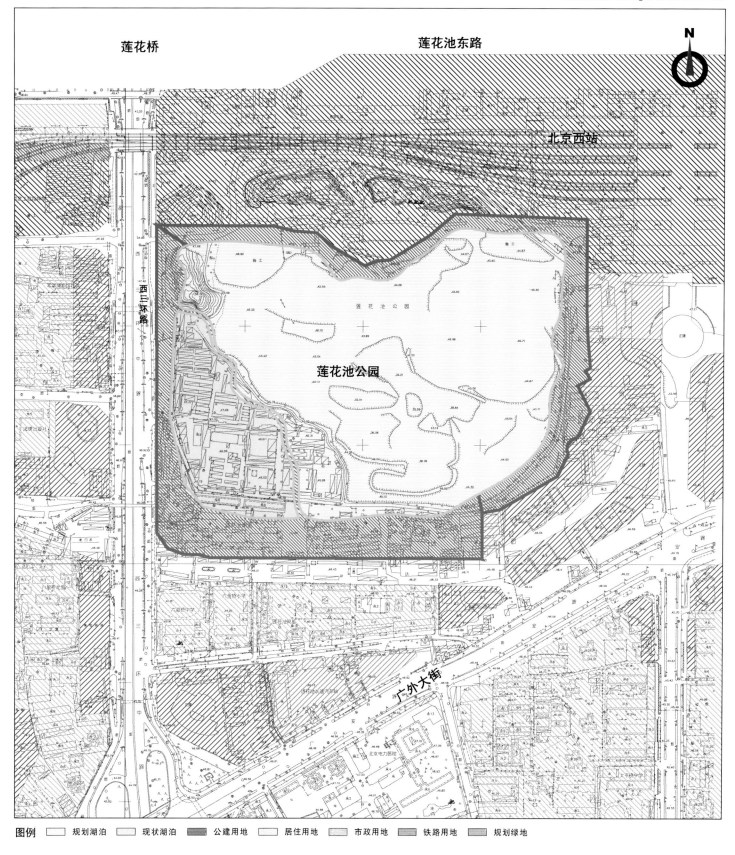

莲花池恢复规划平面示意图
Restoration Planning of Lotus Flower Pond

北京历史文化名城保护规划·规划图纸

图例 规划湖泊 现状湖泊 公建用地 居住用地 市政用地 铁路用地 规划绿地

北京历史文化名城北京皇城保护规划　
Conservation plan for the Historic City of Beijing and Imperial City of Beijing

北京城市中轴线空间结构分析图
Analysis of the Old Axis of Beijing

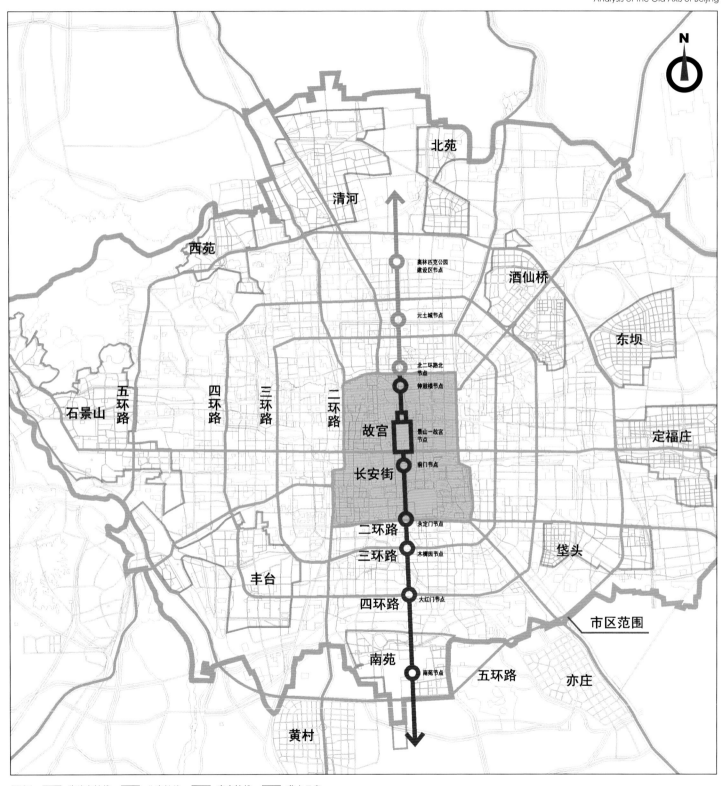

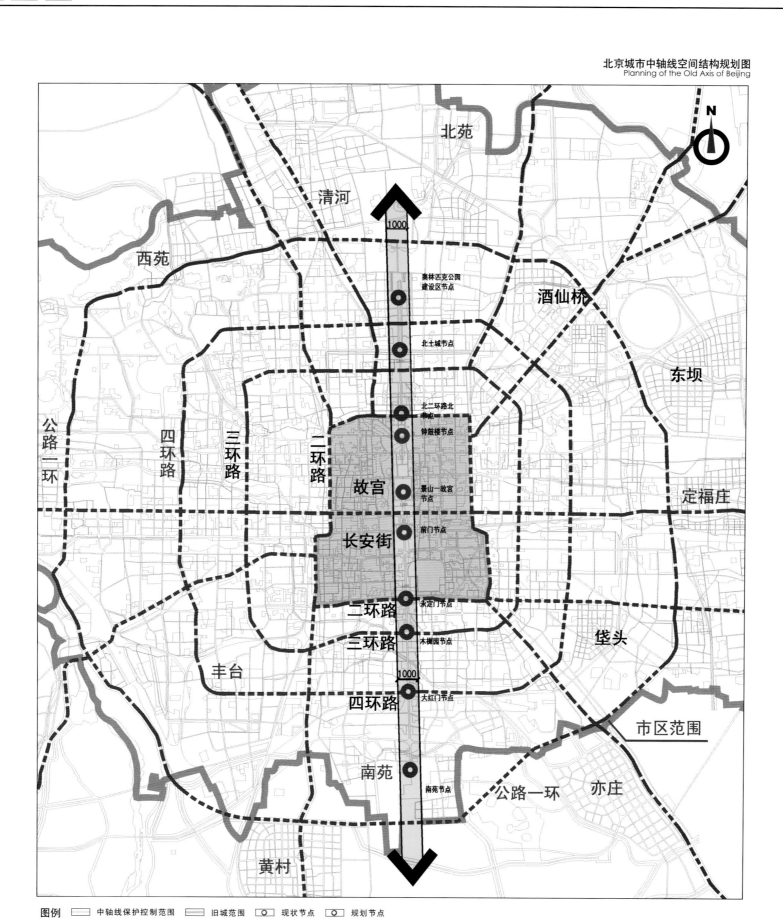

北京历史文化名城北京皇城保护规划 北京名城
Conservation plan for the Historic City of Beijing and Imperial City of Beijing

旧城区原规划主、次干道现状图
Roads System of the Old City

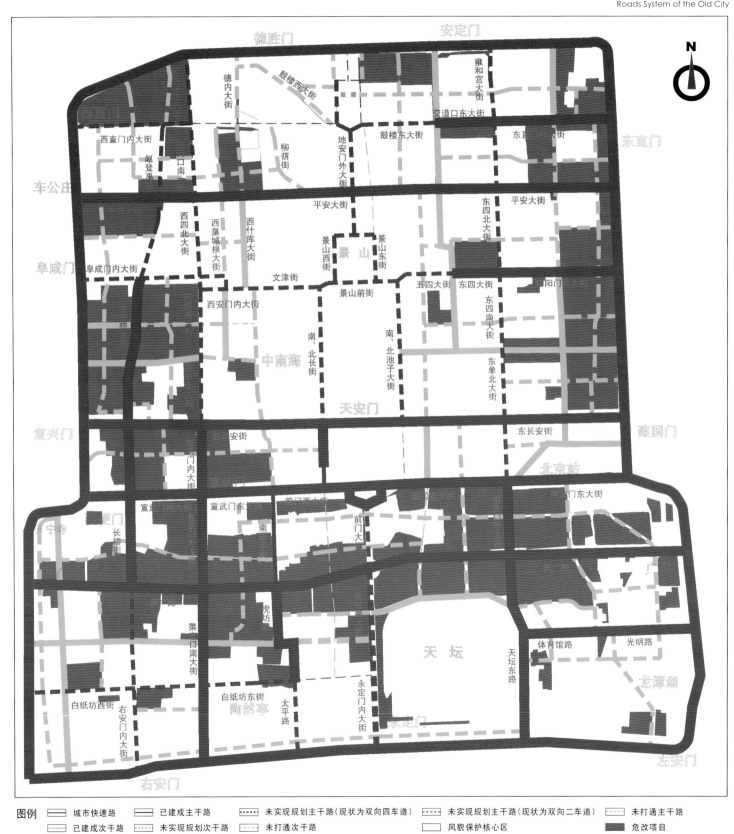

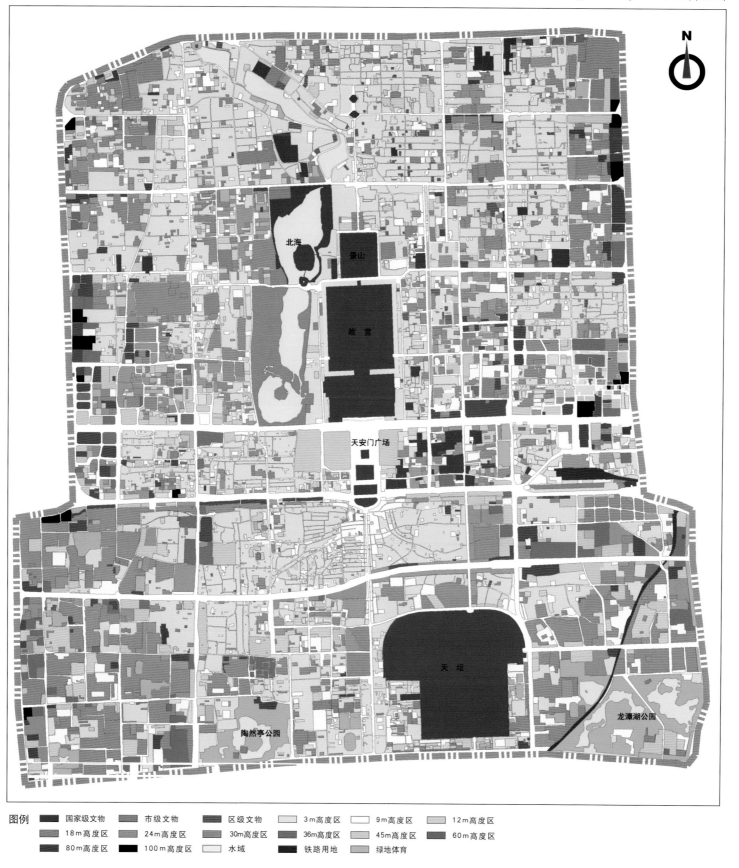

旧城建筑高度现状图（2002年2月）
Height of Buildings in the Old City (2/2002)

北京名城

北京历史文化名城北京皇城保护规划
Conservation plan for the Historic City of Beijing and Imperial City of Beijing

旧城现有高层建筑分布图(2002年2月)
The Distribution of High-Rise Buildings in the Old City (2/2002)

图例　文物　绿地体育　多层建筑区　高层建筑区（高于24m）　旧城城廓

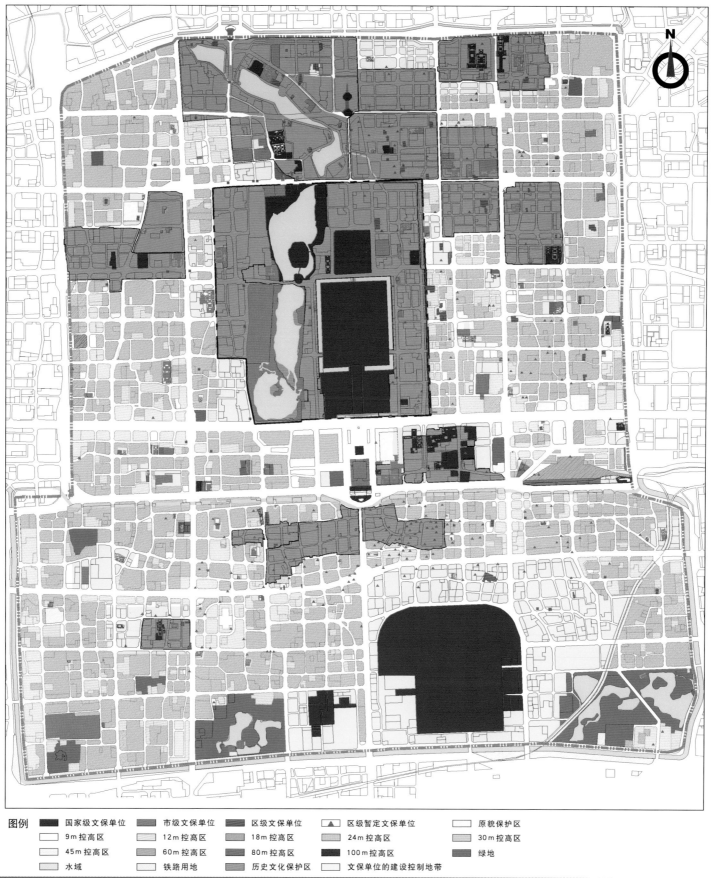

旧城建筑高度控制规划图
Planning of Height Control in the Old City

北京历史文化名城北京皇城保护规划　
Conservation plan for the Historic City of Beijing and Imperial City of Beijing

城市景观线分布图
Landscape Corridors of the Old City

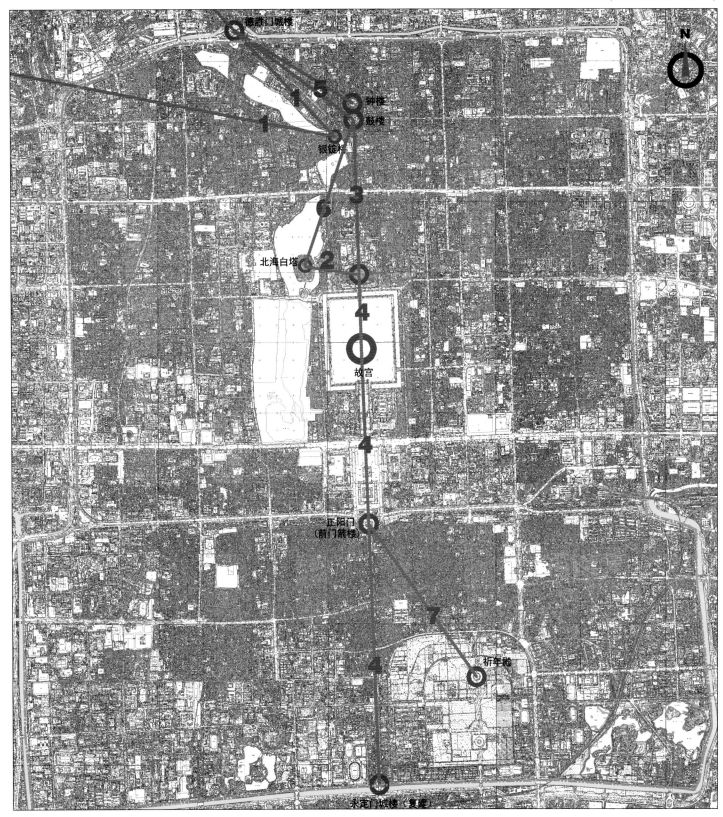

图例　　景观视点　　　城市景观线　　5　城市景观线编号

附件

政协北京市第九届委员会第四次会议党派团体提案

第0658号

案由：关于尽快制定北京历史文化名城总体保护规划的建议
审查意见：建议规划委员会研究办理
内容：

北京是世界著名的古都，是国务院批准的国家级历史文化名城。但在以往的北京发展战略中，历史文化名城保护的地位一直很不稳定，致使北京的古都风貌遭到极大破坏。其原因，一是历史文化名城保护尚未与城市发展战略有机地挂起钩来；二是缺少一个既尊重历史、又从北京城市现状出发的总体保护规划。1982年公布、1992年修订的《北京城市总体规划》中，虽然有有关历史文化名城保护方面的内容，但只是原则规定，并没有具体的规划方案，更谈不上详细规划，这与北京的地位很不相称。

刘淇市长在本次人代会所作的《关于北京市国民经济和社会发展第十个五年计划纲要的报告》中指出："城市建设和发展必须最大限度地保护好历史文化名城风貌。本着'考虑历史、立足现在、着眼未来'的思想，重点保护皇城、城市中轴线、首批25片历史文化保护区、重要文物保护单位和城市水系，维护古城基本格局和原有风貌。"刘市长的讲话明确了历史文化名城保护在北京城市建设中的重要地位，为进一步作好历史文化名城保护工作创造了条件。

为落实"十五"规划纲要提出的目标和任务，真正维护好古城基本格局和原有风貌，应抓紧制定北京历史文化名城保护规划。具体建议如下：

1. 组织专门班子，召开专门会议，对《北京城市总体规划》中有关历史文化名城保护原则的执行情况做出评估，肯定成绩，找出问题，总结经验教训，并以此作为编制"历史文化名城保护规划"的基础。

2. 在上述基础上，抓紧编制"北京历史文化名城保护规划"。在编制"名城保护规划"的过程中，应着重研究解决以下问题：

（1）明确历史文化名城保护的地域范围，摸清保护范围中的文物保护单位和未列入保护单位的历史遗存的现状，提出保护对策。

（2）分层次确定历史文化名城保护的内容，分别对文物保护单位、历史街区、名城传统风貌和城市设计提出具体保护方案；对如何解决危旧房改造与名城保护之间的矛盾，旧城区内新建筑的高度、风格、形式、色调如何与古城的历史风格相协调等问题，做出规划。

3. 鉴于危旧房改造步伐加快，时间紧迫，应规定完成编制规划的时限。

4. 对《北京城市总体规划》作适当的修改，使之与"名城保护规划"相协调。实践证明，现行的《北京城市总体规划》已经不能完全适应当前城市发展与保护的需要。因此建议，在编制"名城保护规划"的同时，对《总体规划》予以修改、调整，使之更加符合实际。

5. 成立一个由多学科、高层次专家学者组成的"北京历史文化名城保护规划课题组"，完成规划文本的编制工作。

提案人：北京市政协文史委员会（T96）

2001年2月5日

政协委员建议抓紧编制《北京历史文化名城保护规划》

近日，市政协文史委员会委员对本市制定"《北京历史文化名城保护规划》"（以下简称《规划》）"工作再次提出建议：

1. "申奥"成功后，城市建设的速度会进一步加快。因此历史文化名城保护工作更加紧迫，必须抓紧编制"《规划》"，并把它当成十万火急的硬任务。

2. 建议由市政府发布相关政府令，对一些问题做出强行规定。

（1）对一些重要地区，采取"先控后放"的方针，目前不可动一砖一瓦。

（2）旧城区的危改绝不允许突破高度限制。

（3）在对25片历史文化保护街区之外的地区进行危改时，必须先进行文物普查，对有价值的历史建筑采取保护措施后再动工。

3. 建议成立由市政府主要领导牵头，包括中央单位和军队同志参加的领导班子，统筹协调有关问题。

4. 编制"《规划》"应突出重点，力争解决难点问题。

（1）皇城保护应作为"《规划》"的一个重点。

（2）对旧城区建筑的控高、历史街区内风貌保护与改善市民生活的矛盾、旧城区道路交通的走向和宽度与文物保护的矛盾等老大难问题，应在此次"《规划》"编制时有所突破。

5. 要解决好"《规划》"与《北京城市总体规划》、"十五计划"、"奥运行动规划"之间的衔接。

6. 编制"《规划》"的工作应与北京历史文化名城保护的立法工作同步进行。

（市政协）
2001年10月27日

关于《北京历史文化名城保护规划》编制工作的报告

市政府:

根据市领导在北京市政协文史委员会提出的"关于尽快制定《北京历史文化名城保护规划》的建议"上的批示,经研究,提出以下工作意见:

1. 规划指导思想

以江泽民总书记年初视察北京时的指示:"北京的城市规划和建设,一定要充分考虑历史、立足现实、着眼未来,在最大限度保护历史文化名城的前提下,加快旧城改造步伐,努力提高城市现代化水平"和1993年国务院"关于北京城市总体规划的批复"为指导,在学术思想和规划上要充分尊重和吸纳各方专家之长。

2. 规划基本思路

工作的思路是:领会中央精神、发掘历史内涵、深入调查现状、认真总结经验、广泛听取意见、提出比较方案。

规划的主要内容是:"三个层次、一个重点、文化融合"。

"三个层次"是:文物保护单位的保护、历史文化保护区的保护、历史文化名城的保护;

"一个重点"是:旧城区;

"文化融合"是:传统文化、商业特色的继承和发扬,丰富历史文化名城的内涵。

3. 编制单位和责任单位

参加编制单位:市规划委、市文物局、市规划院、市商委、市国土房管局、市文物研究所、清华大学、北京建工学院、北京工业大学等。

责任单位:市规划院、市文物局

协调单位:市规划委

聘请专家顾问:由国家文物局、建设部、中国城市规划设计研究院、北京大学、清华大学等单位的著名专家组成。

4. 研究专题

专题一:文物保护单位的保护和利用(包括文物资源的情况,各级文物保护单位、地下文物埋藏区、近现代保护建筑的现状、问题和规划方案,保护和利用的政策意见等)。

专题二:历史文化保护区的保护、整治和更新(包括旧城25片历史文化保护区规划综合、提出市域范围内新的历史文化保护区名单和保护范围、历史文化保护区的保护和更新的实施政策及管理意见等)。

专题三:历史文化名城整体格局的保护与发展(包括中轴线的保护与发展、皇城保护区的保护与整治、旧城城廓遗址的保护、旧城平缓开阔空间特色与建筑高度控制、重要景观视廊和街道对景的保护、古树名木和绿地系统的保护、危旧房改造与历史特色的传承等)。

专题四:道路交通规划与旧城保护(包括旧城道路系统与文保单位、历史文化保护区的关系,主要胡同系统的保留等)。

专题五:水系的保护、整治和恢复(在市区水系整治基础上重点研究旧城护城河保护及恢复的可能性等)。

专题六:北京传统文化和商业的继承和发扬(历史文化名城的保护不仅有建筑、街道、城墙等"硬件",也应有对民俗民风、传统文化、饮食、娱乐等"软件"的继承与发扬)。

5. 时间进度

总进度:2001年6月开始,至2002年6月底全部完成,上报市政府。

(1)2001年6月至7月,准备阶段:制定工作计划,成立协调、编制工作组,落实分工,提出经费方案,报市政府批准。

(2)2001年8月至10月,调查研究阶段:各专题开展现状调查、总结实践经验教训,国内外相关城市学习、比较。

(3)2001年11月,提出各专题报告和图纸。

(4)2001年12月至2002年1月,提出综合报告初稿。

(5)2002年2月至4月,征求人大代表、政协委员、专家学者和各有关部门意见,向市政协文史委汇报。

(6)2002年5月至6月,修改方案和综合报告,完成规划文本、全部图纸和成果。

(7)2002年6月,上报市政府批准。

北京市规划委员会
北京市文物局
北京市城市规划设计研究院
2001年8月9日

附件

政协委员对《北京历史文化名城保护规划》编制工作的意见

2002年1月22日,北京市规划委员会、北京市城市规划设计研究院和北京市文物局向北京市政协文史委员会委员通报了《北京历史文化名城保护规划》(初稿)的编制情况。委员们对《保护规划》的初稿及其编制工作非常满意,一致认为,这份《保护规划》初稿总结了历史经验,吸收了政协委员和社会各界的意见,指导思想明确,思路对头,措施得当,基本成熟。规划的实施,将对北京历史文化名城的保护发挥重要作用。通报会上,委员们还对进一步做好《北京历史文化名城保护规划》编制工作提出了如下建议:

1. 制定《北京历史文化名城保护规划》必须打破常规,边征求各方面意见,边报政府审议。

按照"十五"计划规定,北京市要在五年之内对全市164片、近700万m²的危旧房进行彻底改造,危改在给京城百姓改善住房条件带来希望和福音的同时,也对北京历史文化名城保护形成了巨大威胁。由于大多采取房地产开发的方式,危改出现了"推平头"和摧枯拉朽的态势。致使散布在老街区内一些有人文历史价值或艺术价值的传统建筑,未经甄别就被夷为平地。这种势头有增无减,群众反映强烈。按照常规,《保护规划》还需反复征求意见、修改论证和履行必要的审批程序,所以距公布实施尚需一定的时间。委员们担心,待《保护规划》出台,北京的旧城区已被拆除殆尽。届时即使规划编得再好,也只能是一纸空文。委员们认为,现在关键是名城保护要和危改抢时间、争速度,因此制定《保护规划》必须打破常规,边征求各方面意见,边报政府审议,同时采取紧急抢救措施,使《保护规划》初稿里一些规定及早生效,尽最大的努力减少危改给北京古城保护带来的损失。

2. 立即将皇城公布为北京历史文化名城的核心保护区,制定皇城保护规划,并申报世界文化遗产。

(1)划定皇城历史文化保护区的保护区域,公布保护区域的范围,构建起完整的皇城区域意向,使公众明确地感知皇城的存在,以便对皇城的保护实行监督。

(2)限期制定皇城保护规划,申报世界文化遗产,按申报的要求对皇城的环境进行治理。

(3)立即停止审批修建皇城保护区内两层以上的房屋和与传统风貌不协调的建筑,杜绝"建设性"破坏继续发生。

(4)皇城保护区内的成片危旧房改造必须慎行,应遵循"微循环式"的保护与更新方法,逐步整治、更新保护区环境。严禁随便拆除皇城区内的老建筑,并结合旧城外的土地开发,疏导皇城保护区内的人口,降低其中的人口密度。

(5)皇城保护区内的道路改造应以保护为前提,要尽量减少交通流量,有些地区应设为机动车禁行区(如故宫筒子河)。

3. 采取紧急切实的措施,解决危旧房改造和城市建设中历史文化保护区及文物保护的紧迫问题。

(1)尽早公布新增历史文化保护区名单,研究制定相应的政策。

1)划定第二批历史文化保护区和扩展第一批历史文化保护区时,应尽可能多保留一点胡同和四合院;注意深入挖掘北京的文化底蕴,重视会馆和名人旧居的保护;历史街区应尽量连成片,以更真实地反映北京的胡同格局及四合院民居独特的传统风貌。

2)除《保护规划》初稿中划定的两批历史保护区外,旧城区域内是否还有历史地段可列入保护范围,有关单位应多听取专家和各方面人士的意见。如果这次来不及增加新的历史文化保护区,是否可以把一些目标先定为"缓拆缓建区"。

3)积极研究制定历史文化保护区房屋产权私有化的相关政策,推进私有化的进程。

4)抓紧研究制定鼓励人口外迁政策,是否能在旧城以外划定与危改联动的配套土地或住宅,定向疏散历史文化保护区内的人口。

5)研究制定鼓励产权单位和个人按照保护要求修缮房屋的相关政策。

(2)尽快以"政府令"的形式向社会公布第三次文物普查登记的在册文物名单,对占全市文物资源72%的、没有定级的文物,应明确规定,在危改中视同文物保护单位,不得随意拆除;如确需拆除,必须报市文物局审查批准。

(3)市区文物部门还应立即组织人力,对北京旧城区内即将成片改造的胡同街巷进行详细普查。一方面避免有价值而又没有定为文保单位的建筑因漏查被拆除;一方面保留一份比较详细的资料,作为北京历史文化名城保护的基本依据。

(4)区县文物主管部门应对本区域内尚未定为文保单位的文物进行价值鉴定,分别提出保护措施。对文物暂保单位,尽快报区县政府批准,公布为区县级文物保护单位。在此之前市政府应作出规定,在危改中文物暂保单位应视同正式的文物保护单位,不得随意拆除,确需拆除的,必须报请市文物局审查批准。

(5)对旧城区内地下文物埋藏的情况,要作进一步勘察,对地下文物埋藏比较集中的地区,公布为新的地下文物埋藏区。在这些地区进行建设,应按法定程序进行,以确保地下文物和古遗迹的安全。

《北京历史文化名城保护规划》专家论证会专家意见要点

2002年3月20日,北京市规划委员会、北京市城市规划设计研究院、北京市文物局联合召开了《北京历史文化名城保护规划》(以下简称"保护规划")专家论证会,刘敬民副市长主持会议。

1.关于"保护规划"的总评价

专家们一致认为:北京市政府组织编制《北京历史文化名城保护规划》是一件非常好的事情,对保护北京历史文化名城具有重要意义,并且对全国将起到示范的作用。这次"保护规划"内容翔实、丰富,概念比较清楚,做得也比较细致。无论从理论和实践上都有很大的突破,具有很大的开拓性,是集过去名城保护工作的思想之大成。专家们建议市政府尽快审查批准"保护规划",积极贯彻和组织实施规划。

2.建议和意见

(1) 关于北京城市中轴线的保护

北京城市中轴线非常重要,在北中轴线上特别要注意北二环北节点与钟鼓楼节点的关系。北二环北节点不应建造太高的建筑,应以钟鼓楼为主。(侯仁之、王景慧)

应下决心恢复历史上天坛、先农坛的坛墙,远期将天坛医院等单位搬迁。(傅熹年)

(2) 关于历史文化保护区的保护

有条件的尽可能把历史文化保护区连片,进一步增强历史文化名城的整体保护格局。具体的如平安里以北、北河沿以东等地区可划为保护区。(柯焕章)

历史文化名城保护政府要出钱,北京市政府要出钱,中央也应当出钱。保护历史文化名城不能靠开发商,必须国家跟北京市共同做,这样才能真正保护好。(傅熹年)

历史文化保护区内的改造只能一点一点地改,小规模、逐步地改造,在较长的改造时间里使历史的真实风貌保存下来,不能大片大片地改造。这在日本和欧洲都有好的经验。保护区保护应当采取政府和居民相结合的办法,不能只靠开发商来搞保护。(王景慧、阮仪三)

在实施历史文化保护区保护规划时,要建立专家审议制度,由专家进行把关。(王世仁)

要注意西四"万松老人塔"的保护,这是北京市保存下来的一个元朝的文物,是一个重要的城市景点。(王景慧)

从阜外大街看白塔是一条景观线,白塔寺药店应在适当时候拆除。(王景慧)

(3) 关于皇城的保护

在皇城里可以多搞一点中小型博物馆,把紫禁城展览的功能向外延伸一点,特别是利用南北长街、南北池子的老房屋,可以缓解故宫博物院展览空间不足的矛盾。

对皇城来说要采取"三低"政策,一个是低人口密度,第二是层数要低,第三是交通量要低。主要道路应保持现有道路尺度,不应拓宽。(郑孝燮)

(4) 关于历史河湖水系的保护

关于恢复护城河的问题,除了远期恢复前三门护城河外,还应该恢复东护城河和西护城河,护城河是一个完整的体系。(段天顺)

从后门桥开始到前三门护城河的御河在元代就是大运河的起点,应考虑将御河全线恢复。(侯仁之、段天顺)

建议在规划文本中写上恢复河道的位置、宽度,要控制该范围内的建设。(段天顺)

高梁桥、高梁闸是很重要的文物,要加以保护。(段天顺)

(5) 要强调北京城市的历史文化价值

北京是五朝古都,辽是北京建都之始,这次在规划中没有提到辽代,应补充。

应进一步强调元大都的历史价值。元大都是明清北京的前身,是中国封建社会最后一座,也是惟一的一座按规划新建的街巷制的城市。中国早期是里坊制城市,晚期是街巷制城市。明清北京是在元大都基础上发展,北京现在有很多胡同是从元大都开始就有的。所以必须把胡同以及胡同里的四合院这套体系在历史上的重要性,在世界城市史上的重要性说够,这样才能下决心保护这个城市,保护这些四合院和胡同。(傅熹年)

(6) 关于建筑高度控制

旧城高度控制应有一个总的构思,到底是什么形态,这次规划应该有一个明确概念和要求,做出规定,要不然无所适从。(柯焕章)

对于历史文化名城的高度保护不能仅仅局限在旧城,对城市的西北地区也应有一定的控制要求,特别是对小西山方向。(柯焕章)

(7) 关于交通问题

关于北京市城市交通政策在旧城区、二环路之内,应该建立"交通保护核"的概念,"交通保护核"在国际上就是环保的概念。"交通保护核"内机动车交通量低,人流量不低,这个地方的交通应该以轨道交通作为重要的支撑条件。

关于调整旧城交通政策,是否可以说为了保护中国五千年的历史古城,或者文化也好,必须减少小汽车对古城区的压力,应该将穿越性交通引到外面的道路上去。把这个作为基本原则写到历史文化名城保护规划里。(赵波平)

道路红线的宽度旧城内和旧城外不能采取一样的标准,旧城内只要满足基本交通需要就可以了。(王景慧)

(8) 关于旧城改造问题

对北京旧城应该定一个政策,就是"零发展",就是用减法不用加法。现在这么大规模的开发没有好效果。(吴良镛)

旧城真正核心的问题是人多。旧城区不能再增加人,不能增加建筑面积。从根本上说不能走开发的路子。(李准)

不要把历史街区当成危旧房改造区,历史街区是保护改造更新应用。(王世仁)

附件

北京在有了《北京历史文化名城保护规划》的前提下，还要有更多的战略指导思想。要贯彻毛主席"集中力量打歼灭战"的思想，不要到处"铺摊子"。

北京应建设"文化新城"，就是"人文奥运"，在北郊建新文化城来有效地疏解旧城中的功能和人口，保护历史文化名城。(吴良镛)

目前北京市城市建设蓬勃发展，一方面非常兴奋，但也忧心忡忡于这种干法太快太热，似乎什么都想干、什么都能干，有点失控。现在旧城的破坏已经达到相当的程度了，保护规划是力挽狂澜，也是不可为而为之。(宣祥鎏)

(9) 关于法制化问题

北京在过去的岁月里不断遭到这样那样的破坏，不是没有规划或法规，而是有法不依或者执法不严造成的。应当提醒市领导注意这个问题，我们要逐步走向法制化，解决这些问题。

在北京应当树立一个观念，保护好北京历史文化名城就是行政领导最大的政绩。(阮仪三)

附：《北京历史文化名城保护规划》评审专家工作名单

侯仁之	中国科学院院士、北京大学教授
吴良镛	两院院士、清华大学教授
傅熹年	工程院院士、中国建筑技术研究院研究员
郑孝燮	建设部科技委顾问
阮仪三	同济大学教授
宣祥鎏	首都规划委员会原副主任兼秘书长
李　准	北京市城市规划局顾问总工
王世仁	北京市文物局研究员
王景慧	中国城市规划设计研究院顾问总工 中国历史文化名城保护专家委员会秘书长
柯焕章	北京市城市规划设计研究院顾问总工
段天顺	北京市水利史研究会会长、北京市水利局原副局长
赵波平	中国城市规划设计研究院交通所所长

北京市人民政府常务会议纪要(第48期)

2002年4月2日下午，刘淇市长主持召开第46次市政府常务会议。

会议听取了：市规划院关于《北京历史文化名城保护规划》的汇报。

会议讨论结果如下：

会议原则同意市规划委、市文物局和市规划院提出的《北京历史文化名城保护规划》(以下简称《规划》)，由市规划委等单位根据会议意见进行修改完善后，提交首规委全体会议讨论。

会议决定：

(1)《规划》保护原则中要加入保护古都传统风貌与提高广大人民群众生活质量相结合，与合理开发使用相结合等内容。

(2)《规划》中要明确保护的内容以及如何进行保护，比如，在历史文化保护区内，不是文物古迹的建筑应可以改建，以改善居住条件，但要保护好古都的整体建设格局，也要保持北京民居的原有风貌。

(3)《规划》中要加入保护历史文化名城的有关政策内容，如房改带危改，居民产权改革，历史文化保护区内居民回迁、危改与经济适用住房建设相衔接等政策措施。

(4) 恢复前三门护城河等有争议且又在短期内难以实现的内容暂不纳入规划，但要控制新建项目。

首都规划建设委员会第21次全体会议纪要

2002年4月4日,首都规划建设委员会召开第21次全体会议,会议由刘淇同志主持,贾庆林同志出席会议并作了重要讲话。会议听取并审议通过了《北京历史文化名城保护规划》的专题汇报。现将会议讨论和议定的事项纪要如下:

关于北京历史文化名城保护规划

会议听取了首规委办公室主任单霁翔同志关于《北京历史文化名城保护规划》(以下简称《规划》)编制情况的汇报。委员们在讨论中赞成《规划》提出的指导思想和基本思路、保护内容及实施《规划》的保障措施。

委员们建议,进一步修改完善后向社会公布,接受社会的监督。《规划》一经按程序审定后就具有法律性,要严格执行,要制定控制性的实施措施,不得随意改动。同时,要进一步研究历史文化名城保护工作中的问题,如保护实体与保护风貌的关系、城市水系规划等。

建设部关于《北京历史文化名城保护规划》审查的函

北京市规划委员会:

我部领导在参加首规委第21次全会时获知北京市准备深化历史文化名城保护规划的情况,会后即要求我司依据法定程序积极配合,做好相关工作。现就北京历史文化名城保护规划审批工作提出意见如下:

北京市历史文化名城保护规划是城市总体规划的重要组成部分,对其进行调整,属于对城市总体规划的调整。根据国办【2000】25号文和建设部【2000】76号文的要求,建议由北京市人民政府向建设部提出书面申请,对原保护规划的实施情况、调整或修编的内容、范围和理由进行说明。经我部认定,属于局部调整规划的由市人民政府批准后报建设部备案;属于修编规划的应报总体规划原审批机关审批。

为了保证北京历史文化名城保护规划审查工作的方便与高效,我司成立专门的工作小组设在建设部城乡规划司城市规划处。

建设部城乡规划司
2002年5月17日

建设部关于对《北京历史文化名城保护规划》审核意见的函

北京市人民政府:

你市《北京市人民政府关于报请审查北京历史文化名城保护规划的函》(京政函【2002】44号)收悉。

经与《北京市城市总体规划》核对并商国家文物局,我部认为,《北京历史文化名城保护规划》所提出的调整补充内容,未对总体规划确定的城市性质、规模、发展方向、用地布局等构成影响,属于对城市总体规划深化的性质,并符合北京市历史文化名城保护工作的实际需要。我部原则同意上报的《北京历史文化名城保护规划》。

按照《国务院关于加强城乡规划监督管理的通知》(国发【2002】13号)的规定,《北京历史文化名城保护规划》应当明确规划的强制性内容,请你市按规定组织对规划内容的完善,特别注意对建筑高度的严格控制,并建议采取立法等措施,切实做好保护规划的实施管理工作。

根据《城市规划法》的规定,《北京历史文化名城保护规划》由北京市人民政府负责审批。请你市将批准后的规划报我部备案。

2002年8月13日

关于报审《北京历史文化名城保护规划》的请示

市政府：

根据市领导的指示，我们在市商委、北京建工学院、北京工业大学、北方工业大学等单位的协助下，在开展8个专题研究的基础上，编制了《北京历史文化名城保护规划》。现将规划要点简述如下：

1. 规划指导思想

（1）充分考虑北京悠久的历史和丰富的文物遗存，立足现实，最大限度地保护历史文化名城。

（2）正确处理历史文化名城保护与城市现代化建设的关系，努力提高城市现代化水平，使两者相互协调，相得益彰。

（3）历史文化名城保护要贯彻"以人为本"的思想，改善人居环境。

2. 规划主要内容和建议

根据国内外历史文化名城保护工作的经验和国家建设部的规定，历史文化名城的保护要考虑三个层次，即文物保护单位的保护、历史文化保护区的保护、历史文化名城整体格局的保护。本规划同时考虑了传统文化和传统商业与名城保护的结合。

（1）文物保护单位的保护

1）要在现有各级文物保护单位约1000处的基础上，继续做好申报世界文化遗产（如明十三陵）、国家重点文保单位的工作，继续公布市级、区县级文保单位和地下文物埋藏区名单，继续划定各级文保单位的保护范围和建设控制地带并严格执行。

2）做好文保单位的腾退、修缮和合理利用工作，并整治其周边环境。重点整治"两线"（中轴线、朝阜路）景观，恢复"五区"（皇城、什刹海、国子监、琉璃厂、古城垣）风貌，重现京郊"六景"（西部皇家园林、长城、帝王陵寝、京东运河、宛平史迹、京西寺庙）。

（2）历史文化保护区的保护

1）在第一批25片历史文化保护区的基础上，公布北京第二批历史文化保护区名单并做好保护规划。建议旧城以外10片（西郊清代皇家园林、卢沟桥宛平城、模式口、三家店村、川底下村、岔道城、榆林堡、古北口老城、遥桥峪、小口城堡和焦庄户），旧城区4片（皇城、张自忠路北、张自忠路南、东四南）。此外有4片为第一批历史文化保护区的适当扩展保护控制范围（锣鼓巷、东四北、鲜鱼口、大栅栏和琉璃厂）。旧城区新增和扩展历史文化保护区面积约4.75km²。第二批历史文化保护区的公布，使北京历史文化保护区的类型更为多样丰富，加强了旧城保护的整体性。旧城第一批、第二批历史文化保护区面积合计约15km²，占旧城总面积的24.9%。

2）历史文化保护区的实施和管理难度很大，建议研究探索多种方式进行保护和整治的途径。城四区每年要在历史文化保护区内选择1~2处开展试点工作。同时研究房屋产权、人口疏导、传统建筑修缮、改善市政设施等政策。

3. 历史文化名城整体格局的保护

（1）历史水系的保护

1）把与北京城市发展密切相关的河湖水域列为保护目标，划定保护范围并加以整治（包括护城河、六海、莲花河、长河、通惠河、北运河等）。

2）恢复转河、菖蒲河、御河（上段）及鱼藻池等水域。

3）提出远期恢复前三门护城河的设想，建议2008年前恢复前门两侧约1km的河段。

（2）皇城的整体保护

1）明确皇城作为一个整体加以保护。

2）明令禁止今后在皇城内审批多层以上建筑。

3）考虑制定逐步整治、修缮、保护的计划。

（3）中轴线的保护和发展

1）中轴线分为三段：旧城传统中轴线、北中轴线、南中轴线。三段采用不同的规划思想。

2）传统中轴线要以保护为主，保护与发展、继承与创新相结合；北、南中轴线以发展和创新为主。

（4）旧城道路交通规划与历史文化名城保护相协调

1）研究调整旧城交通政策、规划思想和交通组织方式。大力发展旧城轨道交通和常规公共交通，控制旧城、特别是皇城以内的交通生成吸引量。

2）深入研究旧城道路红线调整的具体措施，妥善处理道路建设与文物保护单位、历史文化保护区的矛盾，景山前街、文津街等历史文化保护区内的主要街道不再拓宽。

（5）旧城建筑高度的控制

1）旧城建筑高度控制是涉及北京历史文化名城整体风貌保护最重要的影响因素之一。旧城高度的不断突破，严重损害了北京历史文化名城的整体性和历史文化价值。

2）关于旧城建筑高度的控制规定，要坚决执行，不再突破。

（6）危改与名城保护

旧城内的危改区不应采用消除历史痕迹的"推平头"做法。危改区规划建设要注意对有一定价值的历史建筑的保留和修缮。保留原有古树、大树，保留原有主要胡同的走向和名称。新建多层住宅宜采用灰色坡屋顶，使老北京街区的历史因素在改造后的新区中得以传承。

4. 保护和发扬传统文化和传统商业

（1）北京具有历史久远、特色鲜明的传统文化和传统商业，应采取积极措施，加以继承、发扬和创新。

（2）努力恢复传统特色的庙会，恢复和合理利用会馆，进一步繁荣以京剧和昆曲为代表的北京传统戏曲，保护和发扬其他传统文化。

（3）大力支持有发展前途的"老字号"商店，保护北京所剩不多的几条传统商业街（前门大街、大栅栏、琉璃厂等）。

北京市规划委员会
北京市文物局
北京市城市规划设计研究院
2002年1月18日

北京历史文化名城北京皇城保护规划
Conservation plan for the Historic City of Beijing and Imperial City of Beijing

关于报请批复《北京历史文化名城保护规划》的请示

市政府：

北京是世界著名的文化古都，是国务院首批公布的国家级历史文化名城。市政府在2001年4月指示规划、文物部门尽快组织编制《北京历史文化名城保护规划》，以切实保护北京历史文化遗产，体现"人文奥运"精神。2002年3月，市规划、文物部门在《北京城市总体规划》和《北京市区中心地区控制性详细规划》的基础上编制出《北京历史文化名城保护规划》，在召开历史文化名城保护专家和市政协委员等评审会的基础上，分别经2002年4月2日第46次市政府常务会议和2002年4月4日首都规划建设委员会第21次全体会议审查，原则通过。经修改，现报请市政府正式批准并按政府规章颁布实施。

1. 编制规划的指导思想

坚持北京的政治中心和文化中心的性质，正确处理历史文化名城保护与城市现代化建设的关系。重点搞好旧城保护，最大限度地保护名城，使历史文化名城在保护中得以持续发展。

2. 保护规划分为三个层次

文物保护单位的保护、历史文化保护区的保护和名城的整体保护。

3. 保护规划的重点

（1）要从整体上保护北京旧城，具体体现在历史水系、传统中轴线、皇城、旧城"凸"字形城廓、道路及街巷胡同、建筑高度、城市景观线和对景、建筑色彩、古树名木、旧城危改等十个方面的内容。

（2）在已批准的旧城25片历史文化保护区的基础上，在旧城和市域范围内新增历史文化保护区。尽快组织编制各保护区的保护规划，报市政府批准后颁布实施。

（3）划定皇城历史文化保护区，编制保护规划并申报世界文化遗产。

（4）加强文物保护单位的升级和普查工作，制定文物保护单位的保护规划。

（5）旧城内的危改项目，位于历史文化保护区内的，应严格按历史文化保护区保护规划实施，采取渐进的保护与更新的方式，以"院落"为单位逐步更新危房，维持原有街区的传统风貌。位于历史文化保护区以外的，要加强对文物及有价值的历史建筑的保护，强调古树、大树、胡同等历史遗存的保护。市规划、文物、园林、房地等相关部门要建立相应的工作程序和制度，切实把好审批关，做好危旧房改造中的保护工作。

（6）历史文化名城的内涵既包括物质形态的城市格局、文物古迹、有特色的建筑等传统城市风貌，也包括文化传统的继承与发扬。加强对继承和发扬名城传统风貌的文化特色的研究。

（7）进一步落实"两个战略转移"的方针，积极推进城市建设重点逐步从市区向远郊区转移，市区建设从外延扩展向调整改造转移的步伐，疏解旧城区人口的功能，为保护历史文化名城创造良好条件。

（8）加强《保护规划》的实施保障工作。制定《北京市历史文化名城保护条例》、《北京市历史文化保护区管理规定》和《北京历史文化保护区保护准则》等，作为法律依据在城市管理工作中严格执行。

北京市规划委员会
2002年4月30日

北京市人民政府关于《北京历史文化名城保护规划》的批复

市规划委：

你委《关于报请批复〈北京历史文化名城保护规划〉的请示》（市规文【2002】1265号）收悉。现批复如下：

1. 原则同意修订后的《北京历史文化名城保护规划》（以下简称《规划》）。《规划》是《北京城市总体规划》的深化，是加强北京历史文化名城保护工作、正确处理城市发展与保护关系的重要保障。市各有关部门、单位和有关区（县）政府要认真贯彻执行《规划》，通过实施《规划》，有效地保护古都风貌，改善人民居住条件，推进城市现代化建设。

2. 要从整体上保护北京旧城。主要体现在历史河湖水系、城市中轴线、皇城、明清北京城"凸"字形城廓平面、旧城棋盘式道路网和街巷胡同格局、旧城建筑高度的控制、城市景观线、街道对景、旧城建筑形态与色彩、古树名木等十个方面。

3. 同意新增的第二批历史文化保护区，要尽快划定其保护和控制范围，并组织编制各保护区的保护规划，报市政府批准后颁布实施。要继续公布文物保护名单，加强升级和普查工作。

4. 北京城市中轴线的保护应遵循"以保护为主，保护与发展、继承与创造相结合"的原则，严格控制城市中轴线保护区域范围内建筑的高度和形态。

5. 同意将皇城整体设立为历史文化保护区。要尽快编制完成《皇城保护规划》，报市政府批准后颁布实施，同时进行申报皇城为世界文化遗产的准备工作。

6. 要采取措施保护明清北京城的"凸"字形城廓、棋盘式道路网骨架和街巷胡同格局。要从历史文化名城保护的角度，组织重点研究旧城区的交通、道路系统，尽快组织编制旧城道路网调整规划。要严格制定北京旧城的建筑高度。

7. 历史文化保护区内的危房修缮改建项目，必须按照《规划》和历史文化保护区具体规划要求，在保护各级别文物及四合院等历史建筑和遗存的前提下，使危房得到改造，公共设施不断完善，居民生活条件得到改善，延续原有街区风貌。改建项目必须先确定保护内容，再做规划和设计。在历史文化保护区以外的危房改造项目，要加强对文物、古树名木以及有保存价值的四合院、胡同等历史建筑和遗存的保护。

8. 要正确处理专家咨询和政府决策的关系。各级政府要加强领导，切实负起历史文化名城保护工作的历史责任。旧城内房屋维修改造方案必须先由有关专家提出咨询方案，在广泛征求意见后，由规划、文物、园林等行政主管部门依法审批。

9. 为做好危旧房改造中的历史文化名城保护工作，市规划、文物、园林、国土房管等相关部门要建立相应的工作程序和制度，并研究制定有利于旧城内人口外迁与疏导的旧城危改政策。

10. 为保证《规划》的贯彻实施，市文物、规划、建设等部门要共同研究，加快制定北京历史文化名城保护管理有关配套文件，依法行政，严格审批，明确责任，规范建设和管理行为。

2002年10月11日

北京市人民政府关于实施《北京历史文化名城保护规划》的决定

各区、县人民政府，市政府各委、办、局，各市属机构：

《北京历史文化名城保护规划》是在党中央、国务院的关怀下，在市委、市政府的领导下，根据《中华人民共和国城市规划法》、《中华人民共和国文物保护法》、《中华人民共和国文物保护法实施细则》、《北京市城市规划条例》、《北京市文物保护管理条例》等法律、法规的规定和国务院批复的《北京城市总体规划(1991年～2010年)》，在广泛征求有关方面专家意见的基础上编制的，是北京历史文化名城保护工作的重要文件。在加快首都城市现代化建设和发展的过程中，保护北京历史文化名城，落实《北京历史文化名城保护规划》(以下简称《规划》)，是本市各级人民政府的重要职责。本市各级人民政府及其部门一定要把历史文化名城保护工作摆在重要位置。为此，特作如下决定：

1. 本市各级人民政府及其规划、文物、建设、园林、市政管理、国土房管等部门必须建立健全制度，严格把好审批关，保证切实执行《规划》确定的保护内容和范围，实现有效保护古都风貌、改善人民生活条件和推进城市现代化建设有机结合的目标。市有关部门和区、县人民政府要积极研究有关政策，支持试点工作，确保《规划》的实施。

2. 坚持依法行政，严格执行《规划》，查处违法行为，切实做好北京历史文化名城保护工作。《规划》依法编制，内容翔实、丰富、具体、细致，坚持了北京的政治中心、文化中心的性质，正确处理了历史文化名城保护与城市现代化建设的关系，重点是搞好旧城保护，最大限度地保护历史文化名城，使其在保护中持续发展。《规划》是今后城市规划建设、旧城区保护和改造的基本依据。本市各级政府行政主管部门必须依法行政，严格执法。对违反有关法律、法规和《规划》的行为，规划、文物等行政主管部门必须依法予以认真查处。对违反规定批准建设的渎职、失职行为，要依法追究其行政部门负责人和直接负责人的行政责任。

3. 必须处理好危房改造与历史文化名城保护的关系。历史文化保护区内的危房修缮改建项目，必须按照《规划》和历史文化保护区具体规划要求，在保护各级别文物及四合院等历史建筑和遗存的前提下，使危房得到改造，公共设施不断完善，居民生活条件得到改善，延续原有街区风貌。改建项目必须在先确定保护内容后，再做规划和设计。在历史文化保护区以外的危房改造项目，要加强对文物、古树名木以及有保存价值的四合院、胡同等历史建筑和遗存的保护。旧城危改政策要有利于旧城内人口的外迁与疏导，通过兴建经济适用住房，先行缓解旧城人口密集的状况。

4. 建设单位和危旧房产权人必须按照《规划》要求，研究和申报危旧房改造方案。旧城内房屋维护改造方案必须先由相关专家提出咨询方案，在广泛征求意见后，由规划、文物、园林等行政主管部门依法审批。危改项目的规划方案必须有历史文化保护专项论述，包括街区的历史沿革、文物保护单位的保护、有价值的历史建筑和遗存的保护、古树名木和大树的保护、对传统风貌影响的评价、环境改善的措施等内容。

5. 要调整旧城区内的交通结构，完善公共设施。旧城区内的交通必须采取以公共交通为主的方式。要从历史文化名城保护的角度，组织重点研究旧城区的交通、道路系统，编制调整旧城区路网规划和修建方式，实施严格的停车管理措施，限制和调节驶入城区的汽车交通量。要抓紧编制旧城区公共设施完善方案并逐步落实，改善居民生活条件。

6. 本市各级人民政府及其部门必须加大宣传力度，鼓励公众参与，加强社会监督。保护历史文化名城既是政府的责任，也是全社会的共同责任。要增强责任意识、法律意识，积极鼓励和支持人民群众为保护历史文化名城工作出谋划策。对通过公众参与、社会监督发现的破坏和影响历史文化名城保护的建设行为，规划、文物等有关行政主管部门要及时坚决查处。

2002年10月16日

北京市城市总体规划（部分）

（1991年~2010年）

44条：

北京历史文化名城的保护，是以保护北京地区珍贵的文化古迹、革命纪念建筑物、历史地段、风景名胜及其环境为重点，达到保持和发展古城的格局和风貌特色，继承和发扬优秀历史文化传统的目的。对于新的建设要体现时代精神、民族传统、地方特色，根据不同情况提出不同要求，使新旧建筑、新的建设与周围环境互相协调，融为一体，形成当代中国首都的独特风貌。

要妥善处理历史文化名城保护与现代化建设的关系。城市现代化建设、社会经济发展，以及市区特别是旧城的调整改造，要与历史文化名城的保护相结合，使北京的发展和建设，既符合现代生活和工作的需求，又保持其历史文化特色。

45条：

各级文物保护单位是历史文化名城保护的重要内容。对公布的文物保护单位，尤其是国家级和市级文物保护单位，包括万里长城、故宫、周口店北京猿人遗址等"世界文化遗产"，必须加强科学保护，合理利用。进一步加强对地面和地下文物古迹的调查、发掘与鉴定，公布新的保护单位；继续划定文物保护单位的保护范围及其周围的建设控制地带，总结经验，不断完善。对地下埋藏区内的建设，坚持先勘探发掘、后进行施工的原则。

46条：

历史文化保护区是具有某一历史时期传统风貌、民族地方特色的街区、建筑群、小镇、村寨等，是历史文化名城的重要组成部分。北京市已确定的国子监街、南锣鼓巷、西四北、什刹海、阜山门街、牛街、琉璃厂、大栅栏、景山前街、景山后街、景山东西街、南北长街、南北池子、东交民巷等25处第一批市级历史文化保护区，要逐个划定范围，具体确定其保护和整治目标。保护区内新建筑的形式和色彩，要与该区原有风貌协调一致，与之不协调的建筑物和其他设施要加以改造。

要继续在旧城区和广大郊区增划各级历史文化保护区。对于历史文化保护区以外的分散的好四合院，在进行城市改建时也要尽量保留，合理利用。

47条：

要从整体上考虑历史文化名城的保护，尤其要从城市格局和宏观环境上保护历史文化名城。

（1）保护和发展传统城市中轴线。必须保护好从永定门至钟鼓楼这条明、清北京城中轴线的传统风貌特点。继续保持天安门广场在轴线上的中心地位，要在扩建改建中增加绿地、完善设施。鼓楼前街和前门大街要建设成为具有传统特色的商业街。

中轴南延长线要体现城市"南大门"形象，中轴北延长线要保留宽阔的绿带，在其两侧和北端的公共建筑群作为城市轴线的高潮与终结，突出体现21世纪首都的新风貌。

（2）注意保持明、清北京城"凸"字形城廓平面。沿城墙旧址保留一定宽度的绿化带，形成象征城墙旧址的绿化环。原城门口的建筑应体现"城门旧址"的标志特点。

（3）保护与北京城沿革密切相关的河湖水系。如长河、护城河、六海等。

（4）旧城改造要基本保持原有的棋盘式道路网骨架和街巷、胡同格局。

（5）注意吸取传统民居和城市色彩的特点。保持皇城内青灰色民居烘托红墙、黄瓦的宫殿建筑群的传统色调。

（6）以故宫、皇城为中心，分层次控制建筑高度。旧城要保持平缓开阔的空间格局，由内向外逐步提高建筑层次，建筑高度除规定的皇城以内传统风貌保护区外，分别控制在9m、12m和18m以下。长安街、前门大街两侧和二环路内侧以及部分干道的沿街地段，允许建部分高层建筑，建筑高度一般控制在30m以下，个别地区控制在45m以下。旧城以外，一般不超过60m。

从生态环境考虑，由市区西北部风景名胜区至东南部，应留出一条"通风走廊"，以保持中心地区良好的大气环境，建筑高度低于相邻地区。市区南部的中轴南延长线两侧，是从景山南望故宫，显示古都传统天际轮廓线的重要背景，建筑高度相对低一些。

市区的东部、北部的适当地段，可按城市设计要求建设个别较高的建筑物，丰富城市轮廓线。

（7）保护城市重要景观线。保护"银锭观山"和从市中心区往西的几条干道遥观西山的重要景观线，以及景山万春亭、北海白塔、妙应寺白塔、钟鼓楼、德胜门箭楼、天坛祈年殿、正阳门城楼和箭楼各景点之间几条主要的传统景观线。景观线保护范围内新建筑的高度，应按测试高度控制，严禁插建高层建筑。

（8）保护街道对景。对于历史形成的对景建筑及其环境要加以保护，控制其前景和背景的建筑高度。对有可能形成新的对景的建筑，要通过城市设计，对其前景和背景建筑的高度、体量和造型提出控制要求。

（9）增辟城市广场。除天安门广场是城市中心广场外，旧城各城门口附近，城市内环路上的各干道交叉口附近，以及重要公共建筑地段，要增辟城市广场，搞好景观设计，增添小品设施，处理好建筑形体与广场、绿化的关系以及广场的交通问题。

（10）保护古树名木，增加绿地，发扬古城以绿树衬托建筑和城市的传统特色。

附件

《北京历史文化名城保护规划》编制工作大事记

在市委、市政府的领导下，从2001年4月至2002年10月，市规划委、市文物局、市规划院作为组织协调和编制单位，在近一年半的时间里编制完成了《北京历史文化名城保护规划》。该项规划是对《北京城市总体规划》的延续和深化，是对历史上关于北京历史文化名城保护经验的总结和发展，是中华人民共和国成立以来北京历史上第一次在市域范围内对北京历史文化名城的保护进行的一次系统而完整的思考和规划，是全市各有关方面集体智慧的结晶。保护规划的编制完成必将对未来北京历史文化名城的保护有着深远而积极的意义。

2001年4月初，北京市政协文史委员会提出了"关于尽快制定《北京历史文化名城总体保护规划》的建议"，市政府主管城市建设的汪光焘副市长（现建设部部长）就此批示，"请规划院组织编制，制定工作方案，限期完成"。

2001年5月15日，根据市领导的指示，规划院拟出《北京历史文化名城保护规划编制工作大纲》，明确了编制要求、规划思路及时间进度。编制协调单位为市规划委，编制负责单位为市规划院、市文物局。

规划总体思路为按照文物保护单位的保护、历史文化保护区的保护、历史文化名城整体格局的保护、传统商业和文化的继承和发扬四个大专题进行工作。

编制工作的具体分工为：

由市文物局负责、市规划院协助完成关于北京市各级文物保护单位的保护和利用，北京传统文化的保护和发扬两个专题。

由市规划委负责完成关于北京历史文化保护区的保护、整治与更新；第二批历史文化保护区名单和保护范围的确定；北京历史文化保护区保护规划实施管理办法三方面内容。

市商委负责、市规划院协助完成关于北京传统商业的继承和发扬专题。

市规划院负责关于历史文化名城整体格局的保护专题，包括历史河湖水系的保护、城市中轴线的保护和发展、皇城的保护、明清北京城"凸"字形城廓的保护、旧城棋盘式道路网和街巷胡同格局的保护、旧城建筑高度的控制、城市景观线和街道对景的保护、旧城建筑形态与色彩的继承与发扬、古树名木的保护、传统地名的保护十个方面内容。同时，规划院负责完成规划综合报告和规划成果的汇总等方面内容。

2001年8月9日，市规划委、市规划院、市文物局联合向市政府报送"关于北京历史文化名城保护规划编制工作的报告（市规文【2001】790号）"，"报告"提出保护规划的基本工作思路、研究内容和工作进度安排。

2001年8月29日，汪光焘副市长批示，"所报意见原则同意，报刘淇同志审示"。

2001年8月30日，刘淇市长批示，"原则同意，进度加快一点。"

2001年9月3日~14日，市规划委和市规划院落实市领导的指示，研究加快编制工作的相关措施。

2001年10月16日，市规划院向政协文史委员会答复其提出的"关于尽快制定《北京历史文化名城保护规划》的建议"的提案。

2001年10月27日，在市政府《昨日市情》特刊第398期上登载市政协提出的关于"政协委员建议抓紧编制北京历史文化名城保护规划"的建议。刘淇市长作了批示："请市规委把此项工作抓紧，争取年内交市里讨论"。

2001年11月7日，市规划委、市文物局、市规划院召开会议讨论文物保护单位保护的规划思路。

2001年12月18~21日，由市规划委、市文物局、市规划院联合召开《北京历史文化名城保护规划》专题阶段性审查会。

2002年1月16日，市规划院完成《北京历史文化名城保护规划》总报告。

2002年1月18日，由市规划委、市文物局、市规划院联合向市政府报文"关于报送《北京历史文化名城保护规划》的请示（城规设发【2002】04号）"，申请市政府审议《保护规划》。同时附上《保护规划》文本、图纸。

2002年1月22日，张茅副市长批示"建议进一步征求有关方面专家意见，请刘淇、敬民同志批示。"刘淇市长对报送《保护规划》的请示作了批示"在征求各界专家意见后，市政府专题会议定"。

2002年1月22日，由市规划委、市文物局、市规划院联合向市政协文史委员会通报了《北京历史文化名城保护规划》的编制情况，获得委员们一致好评。

2002年3月7日，市规划委、市文物局、市规划院到北京市政协文史委员会听取政协文史委"关于《北京历史文化名城保护规划》制定中几个问题的紧急建议"的意见，并将意见加以落实。

2002年3月9日，市规划院向刘敬民副市长报文"关于召开历史文化名城保护规划专家论证会的请示"。

2002年3月20日，根据市领导指示，由市规划委、市文物局、市规划院联合组织召开了《北京历史文化名城保护规划》专家论证会。由刘敬民副市长主持。与会专家12名，参会人员约80名。

2002年3月26日，市规划委、市文物局、市规划院联合向市政府报文"关于报送《北京历史文化名城保护规划》政协委员、专家意见的报告（城规设发【2002】14号）"，同时附上《保护规划》文本、说明、图纸、政协提出的保护区名单及附图，提请市政府对《保护规划》进行审议。

2002年4月2日，市政府第46次常务会议听取了《北京历史文化名城保护规划》的汇报，并原则通过。

2002年4月4日，由市规划委向首都规划建设委员会第21次全体会议汇报《北京历史文化名城保护规划》，获原则通过。

2002年4月30日，市规划委向市政府报文"关于报请批复《北京历史文化名城保护规划》的请示（市规文【2002】538号）"。

2002年5月17日，建设部城乡规划司给首规委发文"关于北京历史文化名城保护规划审查的函"，要求将《北京历史文化名城

保护规划》上报建设部审查。

2002年8月13日，建设部致北京市政府"关于对《北京历史文化名城保护规划》审核意见的函"，并附2002年8月7日国家文物局致建设部城乡规划司"关于《北京历史文化名城保护规划》审核办理意见的函"。原则同意上报的《北京历史文化名城保护规划》。

2002年9月19日，《北京历史文化名城保护规划》文本同时在北京日报、北京青年报等北京十大媒体全文登载，向社会公示。

2002年10月16日，"北京市人民政府关于实施《北京历史文化名城保护规划》的决定（京政发【2002】27号）"下达。"决定"明确必须严格执行《北京历史文化名城保护规划》，依法行政，严格执法。

2002年10月17日，"北京市人民政府关于北京历史文化名城保护规划的批复"（京政函【2002】83号）下达。"批复"原则同意《北京历史文化名城保护规划》。

《北京历史文化名城保护规划》集中了国家、地方各级领导、各方面专家、社会各界人士、规划设计人员的智慧和心血，是北京城市规划史上的重要事件，将编制过程中发生的重要大事如实记录下来，作为史料，对今后北京历史文化名城保护的研究具有重要参考价值。

北京皇城保护规划

北京市规划委员会
北京市文物局
北京市城市规划设计研究院

统一思想,广泛宣传北京名城保护的重要性

认真贯彻中央对北京城市规划建设和管理工作"建设第一流的现代化国际城市和历史文化名城的高促进北京历史文化名城保护"的指示,牢记"把北京态分认真贯彻"的共识。

进一步落实《北京皇帝城总体规划》

在《北京历史文化名城保护规划》基础上,"十五"期间,组织编制落实北京历史文化名城保护规划的具体规划,主要包括:

一、批历史文化保护区保护规划;

二、旧城内道路规划的调整与完善;

三、加快制定北京历史文化名城保护范围和建设控制地带的规定等;

四、组织危旧房改造规划和旧城保护范围内各类项目建设的衔接;

由市文物、规划、建设等部门,共同建立立法的前期调研工作,制定《北京市皇城保护条例》(暂定名),作为法律依据并延伸城市管理工作中最怪修行。

第五批、第六批文物保护单位保护范围和建设控制地带法规,明确修缮。

旧城内危改与建设不宜采用大成片开发的方式,危改中要注意保护原有主要胡同和其它文保护法规。

深入研究新城房屋产权保护机制

稳妥推进旧城历史文化保护区管理体制的改革,并开展试点;

实施政策,各区都要加强历史文化保护区规划实施试点,积累经验。

根据"国家保护为主,动员全社会参与"的主导思想,拓展试点工作,探索文物保护单位和历史文化保护区管理保护、维修、整修、利用的有效途径,建立各项基金保护管理制,除政府财政拨款外,还要通过开发旅游经济,开展使用部份自筹资金,社会捐助等多种渠道,建立保护基金,加强对北京传统风貌和文化特色的继承和发扬的

《北京皇城保护规划》文本

1. 总则

1.1 为加强皇城的整体保护，落实《北京历史文化名城保护规划》，特制定本规划。

1.2 皇城是北京第二批历史文化保护区之一，是北京历史文化名城的重要组成部分。

1.3 在本保护规划范围内进行各项建设活动的一切单位和个人，均应按《中华人民共和国城市规划法》的有关规定，执行本规划。

1.4 本规划未涉及的控制指标和管理规定，应遵循国家及北京市的相关法规、规定执行。

1.5 本规划经北京市人民政府批准后，由市规划行政主管部门负责执行；如有重大调整，须经北京市人民政府批准。

2. 规划依据

2.1《中华人民共和国城市规划法》(1989年12月)
2.2《中华人民共和国文物保护法》(2002年10月)
2.3《北京市文物保护管理条例》(1987年6月)
2.4《北京城市总体规划（1991~2010年）》
2.5 国务院关于《北京城市总体规划》的批复（1993年10月）
2.6《北京旧城历史文化保护区保护和控制范围规划》(1999年4月)
2.7《北京市区中心地区控制性详细规划》(1999年9月)
2.8《北京市文物保护单位范围及建设控制地带管理规定》(1994年)
2.9《北京25片历史文化保护区保护规划》(2001年3月)
2.10《北京历史文化名城保护规划》(2002年4月)

3. 规划指导思想及原则

3.1 规划指导思想

3.1.1 把皇城作为一个整体加以保护，特别是搞好故宫等重点文物的保护。

3.1.2 正确处理皇城的保护与现代化建设的关系，皇城内新的建设要服从保护的要求，保证皇城整体风貌与空间格局的延续。

3.1.3 要用辨证和历史的观点，正确认识和处理皇城的发展和变化，区别和慎重对待已有新建筑。

3.1.4 贯彻"以人为本"和"可持续发展"的思想，努力改善皇城中的居住、工作条件和环境质量。

3.2 规划原则

3.2.1 整体保护与分类保护相结合，最大限度保存真实历史信息的原则。
3.2.2 有重点、分层次、分阶段逐步整治、改善和更新的原则。
3.2.3 逐步、适度疏导人口的原则。
3.2.4 严格控制皇城内建设规模的原则。
3.2.5 文物保护单位的保护与合理利用相结合的原则。

4. 规划范围

皇城的规划范围为东至东黄(皇)城根，南至东、西长安街，西至西黄(皇)城根、灵境胡同、府右街，北至平安大街。规划占地面积约6.8km²，皇城行政区划分属东城、西城两区。

5. 皇城的性质和历史文化价值

5.1 皇城的性质

5.1.1 北京明清皇城以紫禁城为核心，以皇家宫殿、衙署、坛庙建筑群、皇家园林为主体，以满足皇家工作、生活、娱乐之需为主要功能。

5.1.2《北京历史文化名城保护规划》确定皇城为具有浓厚传统文化特色的历史文化保护街区。

5.2 皇城的历史文化价值

5.2.1 北京皇城始建于元代，主要发展于明清时期。现今的皇城基本保持了清末民初的格局与风貌，具有极高的历史文化价值。

5.2.2 惟一性：明清皇城是我国现存惟一保存较好的封建皇城，它拥有我国现存惟一的、规模最大、最完整的皇家宫殿建筑群，是北京旧城传统中轴线的精华组成部分。

5.2.3 完整性：皇城以紫禁城为核心，以明晰的中轴线为纽带，城内有序分布着皇家宫殿园囿、御用坛庙、衙署库坊等设施，呈现出为封建帝王服务的完整理念和功能布局。

5.2.4 真实性：皇城中的紫禁城、筒子河、三海、太庙、社稷坛和部分御用坛庙、衙署库坊、四合院等传统建筑群至今保存较好，真实地反映了古代皇家生活、工作、娱乐的历史信息。

5.2.5 艺术性：皇城在规划布局、建筑形态、建造技术、色彩运用等方面具有极高的艺术性，反映了历史上皇权至高无上的等级观念。

6. 土地分区和特征

6.1 土地分区

6.1.1 根据皇城内土地使用特征及街道的自然分布，按照由南至北、由东至西的原则，将皇城用地划分为10个区域。

6.1.2 01区：皇城内故宫与劳动人民文化宫以东，五四大街以南的区域；

02区：皇城内地安门内大街、景山东街以东，五四大街以北的区域；

03区：由故宫、中山公园和劳动人民文化宫三部分组成的区域；

04区：景山公园所在的区域；

05区：中南海以东，文津街以南，故宫与中山公园以西的区域；

06区：北海以东，景山前街以北，地安门内大街、景山西街以西的区域；

07区：中南海所在区域；

08区：北海所在区域；

09区：文津街以南，中南海以西的区域；

10区：文津街以北，北海以西的区域。

6.1.3 10个区内各类规划地块有572个，在今后的改造中应参照地块编号进行。

6.2 各分区的特征

6.2.1 01、02、05、06、09五个区为传统平房四合院建筑相对集中的地区，胡同和四合院构成街区的主体。

6.2.2 03、04、07、08四个区为重大文物保护单位，是皇家建筑

The Text of "Conservation Plan for the Imperial City of Beijing"

1. General Principles

1.1 This plan was drawn up in order to enhance the overall conservation of the imperial city and implement the "Conservation Plan for the Historic City of Beijing".

1.2 As one of the second group of conservation districts of historic sites in Beijing, the imperial city is an important part of the historic city of Beijing.

1.3 All units and individuals engaged in construction within the range of this plan must implement this plan according to the relevant regulations of the "Urban Planning Law of the People's Republic of China".

1.4 For control index and management rules not mentioned in this plan, relevant laws and regulations of the State and Beijing must be followed.

1.5 Upon approval by the Beijing municipal government, the municipal department responsible for urban planning will supervise the its implementation. Any substantial changes are subject to the approval of the Beijing municipal government.

2. Basis of Plan

2.1 "Urban Planning Law of the People's Republic of China" (December, 1989)

2.2 "Historical Relics Conservation Law of the People's Republic of China" (November 1982).

2.3 "Management Ordnance of Cultural Relics Conservation in Beijing" (June, 1987)

2.4 "Master Plan for Beijing Construction (1991-2010)"

2.5 The Written Reply by the State Council to the "Master Plan for Beijing Construction" (October, 1993)

2.6 "Conservation and Control Scope Plan for the Conservation Districts of Historic Sites in Old City of Beijing" (April, 1999)

2.7 "Detailed Plan for the Control of the Central Area of Beijing" (September, 1999)

2.8 "Scope and Construction Control Regulations for Cultural Relics Sites in Beijing" (1994)

2.9 "Conservation Plan for the 25 Conservation Districts of Historic Sites in Beijing" (March, 2001)

2.10 "Conservation Plan for the Historic City of Beijing" (April, 2002)

3. Guidelines and Principles of Plan

3.1 Guidelines of Plan

3.1.1 Conserving the imperial city as a whole, especially the key cultural relics such as the Forbidden City.

3.1.2 Correctly address the relationship between the conservation of the imperial city and construction of modernization, with the new constructions within the imperial city meeting the requirements of protecting the imperial city, its overall features and spatial layout.

3.1.3 From the dialectical and historical perspectives, correctly recognize and deal with the development and changes of the imperial city, and treat new constructions in a discriminate and cautious way.

3.1.4 Adopting the humanistic and sustainable approach, try to improve the living, working and environmental conditions within the imperial city.

3.2 Principles of Plan

3.2.1 The principle of combining overall and classified conservations, and protecting the authentic historical information to the full.

3.2.2 The principle of rectifying, renovating and updating in a gradual way, and in accordance with order of importance.

3.2.3 The principle of gradually relocating inhabitants in a modest way.

3.2.4 The principle of strictly restricting the construction scale within the imperial city.

3.2.5 The principle of combining the conservation and sensible utilization of cultural relics sites.

4. Range of Plan

The plan scope for the imperial city stretches to the eastern section close to the wall of the imperial palace in the east, to the Dong and Xi Chang'an Avenues in the south, to the western section close to the wall of the imperial palace, Lingjing Allay and Fuyou Street in the west, and to the Ping'an Street in the north. The planned area is about 6.8 square kilometers, and the imperial city administration district is under the jurisdiction of Dongcheng and Xicheng Districts.

5. Nature and Historical and Cultural Values of Imperial City

5.1 Nature of Imperial City

5.1.1 The imperial city of the Ming and Qing Dynasty is centered around the Forbidden City, surrounded by imperial palaces, yamens, temples, altars and imperial gardens, the satisfying the needs of work, life and entertainment of the imperial family.

5.1.2 "Conservation Plan for the Historic City of Beijing" designates the imperial city as a conservation district of historic site that has a pronounced traditional and cultural characteristics.

5.2 Historical and Cultural Values of the Imperial City

5.2.1 Originated from the Yuan Dynasty, the imperial city was finalized in the Ming and Qing Dynasties. The present imperial city has kept the layout and features of the end of the Qing Dynasty and beginning of the Republic of China, thus having a high historical and cultural value.

5.2.2 Uniqueness: Being the only well preserved imperial city of the Ming and Qing Dynasties, the imperial city is the sole largest and most complete imperial architectural complex, forming the cream of the traditional axis of the old city of Beijing.

5.2.3 Completeness: Imperial city is centered around the Forbidden City and situated on the axis. Imperial palaces, gardens, temples, yamens and warehouses are all scattered in the city, demonstrating the complete idea and functional layout of serving the feudal emperors.

5.2.4 Authenticity: Traditional architectural complex of the imperial city such as the the Forbidden City, Dongzi River, Three Seas, Imperial Ancestral Temple, Altar to the God of the Land and Grain, government agencies, warehouses and courtyard houses have been well preserved, providing solid evidences of the historical information on the life, work and entertainment of the royal families.

5.2.5 Artistic Quality: The Imperial City has attained a high artistic quality in planning concept, architectural layout, construction technique, and color application. So it boasts a high historical and cultural value symbolizing the absolute authority of the royal monarch.

6. Zoning and Characteristics of Land

6.1 Zoning of Land

6.1.1 According to the features of land use and the natural layout of streets within the imperial city, as well as the principle of going from south to north, and east to west, the land of imperial city is divided into 10 districts.

6.1.2 01 District: area east of the Forbidden City and the Laboring People's Palace of Culture, and south of Wusi Street within the imperial city;

02 District: area east of Di'anmennei Street, Jingshandongjie Street and Wusibeijie Street;

03 District: area composed of the Forbidden City, the Zhongshan Park and the Laboring People's Palace of Culture;

04 District: area where Coal Hill Park is located;

05 District: area east Zhongnanhai, south of Wenjin Street, and west of the Forbidden City and the Laboring People's Palace of Culture;

06 District: Area east of Beihai Park, north of Jingshanqianjie Street, west of Di'anmennei Street and Jingshanxijie Street;

07 District: area where Zhongnanhai is located;

08 District: area where Beihai Park is located;

09 District: area south of Wenjin Street, and west of Zhongnanhai;

10 District: area north of Wenjin Street, and west of Beihai Park.

6.1.3 There are 572 planned plots of various kinds in the 10 zones, which will be renovated according to their numbers.

群传统风貌的精华，它们对皇城的保护具有决定性作用。

6.2.3　10区基本上为多层建筑区，是新的建设对皇城冲击最大的区域。

7. 土地使用功能规划

7.1　皇城的土地使用功能规划为12类，包括行政办公、商业金融、文化娱乐、医疗卫生、教育科研、宗教福利、普通住宅、一类住宅、中小学和托幼、市政设施、绿地、道路广场。

7.2　依照占地规模大小，主要用地有：绿地226hm²，约占皇城面积的33%；居住用地123hm²，约占18%；文化娱乐用地124hm²，约占18%；道路广场用地85hm²，约占12%；行政办公用地67hm²，约占10%，其他用地占9%。

7.3　规划以调整为主，现有公园绿地、行政办公用地、医疗卫生等用地面积基本不变。

7.4　原则上外迁与皇城性质不符的工业用地和仓储用地。

7.5　将地安门内大街两侧（道路红线50m范围内）规划各10m宽的绿地；景山后街北侧规划5～10m宽的绿地；拆除景山西街东侧、景山公园西墙外侧房屋，规划为绿地，为恢复历史河道创造条件。

7.6　在原御河古河道所在的位置，规划一条宽15～20m宽的绿带，严格控制相关建设，恢复古御河。

8. 文物保护单位的保护和利用

8.1　皇城内拥有各级文物保护单位63个，总占地面积约369hm²，占皇城面积的54%，是皇城保护的核心内容。

8.2　国家级文物保护单位9个，占地面积234hm²，占文物保护单位总面积的63.7%；市级文物保护单位有18个，占地面积约127hm²，占文物保护单位总面积的34%；区级文物保护单位有6个，占地面积约3hm²，占文物保护单位总面积的1%；普查登记在册文物有30个，占地面积约5hm²，占文物保护单位总面积的1.3%。

8.3　皇城内拥有区级以上的文物保护单位33个，其利用状况可分为三类：利用基本合理、利用勉强合理、利用不合理。其中，利用基本合理的文物保护单位有17个，利用勉强合理的文物保护单位有8个，利用不合理的文物保护单位有8个。

8.4　对占用被保护文物的单位但不具备保护和合理利用条件的单位和居民，应采取措施予以外迁、修缮、腾退文物并改善其使用环境。

9. 有历史文化价值的建筑的保护和利用

9.1　指那些尚未列为文物保护单位，但建筑形态或院落空间反映了典型的明清四合院格局或近代建筑特征的，具有真实和相对完整的历史信息的传统建筑。

9.2　皇城内具有一定历史文化价值的建筑或院落约有204个，占皇城院落总数的6.3%；这些建筑或院落占地约21hm²，占皇城面积的3%。

9.3　有历史文化价值的建筑应按照文物的保护要求对其进行保护，并挂牌明示。

9.4　对占用有历史文化价值的建筑但不具备保护和合理利用条件的单位和居民，应采取措施予以外迁、修缮、腾退并改善其使用环境。

10. 城廓、城墙、坛墙

10.1　现状皇城的东、南城廓基本清晰；沿平安大街南侧、西皇城根东侧、灵境胡同北侧结合环境整治规划一条不少于8m的绿化带，作为西、北皇城边界的象征。

10.2　在西安门，地安门所在位置两侧的用地中规划小型绿地，示意皇城城门的所在地。

10.3　皇城内目前存留下来的城墙、坛墙遗存约有41处，大部分只存一些残垣断壁。其中依托于国家级或市级文物保护单位的墙体一般保存较好，如故宫、景山等。对分布于平房区中的府、庙、坛墙，应随着环境的整治加以保护、修缮和展示。

11. 水系

11.1　历史上皇城内的主要水系北海、中海、南海、筒子河、金水河至今保存完好，应严格予以保护。

11.2　御河皇城段、织女河、连接筒子河和菖蒲河的古河道、连接北海和筒子河的古河道，有的已被填埋，有的已改为暗沟。对古河道用地上的建设应加以严格控制，为将来恢复古河道创造条件。

12. 绿化

12.1　皇城内的绿化系统可分为大型公园、小型公园、小型绿地、沿街行道树以及分布在街坊、四合院中的树木。

12.2　大型公园指景山公园、中山公园、劳动人民文化宫、北海公园、中南海（暂不开放），五个公园总占地面积约207hm²。

12.3　小型公园指东皇城根遗址公园和菖蒲河公园，两个公园总占地面积约7.7hm²。

12.4　沿城市主要道路两侧的带状绿地和分布在街区中的小型绿地，总占地面积约11.3hm²。

12.5　沿街行道树和分布在街坊、四合院中的树木，是营造皇城生态环境的重要组成部分，应予以保护。

13. 道路和胡同体系

13.1　与皇城直接关联的道路、胡同体系可分为四类：

13.1.1　皇城周边的城市干道，包括长安街、平安大街、南北河沿大街、西黄城根南北街、灵境胡同和府右街，共计6条。

13.1.2　皇城内部的城市主要道路，包括地安门内大街、景山前街、后街、东街、西街、南北长街、西华门大街、南北池子、东华门大街、五四大街、文津街、府右街、西什库大街、大红罗厂街、爱民街，共计15条，均为城市次干道。

13.1.3　皇城内部街区主要胡同，如陟山门街、沙滩北街、草岚子胡同等，胡同宽度一般大于7m，可承担部分城市支路的功能。

13.1.4　皇城内部街区次要胡同，如吉安所右巷、中老胡同、前宅胡同等，胡同宽度一般小于7m，主要用于组织街区中的非机动车和居民步行交通。

13.2　皇城内部的城市主要道路，应保持其现状道路宽度与尺度基本不变。

13.3　应当尽量保护现有胡同的位置和格局，维持胡同的基本走向，保护并延续传统的胡同名称。

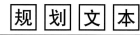

6.2 Characteristics of Each Zone

6.2.1 01, 02, 05, 06, and 09 are areas with concentrated traditional courtyard and single-story housing, and are dominated by allays as well.

6.2.2 03, 04, 07, and 08 are located key cultural relics sites which are the best of imperial architectural complex featuring traditional style. They play a decisive role in protecting the imperial city.

6.2.3 Number 10 is an area with many new multi-storied buildings, having the most negative impact on the imperial city.

7. Plan of Land Use Functions

7.1 Land use functions in the imperial city is divided into 12 categories, including office, business and finance, cultural and entertainment, medicare and health, education and research, religion and welfare, ordinary housing, first-grade housing, daycare centers, primary and secondary schools, municipal installations, greenbelt, and roads and squares.

7.2 Based on the land covered, they are: 226 hectares of greenbelt accounting for 33% of the area of the imperial city; 123 hectares of residential quarters accounting for 18%; 124 hectares of recreational land accounting for 18%; 85 hectares of roads and squares accounting for 12%; 67 hectares of office buildings accounting for 12%, and others accounting for 9%.

7.3 The plan gives priority to adjustment with areas of present park greenbelt, office buildings and land for medical facilities unchanged.

7.4 In principle, land for industrial use and warehouse that are not in tune with the nature of imperial city will be moved.

7.5 A 10-meter wide greenbelt along both sides of Di'anmennei Street (within 50 meters of the road red line) will be planned; another 5 to 10 meter greenbelt in the area north of Jianghshanhou Street will also be planned; the houses east of West Jiangshan Street and outside the west wall of the Coal Hill Park will be demolished and replaced by a greenbelt, paving the way for recover the historical river course.

7.6 A 15 to 20-meter-wide greenbelt will be planned at the original imperial river, where construction shall be strictly restricted with a view of recovering the imperial river.

8. Conservation and Utilization of Cultural Relics Sites

8.1 Cultural relics sites of various levels within the imperial city total 63, covering an area of 379 hectares, accounting for 54% of the area of the imperial city. They are the key to protecting the imperial city.

8.2 Nine sites are under the State conservation, covering an area of 234 hectares and accounting for 63.7% of the total area of cultural relics sites; 18 are under the municipal conservation with an area of 127 hectares, accounting for 34% of the total area of cultural relics sites; 6 are under the conservation of districts with an area of 3 hectares, accounting for 1% of the total area of cultural relics sites, and 30 are registered in the general survey with an area of 5 hectares, accounting for 1.3% of the total area of cultural relics sites.

8.3 There are 33 cultural relics sites at or above the conservation of the district levels, which can be divided into 3 categories in terms of uses: sensible, barely sensible and insensible. 17 belong to the first group; 8 belong to the second and 8 belong to the third.

8.4 For units and individuals that are occupying the sites but have no ability to protect and utilized them in a sensible way, measure must be taken to ask them to relocate, renovate and vacate the places so that the environment there can be improved.

9. Conservation and Utilization of Architecture of Historical and Cultural Values

9.1 Referring to traditional architecture that have not been listed as cultural relic sites under conservation, but their shapes and courtyard spaces reflecting typical courtyard layout of the Ming and Qing Dynasties and modern architectural styles, as well boasting authentic and comparatively complete historical information.

9.2 There are 204 architecture or courtyards of historical and cultural values within the imperial city, accounting for 6.3% of the total courtyards of the imperial city. They cover an area of 21 hectares, which equals 3% of the area of the imperial city.

9.3 Architecture that have historical and cultural values should be in line with the conservation requirements of cultural relics, and be marked as such.

9.4 For units and individuals that are occupying architecture of historical and cultural values, yet have no ability to protect and use them in a sensible way, measures should be adopted to relocate them, and the vacated architecture and their surroundings should be renovated and improved.

10. City Contour, Walls and Altar Walls

10.1 The east and south wall contours of the imperial city are clear-cut; a no less than 8-meter-wide greenbelt will be planned south of Ping'an Street, east of sections close to the west wall of the imperial palace, and north of Lingjing Allay to make it the symbol of the western and northern borders of the imperial city.

10.2 Smaller greenbelts in areas on both sides of Xi'anmen Gate and Di'anmen Gate will be planned to indicate the sites of imperial gates of the imperial city.

10.3 There are 41 broken walls in the imperial city. Among them those under the State and municipal conservations are well preserved such as the Forbidden City, the Coal Hill Park, etc. For those temple, altar and residential walls that are scattered amidst single-storied housing should be protected, updated and on display along the rectification of environment.

11. River System

11.1 The main ancient rivers systems within the imperial city such as Beihai, Zhonghai, Nanhai, moat, and Jinshui River are well preserved, and must be under strict conservation.

11.2 Some ancient river courses have either been filled up or turned into covered sewerage, such as the imperial city section of the Imperial River, Zhinu River, ancient river linked with Dongzi and Changpu Rivers, ancient river linked with Beihai and Dongzi Rivers, etc. Construction on these river courses must be strictly restricted to make room for recovering the ancient rivers in the future.

12. Afforestation

12.1 The afforesting system within the imperial city can be divided into large parks, small parks, small greenbelt, sidewalk trees, and trees in the neighborhoods and courtyard housing.

12.2 Large Parks refer to the Coal Hill Park, Zhongshan Park, Laboring People's Palace of Culture, Beihai Park, Zhongnanhai Park (not open at the moment), covering altogether area of 207 hectares.

12.3 Small Parks refer to the Ruins Park at sections close to the east wall of the imperial palace and Changpuhe Park. The two cover an area of 7.7 hectares.

12.4 The greenbelt on both sides of the main roads of the city and small green areas scattered in the neighborhoods cover an area of 11.3 hectares.

12.5 The sidewalk trees and those in neighborhoods, courtyard housing must be protected because they constitute a vital part in creating an ecological environment in the imperial city.

13. Road and Allay System

13.1 The road and allay systems closely related with the imperial city can be divided into 4 categories:

13.1.1 The 6 main roads around the imperial city: Chang'an Avenue, Ping'an Avenue, Nan and Bei Heyan Street, Nanbei Street at Xihuangchenggen, Lingjing Allay and Fuyou Street.

13.1.2 The 15 main roads within the imperial city: Di'anmenei Street, Jingshanqian Street, Jingshanhou Street, Jingshandong Street, Jingshanxi Street, Nan and Bei Chang Street, Xihuamen Street, Nanbeichizi Street, Donghuamen Street, Wusi Street, Wenjin Street, Fuyou Street, Xishiku Street, Dahongluochang Street, and Aimin Street. They are all secondary roads.

13.1.3 Main allays located in the streets within the imperial city: including those in Zhishanmen Street, Shatanbeijie Street and Caolanxi Street. Allays that are wider than 7 meters can assume some of the side road functions.

13.1.4 Minor allays within the imperial city such as Ji'ansuoyouxiang, Zhonglao, Qiangzhai, etc. Allays that are narrower than 7 meters usually serve non-motor vehicles and as pedestrian roads.

13.2 The main roads within the imperial city should remain as they are in terms of width and sizes.

13.4 根据居民出行或改善街区市政设施条件的要求,允许适当打通或拓宽个别胡同,但总体上要满足保护的要求。

14. 建筑高度控制

14.1 皇城是北京旧城保护的重要内容,必须严格保护其传统的平缓、开阔的空间形态。

14.2 在皇城内,对现状为1～2层的传统平房四合院建筑,在改造更新时,建筑高度应按照原貌保护的要求进行,禁止超过原有建筑的高度。

14.3 对现状为3层以上的建筑,在改造更新时,新的建筑高度必须低于9m。

14.4 必须停止审批3层及3层以上的楼房和与传统风貌不协调的建筑。

15. 保护与更新方式

15.1 综合现状建筑质量和建筑风貌的评估,皇城内建筑的保护与更新方式可分为6类。

15.2 文物类建筑,指国家级、市级、区级以及普查登记在册文物,这类建筑必须严格按照国家或北京市文物保护的相关法规进行保护和管理。

15.3 保护类建筑,指有一定历史文化价值的建筑或院落,这类建筑应参照国家或北京市文物保护的相关法规进行保护和管理,以修缮、维护为主。

15.4 改善类建筑,指街坊中一般的平房四合院建筑,这类建筑应以修缮和按原貌翻建为主,对具有传统四合院空间形态的建筑,应保持四合院的空间肌理。

15.5 保留类建筑,指街坊中和传统风貌较协调的新建筑,这类建筑大部分为新建的仿四合院建筑,与传统风貌较协调,应予以保留。

15.6 更新类建筑,指建筑质量差又没有传统风貌特征的破旧平房或对皇城传统空间具有破坏作用的多层或高层建筑。对破旧平房可采用改造的方式,恢复传统四合院的形态;对破坏性的多、高层建筑有条件时应予以拆除。

15.7 整饰类建筑,指一时无条件拆除的多、高层建筑或沿主要街道、胡同分布的一些被广告牌、墙面砖、色彩所充斥的平房。对这类建筑应通过统一的设计规则,对屋顶、墙面材料、色彩等要素统筹考虑,进行整饰。

16. 市政设施规划

16.1 排水设施:现状皇城街坊内排水基本上以合流制为主。规划皇城内的排水,在有条件的地区实行雨污分流,不具备条件的地区实行雨污合流的排水体制。

16.2 供水设施:随着房屋改建和道路系统的建设,逐步完善供水干、支线系统。

16.3 供热设施:供热应以清洁能源为主,能够利用城市热力管网的地区应充分利用城市集中热力系统;热力管道无法覆盖的地区可以采用燃气采暖或电采暖。

16.4 供气设施:随着房屋改建和道路系统的建设,逐步完善供气干、支线系统和中低压调压设施。

16.5 通信设施:随着房屋改建和道路系统的建设,逐步完善通信线路干、支线系统和电信局所。有线广播电视网、宽带网络应与通信线路统一规划、建设。

16.6 供电设施:随着房屋改建和道路系统的建设,逐步完善供电干、支线系统和开闭站、配电室;有条件地区应逐步将架空线入地。

16.7 市政管道布置:布置市政管道时应统筹规划。局部路段市政管道设置如不能满足规范要求,应采取特殊处理方式。

17. 环境整治的实施

17.1 重塑皇城边界,特别是皇城的西、北边界,展示皇城墙的位置。

17.2 强化中轴线沿线的绿色景观。沿景山前街、后街、西街、东街的一侧或两侧,结合环境整治规划一条连续的绿化带。保护南北长街、南北池子街道两侧的行道树木。

17.3 加强皇城内文物保护单位的保护、修缮、腾退和合理利用,加快对皇城内文物保护单位利用不合理情况的调整和改善,首先改善大高玄殿、宣仁庙、京师大学堂、智珠寺、万寿兴隆寺等文物的环境。

17.4 加强对有历史文化价值的建筑或院落的保护和修缮,先期选择2～6片有价值院落(真如镜胡同、张自忠故居、互助巷47号、会计司旧址、帘子库、北河等周边地区)比较集中的区域进行房屋修缮、腾退居民、拆除违章建筑、改善居住环境的试点,为普通民居的保护积累经验。

17.5 对在皇城环境整治过程中发现的,经考证为真实的历史建筑物或遗存,必须妥善加以保护,并加以标示。

17.6 采取措施分阶段拆除一些严重影响皇城整体空间景观的多、高层建筑。近期需拆除的建筑有:陟山门街北侧的6层住宅楼、景山公园内搭建的圆形构筑物、京华印刷厂烟囱、欧美同学会北侧的市房管局办公楼、北池子大街北端的北京证章厂办公楼等5处;中期应重点整治主要道路两侧的多、高层建筑。

17.7 对建筑质量较好且与皇城风貌不协调的多层住宅,应予以整饰。对于现状为平屋顶的多层建筑,特别是居住建筑应采取措施将平屋顶改造成坡屋顶,坡屋顶颜色以青灰色调为主。对新审批的建设项目,均应严格控制建筑高度,采用坡屋顶(青灰色调)的形式,并不得滥用琉璃瓦。

17.8 结合住房产权制度的改革以及旧城外新区的土地开发,逐步向外疏解皇城内的人口,降低人口密度,使居住人口达到合理规模。

17.9 加大拆除皇城内违法建设的力度,不在皇城内新建大型的商业、办公、医疗卫生、学校等公共建筑。

17.10 在环境整治的基础上,进一步宣传皇城作为历史文化保护区的重要意义,提高全民对皇城整体保护的意识,加快立法工作,制定《北京皇城保护规划》实施管理办法。

2003年4月18日

13.3 The location and layout of present allays, as well as their directions should be retained, and their traditional names protected and kept unchanged.

13.4 Some allays can be connected in line with the convenience of inhabitants and improvement of municipal installations. But the requirements of conservation as a whole must be met.

14. Control of Architectural Height

14.1 As the imperial city is the key component of protecting the old city of Beijing, its traditional flat and open spatial layout must be strictly preserved.

14.2 During renovation, the height of traditional single- and two-storied courtyard houses within the imperial city must be kept according to their original shapes. Height exceeding the original one will be prohibited.

14.3 During the renovation, the height of buildings of or above three stories shall not exceed 9 meters.

14.4 Buildings of or above three stories and those that are no in agreement with traditional features shall not be approved.

15. Means of Conservation and Updating

15.1 In accordance with the quality and styles evaluations of present architecture, the means of protecting and updating the buildings within the imperial city can be divided into 6 categories.

15.2 Architecture of cultural relics. These refer to cultural relics under the State, municipal and district conservations and are registered in the general survey. They must be protected and managed strictly according to the relevant regulations of the State and Beijing municipal government.

15.3 Architecture that should be protected. They refer to buildings or courtyards that have certain degree of historical and cultural values, which must be protected and managed according to the relevant regulations of the State and Beijing municipal government. Priority should be given to renovation and maintenance.

15.4 Architecture that should be renovated. They refer to ordinary single-storied and courtyard structures in the neighborhood, which should be mainly renovated and rebuilt according to their original style. For those that boast spatial shapes of traditional courtyard, the courtyard space should be retained.

15.5 Architecture that should be kept. It refers to new buildings that are in harmony with traditional features, most of which are pseudo courtyard housing. As they are in tune with traditional features, they should be retained.

15.6 Architecture that needs renovation. It refers to dilapidated single-storied houses of poor quality which are not compatible with traditional character. It also includes multi-storied buildings or high-rises that have destructive effect on the traditional space of the imperial city. For the former, renovation should be done in order to salvage the shapes of the traditional courtyard; for the latter, they could be demolished if needed.

15.7 Architecture that needs rectifying. Referring to the multi-storied buildings and high-rises that can no be torn down at the moment, as well as single-storied house along the streets and allays which are covered with billboards, facing tiles and colors. These kinds of architec.

16. Pan for Municipal Installations

16.1 Drainage facilities: Combined system is the dominant way of drainage in the neighborhood of the imperial city. In areas where conditions allow, separate drainage system should be installed in the plan, while combined system has to be set up in areas where conditions are not available.

16.2 Water supply facilities: Gradually perfect the main and branch lines of water supply alongside the housing renovation and road construction.

16.3 Heating facilities: Priority must be given to clear energy in heating supply. Areas that can make use of the urban heat supple network should fully utilize the central heating system. Areas that cannot be covered by heating pipes can use heating generated by gas or electricity.

16.4 Gas supply facilities: Gradually perfect the main and branch line system of gas supply and medium and low pressure adjusting facilities along with the housing renovation and road construction.

16.5 Communication facilities: Gradually perfect the main and branch line systems of communication and telecommunications bureaus along with the housing renovation and road construction. Cable TV and Broadband networks should be planned and constructed with communication lines.

16.6 Electricity supply facilities: Gradually perfect the main and branch line systems of electricity supply and switching stations and distribution rooms along with the housing renovation and road construction. Areas where conditions allow, aerial lines should gradually be buried underground.

16.7 Installation of municipal piping: Overall planning is needed in installing municipal pipelines. Specific measures should be taken to deal with sections where installations of piping cannot meet the requirements of the plan.

17. Implementation of Environmental Rectification

17.1 Restore the border of the imperial city, especially the borders on the west and north so as to demonstrate position of the walls of the imperial city.

17.2 Highlight the green landscape along the axis. A continuous greenbelt will be planned with the revamping of environment on one or both sides of Jiangshanqian Street, Jiangshanhou Street, West Jiangshan Street, and East Jiangshan Street. Roadside trees on both sides of North and South Street and Nanbeichizi Street must be protected.

17.3 Strengthen the conservation, facelift, vacating and reasonable utilization of the cultural relics in the imperial city, quicken the adjustment and improvement of reasonable use of relics. The first step should be to improve the environment of cultural relics such as Gaoxuan Hall, Xuanren Temple, Capital University, Zhizhu Temple, Wangshoulong Temple, etc.

17.4 Enhance the conservation and renovation of architecture and courtyard houses that have historical and cultural values. First select 2 to 6 areas with concentrated valuable courtyards (areas around Zhenrujing Allay, former residence of Zhang Zizhong, No 47 of Huzhuxiang, old site of Accounting Department, Lianziku, Beihe River) to conduct revamping, vacating inhabitants, demolishing unauthorized buildings, improving living environment, and accumulating experience in protecting ordinary residences.

17.5 With regard to architecture or relics that are discovered and found as a result of research to be authentic relics in the process of improving the environment of the imperial city, measures should be taken to protect and label them as such.

17.6 Take measures to tear down by stages some of the multi-storied buildings or high-rises that have seriously impacted the overall spatial landscape of the imperial city. The ones that must be demolished recently include: the 6-storied apartment north of Zhishanmen Street, the round-shaped structure within the Coal Hill Park, the chimney of the Jinghua Printing House, the office building of the municipal Housing Administration north of the Union of Western Returned Scholars, and the office building of the Beijing Badge Plant north of the Beichizi Street. During the middle stage, priority should be given to rectify the multi-storied and high buildings along the sides of main roads.

17.7 With regard to buildings that are of good quality but not in agreement with the character of the imperial city, revamping should be done. Measures must be taken to change the flat-roofs of high-rises, especially that of the residential architecture, into sloping roofs. The dominant color of the sloping roof should be greenish gray. Strictly control the newly approved constructions, and sloping roofing (greenish gray color) must be applied, without overusing glazed tiles.

17.8 With the help of property rights reform and land development of new areas outside the old city, the relocation of population from the imperial city should be conducted in order to cut the population density and keep the number of inhabitants in the old city on a reasonable scale.

17.9 Strengthen the measure of demolishing buildings that are in defiance of rules and regulations, and prohibit construction of large public buildings for business, office, medical and educational uses within the imperial city.

17.10 On the basis of environmental rectification, further promote the important significance of regarding the imperial city as a conservation district of historic city, raise the public awareness of protecting the imperial city in an overall manner, quicken the legislative work, and work out the methods of implementing and managing the "Conservation Plan for the Imperial City of Beijing".

April 18, 2003

北京皇城

北京历史文化名城北京皇城保护规划
Conservation plan for the Historic City of Beijing and Imperial City of Beijing

皇城鸟瞰（二战时期拍摄）
（资料来源：国家历史博物馆遥感与航空摄影考古中心）
Bird's eye view of the Imperial City (taken in period of WW II)
(Reference offered by: Archeological Center of Remote and Aerophotography of the National History Museum)

《北京皇城保护规划》说明

1. 任务背景

北京古城，曾经是"世界上现存最完整、最伟大之中古都市，全部为一整体设计，对称均齐，气魄之大举世无比"。1993年由国务院批准的《北京城市总体规划》(1991年至2010年)明确指出，北京作为历史文化名城必须保护古都的历史文化传统和整体空间格局。同时明确了历史文化保护区概念，提出在现有基础上继续划定保护区范围、划定文物保护单位的保护范围和建设控制地带，制定保护管理方法。

根据《北京城市总体规划》的要求，1999年，北京市城市规划设计研究院主持完成了在北京旧城区范围内对25片历史文化保护区的重新认定工作，并划定了保护范围和控制范围，经北京市政府批准，正式确立了25片历史文化保护区在北京传统风貌整体保护中的地位和作用。2000年，北京市完成了旧城25片历史文化保护区保护规划的编制工作，2001年初由市政府批准实施。

在25片历史文化保护区中，南北长街、西华门大街、南北池子、东华门大街、景山八片等14片历史文化保护区位于"世界文化遗产"故宫（紫禁城）周边地区，是历史上北京皇城的组成部分。皇城内以故宫为中心的庞大宫殿群、古寺庙、街巷胡同、四合院民居等至今保存较完整，是北京旧城的核心和精华。鉴于此，部分市政协委员提出设立"皇城原貌保护区"的提案，认为"皇城保护区"内的绝大部分地区应划为永久保护地段和暂不开发地段。对新批建筑的用途、高度、形式、色调等要严加限制；对一些严重损害保护区历史风貌的现存建筑物，要下决心有计划地予以拆除；对不合理占用和使用文物保护单位的机构、部门，应有计划地尽早腾退。

2001年，根据北京市政府要求，由北京市规划委员会、北京市文物局、北京市城市规划设计研究院组织相关部门、部分科研单位、高校共同编制了《北京历史文化名城保护规划》，提出以故宫为核心的北京旧皇城地区应作为一个整体进行积极保护，明确将皇城列入北京市第二批历史文化保护区名单，并将其作为"旧城整体格局的保护"中的一个重要专题进行了初步研究。

2002年，市政府46次常务办公会和首都规划建设委员会第21次全体会议审议并通过了《北京历史文化名城保护规划》，并上报建设部备案。与此同时，市政府明确要求，在"25片历史文化保护区的基础上，研究制订北京皇城保护规划，并征求专家意见后，报市政府"。根据市政府决定，由市规划委牵头，市文物局、市规划院、东、西城规划局等单位共同研究参与，深入开展了规划编制工作，通过半年多的紧张工作，编制完成了《北京皇城保护规划》。2002年10月，由市政府142次常务办公会和首都规划建设委员会第22次全体会议审议并通过了《北京皇城保护规划》。

2003年4月，北京市政府京政函[2003]27号下达了关于北京皇城保护规划的批复，要求市、区政府相关部门认真贯彻执行。

2. 规划范围确定

皇城历史文化保护区位于北京旧城的核心部位，其规划范围是在明清北京皇城城廓平面基础上，结合北京城市发展的现状，加以确定的。

明清北京城以皇城墙作为皇城区域的限定，其城墙大致位置东以皇城根遗址公园为界，南接长安街（南河沿大街—府右街段）北侧红墙（历史遗存，红墙南侧千步廊无存）。北以今平安大街为界，城墙位置在道路红线内。西边界大致以今西黄（皇）城根南北街为界，南至灵境胡同折向东，至府右街折向南，其中，府右街灵境胡同至长安街仍保留了皇城墙。皇城墙在西黄（皇）城根与灵境胡同转折处已挖掘出城墙遗迹，显示了原城墙的走向。

故此，规划范围四至边界确定如下：东至南北河沿大街东侧道路红线中心线，南至现存红墙，北至平安大街道路红线中心线，西至西黄城根南北街道路红线中心线、灵境胡同道路红线中心线、府右街道路红线中心线。

规划总用地约683hm²。

（注：民国时期，皇城根这类带有封建君主色彩的地名也被改为黄城根，解放后，该地名一直延续下来。在道路定线中，东黄（皇）城根南北街和南北河沿大街统称为南北河沿大街。）

北京，不仅是新中国的首都，也是辽燕京、金中都、元大都和明清的京都。不难想见，在这块土地上，书写的是什么样的历史，上演的是什么样的活剧，集聚的是什么样的人物，积淀的又是什么样的文化。这里的每一个街区、每一条胡同、每一座旧宅，甚至每一棵古树，差不多都有一个甚至几个值得细细品味慢慢咀嚼的故事。那些毫不起眼的破旧平房，可能是当年的名流住宅；那些杂乱不堪的荒园大院，也可能是昔日的王府侯门。更遑论闻名遐迩的故宫、景山、天坛、颐和园……即便是那些民间的东西，比如老北京的五行八作、时令习俗、工艺制作、风味小吃、儿歌童谣，也都是一本本读不完的书。——易中天《读城记之北京城》

3. 历史沿革

北京是世界名城，也是全国六大古都之一。作为历史文化名城，北京的历史文化价值和传统建筑艺术魅力主要体现在古都的规划格局、

Introduction to the "Conservation Plan of Imperial City of Beijing"

1. Mission Background

The old city of Beijing used to be the "best preserved and greatest capital city of the middle ages in the world. Holistically designed and symmetrical, it has no rivalry in grandeur the world over." The Master Plan for Beijing (from 1991-2010) approved by the State Council in 1993 stipulates that as Beijing is a historic city, its historical and cultural features and organic spatial layout must be protected. At the same time, the document states that the concept of conservation districts of historic sites was also clarified, the range of conservation areas will be further ascertained on the present basis, the protection scope and construction control area of the cultural relic sites will be established, and conservation management methods will be worked out.

According to the requirement of the this document, in 1999, the Beijing Urban Planning Designing Institute completed the establishment of 25 conservation districts of historic sites within the old city of Beijing, as well as their protection scope and construction control areas. Approved by the Beijing municipal government, the institute formally validated the status and role of the 25 sites in the protection of the overall traditional features of Beijing. In 2000, Beijing completed the compilation of the conservation plan for the 25 sites which was approved by the Beijing municipal government in early 2001.

Among the 25 sites, 14 (including Nanchang Street, Beichang Street, Nanchizi Street, Beichizi Street, Donghuamen Street, and 8 areas in Jingshan) are around the Forbidden City, a "world cultural heritage", thus forming the integral part of the Beijing Imperial City in history. The intact compounds of palaces, temples, streets and hutongs, as well as quadrangle courtyards around the Forbidden City are the core and cream of the old city of Beijing. Because of this, some members of the Beijing Political Consultative Conference put forward the proposal of setting apart a "conservation district of original Imperial City", saying that most part in the Imperial City conservation area should become permanently protected areas which are off limits to development. Strict measures should also be adopted to limit the function, height, form, color of newly approved architecture; systematic demolition of some buildings that have done harm to the traditional features in the conservation districts must be carried out in a determined way; and evacuation of institutions, agencies that are using and occupying cultural relic sites in a unreasonable manner must be implemented as soon as possible.

In 2001, based on the instructions of the Beijing municipal government, the Beijing Urban Planning Commission, Beijing Administration of Cultural Heritage and Beijing Urban Planning Designing Institute appointed relevant departments, some research units and universities to jointly compile the Conservation Plan for Historic City of Beijing, stipulating that the Imperial City of Beijing with the Forbidden City as the core should be protected as whole. They also filed the Imperial City on the list of the second group of cultural relic units under protection in Beijing, and undertook initial research on it as an important special topic of the "protection of the overall layout of the old city".

In 2002, the 46th routine working conference of the municipal government and 21st plenary session of the Capital Planning Commission jointly deliberated and approved the Conservation Plan and submitted it the Ministry of Construct for record. Meanwhile, the municipal government demanded to "draft a conservation plan for the Imperial City of Beijing on the basis of the 25 conservation districts of historic sites, and submit it to the municipal government after soliciting the opinions of experts." Thus, the Beijing Administration of Cultural Heritage, Beijing Urban Planning Designing Institute, and Planning Bureaus of the Dongcheng and Xicheng Districts and other units, led by the Beijing Urban Planning Commission, were involved in the compilation of the Conservation Plan of Imperial City of Beijing. After 6 months' intense work, it was completed on October 2002, and approved by the 142nd routine working conference of the municipal government and the 22nd plenary session of the Capital Planning Commission.

On April 2003, a rely was made by the municipal government in the form a document (no. 27, 2003) concerning the conservation plan for the Imperial City, requesting relevant departments at municipal and district levels to earnestly implement the plan.

2. Establishment of Plan Scope

The conservation district of Imperial City is located at the heart of the old city of Beijing. And the establishment of the plan scope for its protection is based on the contour plane of the Imperial City of Beijing during the Ming and Qing Dynasties coupled with the current development of Beijing.

The boundary of the Imperial City of the Ming and Qing periods was prescribed by the imperial walls, which stretches to the Huangchengen Ruins Park in the east, and red wall (historical relics; the Thousand-Step Corridor south of the red wall is gone) north of the Chang'an Street (Nanheyan Street-Fuyou Street section) in the south. It goes to Ping'an Street in the north with the original wall located within the red line of the road. Its western boundary is at the Nan and Bei Streets at West Huangchengen, with its southern part reaching Lingjing Alley, then turning to the east till Fuyou Street. The imperial walls still remain at Fuyou Street, Lingjing Alley and Chang'an Street, and the ruins of imperial walls at the corner of West Huangchenggen and Lingjing Alley have been excavated which indicated the their direction.

As a result, the four boundaries are: the central line of the red line on the road east of Nanheyan and Beiheyan Streets in the east; the extant red wall in the south; the central line of the red line of Ping'an Street in the north; and central line of red line of Nan and Bei Streets of West Huangchenggen, central line of road red line Lingjing Alley and central line of red line of Fuyou Street in the west.

The total planned land is about 683 hectares.

Note: During the period of the Republic of China, place names with feudal implications were changed. For example, Huangchengen (literally foot of the imperial wall) was changed into Huangchenggen with the first character "huang" (imperial) changed into "yellow", although the sounds are the same. This practice was retained after new China was founded. In the demarcation of boundaries, the Nan and Bei streets at East Huangchenggen and Nanhueyan and Beiheyan Streets are referred to as Nanheyan And Beiheyan Streets in a generic way.

Beijing is not only the capital of new China, but also the capitals of the Liao, Jin, Yuan, Ming and Qing Dynasties. Therefore, it is not difficult to imagine what kind of history, drama, people and culture were and are being produced and performed on this piece of land. Nearly every block, hutong, old house, even every old tree in Beijing has one or some stories worth pondering over slowly. The insignificant and shabby bungalows were perhaps mansions of VIPs; maybe the disorderly and desolate courtyards used to be the mansions of nobility in the past, let alone the world famous Forbidden City, Coal Hill Park, Temple of Heaven, the Summer Palace... Even every sort of trade in old Beijing, such as folk customs, handicrafts, dishes of local flavor, children's songs are all books of endless length

From City of Beijing, A Series of Cities with Walls by Yi Zhongtain

3. Historical Development

Beijing is not only a world famous city, but also one the six old capitals in China. As a historic city, Beijing's historical and cultural values as well as the charm of its

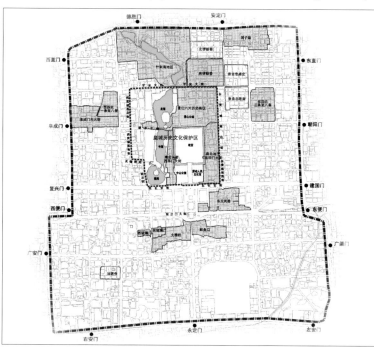

25片历史文化保护区分布图
The distribution map of the 25 conservation derisricts of historic sites

城市的空间塑造和重要历史建筑群的营造上，而处在北京旧城核心的皇城，则代表了中国封建社会在都城建设上的最高成就，集中体现了中华民族的悠久历史和文化古都的恢弘气魄。

北京城最初叫做"蓟"或"蓟城"，曾是燕国的国都，其址位于现在北京的西北一隅（一说位于白云观北一带）。公元前221年被秦始皇的军队所毁。隋文帝统一中国后，北京地区通称幽州，隋炀帝改称涿郡，公元605至609年，修筑了西到榆林长1500km的驰道和南通杭州的大运河，使北京与中原、江南联系起来，遂发展成军事重镇。公元938年，辽太宗耶律德光将幽州升格为"南京"，又叫"燕京"，作为他的四大陪都之一。金贞元年（公元1153年），击败了辽国的金人，将燕京定为他们占领的北部中国的政治中心，是为金中都。又过了一百多年，统一了中国的元世祖忽必烈，决定将北京地区作为他庞大帝国的中枢所在，他废弃了金中都，而以其东北的琼华岛离宫为中心，于至元元年（公元1264年）着手大规模的建设，建成了闻名于世的元大都。明清时期，在元大都基础上进行了改建和扩建而发展成为明清北京城。

蓟城、幽州、辽燕京、金中都和元大都，这些辉煌一时的巍峨城阙，早就已经"被西风吹尽，了无陈迹"，现在人们能够记起说起的，实际上是明清时代的北京城；而明清时代的北京城，则是由里外三层的"城"构成的"城之城"，分别是宫城、皇城、京城（内城、外城）。这个"城之城"的里圈，是通常称为"紫禁城"的"宫城"，城墙周长3km，开有四门，即午门、神武门、东华门、西华门。中间一圈就是"皇城"，城墙周长9km，也开有四门，即天安门、地安门、东安门、西安门。

3.1 皇城的形成和历史变化

历史上的皇城，是与北京城同时代修建的一座拱卫紫禁城，为帝王统治服务的皇宫外围城，建设始于元代，发展于明清时期。据文献记载，元大都即在宫城外建有皇城。明代皇城是在元大都萧墙旧址上改建的，其城墙较元代向外有所扩展。清代皇城是在明代基础上继续发展的。民国消灭帝制后，皇宫禁地被打破，并根据交通需要拆除和改造了原来封闭的皇城墙。现今，皇城内的主体建筑、园林体系、民居和街巷基本保持了清末和民国初年的面貌。

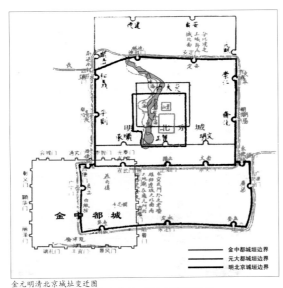

金元明清北京城址变迁图
Maps of the changing sites of Beijing in Jin, Yuan, Ming and Qing Dynasties

"你们必须知道，那里有一个又大又繁华的古城叫做汗八里，在我们的话说起，就是'大汗之城'的意思"，"城是如此的美丽，布置的如此巧妙，我们竟是不能描写它了。——《马可·波罗游记》

3.1.1 元代

元大都是中国古代都城规划史上的经典之作。元大都在兴建时，充分利用了原有条件和地理特点，按照《周礼·考工记》上所记的"国中九经九纬，经涂九轨，前朝后市，左祖右社"的王城规划制度，进行了周密设计和整体布局，在我国城市建设史上占有重要地位，奠定了后来北京城的发展基础。

《三礼图》中的周王城图（资料来源：刘敦桢主编《中国古代建筑史》）
Imperial City of Zhou Dynasty in Tri-Rites Painting (from The History of Chinese Ancient Architecture ed. by Liu Dunzhen)

新建的大都城由大城、皇城和宫城三部分组成，并按一定比例分区布置。它是当时世界上最大的城市，也曾是世界政治的中心。

元大都的宫苑在全城中央偏南，它的市场集中在宫苑之北的海子沿岸。太庙在宫苑的左侧齐化门内，社稷坛在宫苑的右侧和义门内。

大都的皇城在城市南部的中央地区，它的东墙在今南北河沿的西侧，西墙在今西黄（皇）城根，北墙在今地安门南，南墙在今东、西华门大街以南，其主要作用是将宫城、太液池、兴圣宫和隆福宫包围起来，加筑一道防御的墙垣。皇城的城墙，称为萧墙，也叫阑（拦）马墙，周围约10km。"阑（拦）马墙临海子边，红葵高柳碧参天"。阑（拦）马墙外密密麻麻地种植着参天的树木，更增添了皇城威严的气氛。皇城城门都用红色，又称红门，"人间天上无多路，只隔红门别是春"。红门内外，景色完全不同。皇城南墙中门为棂星门，其位置大致在今午门附近。它的南面，就是大都城的丽正门。在丽正门与棂星门之间，是宫廷广场，左右两侧为千步廊，设置中央官署。在元代以前，宫廷广场一直处于宫城正门的前方，大都城却把它安排在皇城正门之前，这在都市规划上是一个极大的变化，它加强了从大都城正门到宫城正门之间在建筑上的层次和秩序，从而使宫阙的布置更为突出，门禁更为森严。

皇城之内，以太液池为中心，围绕着3组大的宫殿建筑群，即宫城、隆福宫和兴圣宫，每组宫殿群分别配有皇家宫苑（御苑、前苑和后苑），分别供皇帝、太后和太子居住。宫城位于皇城南部偏东，因西临太液池而没有位置居中，呈长方形，周长4.5km，是皇城内规模最大、形制最严整的宫殿群。宫城有四门，东为东华门，西为西华门，北为厚载门，南为崇天门。崇天门东西长约51m，南北深约16m，高约26m，门11间，门上有楼，左右两观，下开5门。宫城内的主要建筑分为南北两部分，南面以大明殿为主体，北面以延春阁为主体。

其中大都城以内萧墙以外，街道纵横整齐，互相交错，把全城划分为50个坊，这是居民的住所，至今仍可看出旧日的痕迹来。而且，"胡同"据说就是元代对街道的称语。

在新建大都城的同时，为了解决大都城的供水及漕粮北运的问题，元朝还进行了大规模的水利工程建设。当时大都城内有两条水系，一是由高梁河、海子、通惠河构成的漕运系统；南来船舶可由通州，沿通惠河经皇城墙东侧直达大都城内，积水潭内"舳舻蔽水"，这条水系一直保存到明清。另一条是由金水河、太液池构成的宫苑用水系统；明初金水河系湮废，改由积水潭供水，经历后世的不断疏浚，一直沿用

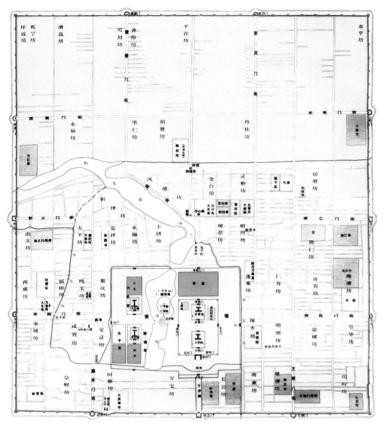
元大都平面图（资料来源：侯仁之主编《北京历史地图集（一）》）
Plane figure of the capital of the Yuan Dynasty (from *A Collection of Historical Atlas of Beijing (Vol.1)* ed. by Hou Renzhi)

traditional architecture are embodied in the planning layout, creation of urban space and important architectural compounds; while the Imperial City, located at the heart of the city, represents the highest achievement of the capital construction in the Chinese feudal society, the long-standing history of the Chinese nation and the grandeur of the ancient cultural capital.

Formerly called "Ji" or "city of Ji", Beijing used to be the capital of Yan Kindom in the northwestern part of present day Beijing (some say it was in the area of the White Cloud Temple). In 221 BC, it was destroyed by the army of emperor Qinshihuang. After emperor Wendi of the Sui Dynasty unified China, Beijing was referred to as Youzhou, which was changed to Zhuo Prefecture during the reign of emperor Yandi of the same period. From 605 to 609, a 3,000-mile royal road was built extending to Yunlin in the west, and the grand canal running to Hangzhou in the south, thus connecting Beijing with middle and southern parts of China and making it a place of strategic importance. In 938, emperor Yeludeguang of the Liao Dynasty elevated Youzhou (a prefecture) to "Nanjing", or "Yanjing" as one of his four secondary capitals. In 1153, the Jin Dynasty defeated Jiao and made Yanjing the political center of the North China they had occupied. They called it as Zhongdu. More than a hundred years later, the first emperor of the Yuan Dynasty Kublai Khan unified China and decided to make Beijing as the center of his huge empire. Abandoning Zhongdu and choosing the provisional imperial palace on the Jade Islet as the center, he started large-scale construction in 1246 and built the world renowned Dadu, capital of the Yuan period. During the Ming and Qing periods, enlargement and renovation were done on the basis of Dadu, which resulted in the coming into being of the city of Beijing.

City of Ji, Youzhou, Yanjing, Zhongdu and Dadu, all of these once magnificent palaces have gone with the wind. What people can talk about is actually the Beijing of the Ming and Qing Dynasties, which was "the city within city" composed of three layers of "cities". They are the palace city, Imperial City and capital city (inner and outer cities). The first refers to the Forbidden City, the wall of which is 6 miles in perimeter. It has 4 gates: Meridian Gate, Divine Prowess Gate, Donghua Gate and Xihua Gate; the second refers to the Imperial City which has a 18-mile long city wall and also 4 gates: Heavenly Peace Gate, Di'an Gate, Dong'an Gage and Xi'an Gate.

3.1 Formation and Historical Evolution of Imperial City

The Imperial City in history was constructed, together with the city of Beijing, in the Yuan Dynasty and further extended in the Ming and Qing Dynasties. Built as a defense, it served as an outer city for the rule of emperors by surrounding the palace. Ancient documents revealed that even Dadu had Imperial City outside the palace city, so the Imperial City of the Ming period was built on the former site of the "screen wall" of the Yuan period with its walls extended further. The Imperial City of the Qing Dynasty was modeled after that of the Ming. The royal system was dicarded in the Republic of China period, so the palace was open to the public with the result that the once closed Imperial City walls were pulled down and renovated in line with the transportation needs. Today, the main buildings, garden system, residential quarters and roads basically retain the style of the periods of the end of the Qing Dynasty and the early yeas of the Republic of China.

"You must know there exist a huge and prosperous ancient city called Hanbali. To use our words, it means the 'city of Great Khan'. It is so beautiful, its layout so ingeniously created that it is ineffable."
From The Travels of Marco Polo

3.1.1 Yuan Dynasty

The Dadu of the Yuan Dynasty is a classic work in the history of ancient capital planning. The construction of Dadu fully made use of the conditions at the time and geographical features. Based on the planning concept for the kingdom city prescribed in the Book of diverse Crafts that "palace in the front and market at the rear, while ancestral temples on the left and altars to gods on the right", the design was meticulous and layout was considered in an organic way, which played an important role in the urban construction history of China, and laid a solid foundation for later development of the city of Beijing.

Composed of Greater City, Imperial City and palace city, and divided into parts according proportionate ratio, Duda was the biggest city in the world, as well as the political center of the globe.

The palaces of Duda were concentrated a little bit south to the heart of the city whereas its markets were on the bank of Haizi Lake north of the palace compounds. The imperial Ancestral Temple was in the Qihuamen on the left of palaces, while the Altar to the God of the Land and Grain was in the Yimen on the right of the palaces.

The Imperial City of Dadu was located in the center of the southern part of the city, with its east wall ending at the western side of present day Nanheyan and Beiheyan; its west wall at today's Huangchenggen; its north wall at the southern side of Di'anmen; and its south wall at southern side of Donghuamen and Xihuamen Streets. Its main function was to serve as a defense system surrounding the palace city, Taiye Lake, Xingsheng Palace and Longfu Palace. The wall of Imperial City was called screen wall, or horse-blocking wall, with a length of 10 kilometers. Tall trees soaring into the skies were planted outside of the horse-blocking wall adding to the solemnities of the Imperial City. As the gates of the Imperial City were painted red, they also called "red gates". The scenes inside and outside those red gates were totally different. The gate in the middle of southern wall was called Lingxing Gate, which was near today's Meridian Gate. To its south was the Lizheng Gate of Dadu. In between these two gates was the royal square flanked by Thousand-Step Corridors where government agencies were located. Prior to the Yuan period, royal square was placed in front of the main gate of the palace, not in front of the main gate of Imperial City as the case in Dadu, so the change is a significant one in urban planning in that it enhanced the gradation and sense of order of architecture between the two main gates, so that the layout of palaces were rendered more conspicuous and the gates were more heavily guarded.

Inside the Imperial City were three large compounds of palaces around the Taiye Lake, namely, the Palace City, Longfu Palace and Xingsheng Palace, each equipped with royal mansions (imperial mansion, front and rear mansions) as residential areas for emperor, empress and crown prince. The Palace City, located on the eastern side of the southern part of the Imperial City is not in the center because it borders the Taiye Lake on the west. Rectangular in shape and 4.5 kilometers in circumference, it is the largest and most standard among palace compounds. The four gates of the Palace City were Donghua Gage in the east, Xihua Gate in the west, Houzai Gate in the north and Chongtian Gate in the south. Chongtian Gate is 51 meter long from east to west, 16 meter deep from north to south, and has a height of 26 meters. It has 11 doors, on which was a tower with 5 doors right below it. The main buildings within the palace were divided into two parts with Daming Hall as the primary structure in the south, and Yanchun Pavilion as the main structure in the north.

至今，而金水河故道皆已变为胡同，尚余清晰遗痕。

另外，除了街道和水系布局之外，许多元代建筑遗迹，如中书省（明清皇城内太庙的占地范围）、御史台、枢密院、太史院、国子监、大都路总管府、社稷坛、太庙、孔庙，以及大天寿万宁寺、大圣寿万安寺（今白塔寺）、大庆寿寺（已毁）等，也都历历在目地保留在北京旧城内。

明清两代对北京城的改建和扩建，典型地体现了我国封建帝王之都的设计理想，奠定了北京城的规模和格局，在北京城的发展史上是一个极为重要的阶段。

3.1.2 明清时期

徐达攻克大都后，遂将大都改为北平。明初，由于大都城北人烟稀少，市况衰落，土地荒芜，加之元顺帝虽退走蒙古草原，但时刻伺机图复。明军为了便于防守，进城后首先将元大都的北城墙南移2.5km，从而奠定了明代北京城的北界。至于东西城墙，基址没有改动，门位也没变更，只是名称有所改换。到了明成祖（朱棣）永乐十七年（公元1419年），才将原大都南城墙南移1km。为了"压胜"，除掉元朝"王气"，明军把元朝的宫殿尽行销毁。

明成祖朱棣即位后，鉴于北平形势踞形胜，为"龙兴之地"，更为了有效地抗击蒙古人的南侵，进一步控制东北地区，明成祖决定迁都北平。永乐元年（公元1403年）正月，他改北平为北京，这是"北京"名称的最早出现。永乐四年（公元1406年），明成祖下诏迁都北京，并开始大规模地重新营造北京城。在元大都的基础上，吸取了历代都城规划的优点，布局整齐，设计精巧，南北中轴线十分突出。这条中轴线南起永定门，向北穿过正阳门、承天门、端门、午门、紫禁城的正中心和万岁山主峰，最后止于鼓楼和钟楼，全长约8km。外城、内城、皇城和宫城，都以这条中轴线对称展开，从而形成了完整和谐、举世无双的巨大建筑群。

明北京城的建设重点是改建宫城和皇城，突出宫城位置，把五府六部摆到宫城前面。修筑外城以后，中轴线从永定门经正阳门、故宫、景山到钟鼓楼，纵深加大，中轴线两侧大建筑物明显对称，全城拱卫宫城。把为皇家服务的内官系统全集中到皇城内，开挖南海，扩大西苑，皇城变成北京城的中心，使东西城之间交通阻隔，造成北京城内居民在心理上总感到有皇帝存在的效果。从城市规划上体现了"皇权至尊"的理想，利用城市规划为政治服务达到了一个新的高度。

（1）明皇城

明皇城位于北京内城中部，是在元大都萧墙旧址上改建的，其南、北、东三面城墙较旧址稍有开拓，南城墙在今东、西长安街北侧，北城墙在今地安门东、西大街南侧，东城墙在今东黄（皇）城根，西城墙在今西黄（皇）城根。设四门，南为承天门，北为北安门，东为东安门，西为西安门。承天门内经端门至宫城之间的空间联系，还收到了平面布局上更为突出的艺术效果。

承天门前有外金水桥和宫廷广场，东、西、南三面绕以红墙。广场东西两侧分别筑长安左门和长安右门，正南一面筑大明门。大明门内东、西红墙内侧各有连檐通脊的千步廊，外侧则是直接为朝廷服务的中央衙署所在地，按"左文右武"的规制，对称地排列着中央政府的主要官署，从而改变了元大都城内中央衙署分散的局面。

宫城之北，在元朝后宫延春阁故址上利用挖南海的土在玄武位堆筑土山，高十四丈七尺（约合46.67m），命名为万岁山。山上五峰并峙，主峰恰当全城中轴线上。因此，万岁山不仅作为全城的制高点标志着明北京内城的几何中心，而且还有压胜前朝的用意，故亦称"镇山"。登临山巅，足以俯瞰全城。万岁山后有寿皇殿等建筑。

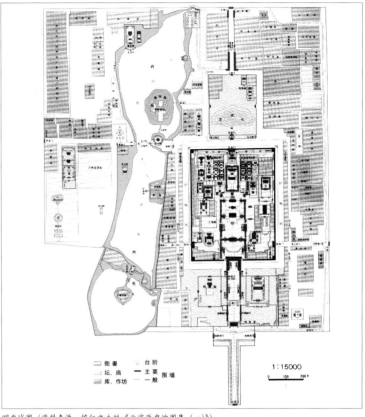

明皇城图（资料来源：侯仁之主编《北京历史地图集（一）》）
Imperial City of Ming Dynasty (from *A Collection of Historical Atlas of Beijing* (Vol.1) ed. by Hou Renzhi)

宫城之西为西苑，相沿为皇城内宫苑区，琼华岛和瀛洲依旧屹立太液池中，瀛洲岛上新建承光殿，昔日岛上有木桥，连接太液池东西两岸，明代废东桥，填为平地，又改西桥为石桥，曰玉河桥，两端各筑牌坊，西曰金鳌，东曰玉𬯀，因有"金鳌玉𬯀桥"之称，明代又在太液池南端开拓湖面，是谓"南海"，而金鳌玉𬯀桥南北湖面别称"中海"与"北海"，是谓"三海"，太液池北岸有嘉乐殿、五龙亭、雷霆洪应殿等；西岸有玉熙宫、紫光阁、寿明殿，东岸有崇智殿、船屋等；南部有昭和殿、台坡等；太液池沿岸垒石堆山，遍植花木，湖光山色，殿堂亭阁，与古木繁花交织成一幅优美的园林画卷，遂使西苑在明代成为皇城内风景优美的皇家园林。

西苑以西的皇城西南部，即元代隆福宫故址一带，有万寿宫、大光明殿、兔儿山、旋坡台等。西苑以西的皇城西北部，即元代兴圣宫故址一带，有清馥殿及内府诸库。

宫城以东为东苑，离宫东南部系由重华宫、崇质殿、玉芝宫等主体建筑组成的小南城；又名"南城"、"南内"，其西南隅是皇史宬，即珍藏实录典籍的所谓"金匮石室"。

皇城东北部，西北部以及宫城东西两侧，多是内宫衙署，如各种监、局、厂、作、房、库等，直接为皇室服务，亦是官办手工业的集中地区。

道路交通形态则呈现为连接和划分这些建筑及机构组织的直接结果，即道路网络不同于方正的里坊肌理，而呈现为较为自然的特征，这也是其形态构成的独特之处。主要的外部道路系统如后来清代的东华门大街、西安门内大街、地安门内大街、景山东西街、景山后街、北池子、北长街等骨架结构此时均已形成。

（2）清皇城

清朝（公元1644～1911年）定都北京，清代的京师建设完全沿袭了明

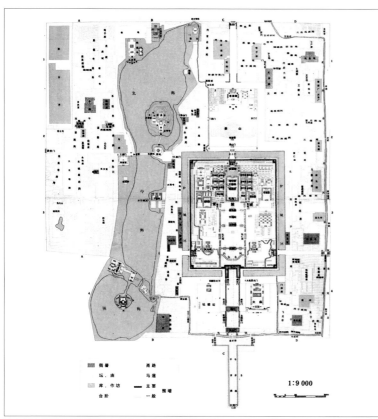

清皇城图（资料来源：侯仁之主编《北京历史地图集（一）》）
Imperial City of Qing Dynasty (from *A Collection of Historical Atlas of Beijing (Vol.1)* ed. by Hou Renzhi)

Between the Dadu City and the screen wall crisscrossed streets and roads, and the whole city was divided into 50 fang which were residences for inhabitants. The vestiges of the structure are still visible even today. It is also said that "hutong" was used first during the Yuan Dynasty to refer to street.

While constructing Dadu, large-scale irrigation works were also undertaken to solve the problem of water supply and transporting grain by water to the north. There were two water systems at the time, the first being the grain transporting system through Gaoliang, Haizi and Tonghui Rivers; boats from the south could reach Dadu by way of Tongzhou, Tonghui River and east side of the Imperial City. So Jishuitan area at the time was full of ships. This system was practiced also in the Ming and Qing periods. Another system was water supply for use in the palace through Jinshui River and Taiye Lake. Jinshui River was in disuse in the Ming Dynasty and replaced by Jishuitan for water supply which, being dredged by later generations, is still used today; whereas Jinshui River was replaced with hutongs, although its watercourse is still visible.

Apart from water systems and streets, many other architectural relics of the Yuan Dynasty still exist in Beijing such as Supreme Military Council, Board of Historians, Imperial College, Chief Steward Mansion at Dadu Road, Imperial Ancestral Temple, Confucius Temple, Altar to the God of the Land and Grain, Datianshou Wanning Temple, Dashengsi Wan'an Temple (today's White Pagoda Temple), and Daqing Temple (being destroyed).

The extension and renovation of city of Beijing by the Ming and Qing Dynasties constitute an extremely important stage in the development history of the city because it typically manifests the design concept of the capital of feudal kingdom and lays down the scale and layout of the city.

3.1.2 Ming and Qing Dyasties

When Xuda conquered Dadu, he changed its name to Beiping. In the early years of the Ming period, as the population in the northern part of Dadu was scarce, the market languished and land was laid waste, coupled with the fact that emperor Shundi of the Yuan Dynasty, after pushed back to the prairie of Mongolia, was always ready to make a comeback, the Ming troops moved the northern wall of the Dadu 5 miles south for the sake of better defense. This also laid down the boundary of the city of Beijing of Ming Dynasty. The sites of east and west walls remained unchanged, although the names of the gates was changed. It was not until the year 1419 during the reign of emperor Zhudi of the Ming period that the southern wall of the former Dadu city was moved south for 2 miles. In order to eliminated the "peremptory air" of the Yuan Dynast, all palaces of the Yuan period were destroyed.

Upon coming to the throne, emperor Zhudi believed that Beiping commanded a vantage point that could help him realize his ambitions. And in order to effectively resist the invasion of the Mongols and exercise control over the northeastern area, he ordered to move the capital to Beiping. In 1403, he changed Beiping to Beijing, the first time this name appeared. Next year, the resettlement of capital started which resulted in large-scale reconstruction of Beijing. Based on the modal of Dadu, and absorbing the strengths of capital cities of various periods, the design of the Ming period was exquisite, the layout more neat, the north and south axis more conspicuous. This axis, starting from Yongdingmen in the south, goes north through Zhengyangmen, Chengtianmen, Duanmen, Meridian Gate, the center of the Forbidden City and the highest peak of the Longevity Mountain and ends in the Drum and Bell Towers, with a total length of 8 kilometers. The inner and outer cities, Imperial City and palace walls are all symmetrically arranged around this axis, thus forming a harmonious and unparalleled huge architectural compounds.

The priority of constructing Beijing in the Ming Dynasty was on the renovation of the palace city and Imperial City, in order to accentuate the location of the palace city and place the central government in front of it. After outer city was constructed, the axis was further extended by going through Zhengyangmen, the Forbidden City, Jingshan Park and the Drum and Bell Towers, along each side of which big architecture is obvious arranged in symmetrical order with the result that the whole city surrounds the palace walls. In order to concentrate all service for the emperor in the Imperial City, the Nanhai (south lake) was excavated, the west park was expended, with the result that the Imperial City became the center of Beijing. Because of this, the flow of east and west sides of Beijing were interrupted with the presence of emperor heavily weighed on the subconsciousness of the local inhabitants and the idea that "imperial power is most revered" fully demonstrated in the urban planning. Thus, to use urban planning to serve politics reached another climax.

(1) Imperial City of the Ming Dynasty

Located in the center of the inner city of Beijing, the Imperial City of the Ming period was renovated on the site of the screen wall of the Yuan period. Its southern, northern and eastern walls were a little bit extended than the original sites, with its southern wall at the north side of the today's east and west Chang'an Avenues; its northern wall at the south side of today's Dong and Xi Di'anmen Streets; its eastern wall at today's east Huangchenggen and its western wall at today's west Huangchenggen. The four gates of the Imperial City are: Chengtianmen in the south, Beianmen in the north, Dong'anmen in the east and Xi'anmen in the west. The spatial link between Duanmen and palace city within Chengtianmen creates a conspicuous artistic effect in plane layout.

In front Chengtianmeng are the Golden Water Bridge and imperial Square which was enclosed on the east, west and south sides by red walls. The east and west sides of the square are equipped with Chang'anzuo Gate and Chang'anyou Gate respectively, and the south side has the Daming Gate. Within the Daming Gate, and on the inside of the red walls are thousand-step corridors, and on the other side are government agencies serving the court. This agencies are arranged with the rule of "those in charge of military are on the right and those in charge of civilian affairs on the left", thus transforming the scattered layout of central government practiced in the Yuan Dynasty.

On the northern part of the palace, a hill of 46.67 meter high was built on the site of Yanchunge, the Yuan Dynast imperial harem with the earth excavated from Nanhai. Named Longevity Hill, it has 5 peaks with the highest peak right on the axis. As the highest place in Beijing, the hill also carries the intention of conquering the previous dynasty, hence also the name the "Hill of Conquest". One can overlook the whole town on the top of the hill. Behind the hill are found other architecture such as the Shouhuang Hall.

The West Park lies to the west of the palace and within the Imperial City. The Qionghua (Jade) Islet and Jingzhou are still in the Taiye Lake, on which Chengguang Hall is newly built. There used to be wooden bridges on the Islet linking the east and west banks, but the east section was dismantled in the Ming Dynasty and filled up;

代的北京城,没有多大的变化,只是做了一些重修和小范围的改建,如清中期曾对皇城进行了一次较大规模的修缮。清军进驻北京后,清朝统治者最关心的莫过于"和"、"安"二字,要把各族协和在统治之下,以求长治久安。于是他们在城市建设、居民安排等方面都突出体现了这一政策。

顺治二年(公元1645年),清朝重新修建宫城内的皇极殿、中极殿、建极殿后,依次将其名改为"太和殿"、"中和殿"、"保和殿",重点突出了一个"和"字。清皇城同明皇城,仍设四门。顺治八年九月,重修承天门竣工后,改承天门为"天安门",次年七月,改皇城后门名为"地安门",再加上皇城原有的东安门与西安门,这样,皇城的东、南、西、北四门都突出了一个"安"字。天安门外广场及长安左门、长安右门仍如旧,惟大明门改称大清门。天安门内由端门至午门以及太庙与社稷坛,俱仍旧制。

还有紫禁城各城门的匾额,都是以满文与汉文合璧书写;太庙祭祖、天坛祭天时读祝文,也是既读满文,又读汉文。所有这一切都体现着清初"协和求安"的基本国策。

同时,清代皇城里的楼堂塔寺建筑,也集中地反映出北京是统一的多民族国家的政治中心。如满人信奉萨满教,清刚定都北京时,就在御河桥东和路南,建造堂子,立杆祭天。蒙古等族信奉喇嘛教,康熙时将皇城东南隅明代小南城(南内)重华宫改建为玛哈噶喇庙(附近之明建皇宬,保留如旧),以笼络和安抚蒙古王公贵族。此外,在北长街东侧临筒子河边修建了一座喇嘛庙,至今该用地仍为班禅的宅邸。乾隆二十三年,相传为新疆维吾尔族的香妃修建了"宝月楼"(今中南海新华门),外设回回营,建礼拜寺,以为望乡之阁。同时,随着天主教的传入,在蚕池口修建了天主教堂。

地安门内明代万岁山改名景山,俗名煤山。山上五峰并列。乾隆十五年(一说乾隆十六年),五峰顶上各建一亭。中峰一亭名万春,适当全城中轴线上,形制特大;其左为观妙、周赏,右为辑芳、富览,东西四亭,两相对称。皇城西苑内若干宫殿,如玉熙宫、清馥殿、万寿宫、兔儿山和旋坡台等,或废除,或改建,变化颇大。

与明代不同,清皇城东安、西安、地安三门内均允许八旗居民迁往。前明内宫各衙署所在地,服务机构数量减少,用地规模也减小,大部分转变为居民胡同,增加了许多内务府太监和家属的住所,如内官监胡同、织染局胡同、酒醋局胡同、惜薪司胡同等。用地进一步细化,胡同肌理变得细密且呈自由生长的趋势。

上述变化的产生有其深刻的社会、历史原因。明朝中后期,官营手工业已有走向衰落的趋势。据宪宗威化二十一年(公元1485年)统计,内织染局已从明初拥有三十二种行业工匠的规模缩减到工匠一十人。大量的工匠走向民间,出现了"若闾里之间,百工杂作,奔走衣食者尤众"的现象。到了清代,官营手工业分隶于内务府和工部,生产规模远不及明代。内务府在北京主要设有内织染局,广储司七作(银作、铜作、染作、衣作、绣作、花作、皮作)和营造司三作(铁作、漆作、炮作)。统治阶级的大部分需求开始转向市场。清代民间商业繁荣,"汇万国之车书,聚千方之玉帛","帝京品物,擅天下以无双",私营作坊的扩大,是商品经济发展的必然结果,也是导致皇城中官办服务机构锐减的根本原因;更直接导致了皇城内用地性质的变化。同时,清朝以明朝宦官之祸为戒,也收缩了内官署,更加剧了官营手工业的衰落。

清代虽然没有对北京城池大兴土木,但在开发和营造皇家园林上则投入了巨大的财力,役使了难以计算的匠师和夫役,清代在"西苑"的基础上,对"三海"进行了大规模的营建,三海周围相继修建,园林景色益加秀丽多姿,使其成为风景优美的皇家园林。其中,南海自宝月

北海、中海(1965年摄)
Beihai and Zhonghai (shot in 1965)

楼至蜈蚣桥,最主要的建筑是瀛台。它在明时称"南台",清顺治二年(公元1645年)进行了扩修,并改名为"瀛台",其名取法于传说中的瀛洲仙境。这里是清帝夏季避暑、冬季赏雪的主要场所,也是戊戌变法失败后,慈禧太后囚禁光绪的地方。

中海位于蜈蚣桥和金鳌玉蝀桥(今北海大桥)之间,主要建筑有"勤政殿"、"丰泽园"、"海晏楼"、"怀仁堂"、"紫光阁"和"水云榭"等。勤政殿与瀛台隔海相望,为清帝驻苑听政之所。殿西的"丰泽园"为康熙帝所建,园内有"颐年殿"。殿东为"菊香书屋",殿西为"澄怀堂",园中有稻田0.67hm²。每年春季,帝王都要到这里举行"演讲"仪式,以示勤劳和劝人耕作。中海的"海宴堂"(后改为居仁堂),是慈禧当权时为讨好洋人在"仪鸾殿"的废墟上修建的,它是一组中西合璧式的建筑群。

北海在金鳌玉蝀桥以北,辽时这里有瑶屿行宫,金时筑有"琼华岛",元称"太液池",并以此为中心营造大都宫殿。明称"西苑",并在圆坻(团城)上修复了"仪天殿"。同时在北海北岸、东岸和西岸又兴建了"太素殿"、"凝和殿"和"迎翠殿"等建筑。到了清代,北海得到进一步修缮。顺治八年(公元1651年),清在广寒殿的旧址上建立了"白塔寺",又称"永安寺"。乾隆年间又在北海的西北岸、东岸和琼华岛上,增建、改建和扩建了许多建筑,今天我们见到的大部分建筑都是乾隆时遗留下来的。光绪十一年至十四年(公元1885~1888年),慈禧挪用海军经费,重修北海,在西岸和北岸沿湖的地带铺设了小铁轨,称为"紫光阁铁路",并同时增建了部分建筑,供其享乐。至此,"三海"地区不仅成为清朝著名的皇家园林,同时也是清帝听政赐宴的重要场所,成为清朝的另一个政治中心。

3.1.3 民国时期

随着岁月的流逝、时代的交替。北京这座古老城市的城垣、宫殿、坛庙、街道和苑囿等都发生了极大的变化。

首先是在辛亥革命之后,虽然还允许溥仪等在紫禁城内廷居住,但外朝于1913年改为故宫博物院,并先后开放了太和、中和、保和等三大殿。溥仪被驱逐出宫后,1925年,故宫各部分开放,供人参观。自紫禁城开放之后,北京城内外的皇家苑囿以及过去神圣的殿坛宗庙,也陆续开放,改为他用。1914年,天安门西侧的社稷坛辟为中央公园(现名中山公园);太庙于1924年改为和平公园。1926年北海开放,名为北海公园;1929年中南海开放;1928年景山也成为公园。

随着城市的发展,原来北京城所具有的皇城、内城、外城层层包围的高大城墙以及城门、瓮城等层层设防的旧建筑,已构成市内交通

while the west section was replaced with a stone bridge called Yuhe Bridge. Two ceremonial arches were erected on both sides of the bridge, the one the west being called Golden Sea-Turtle, the other called Rainbow, hence the name "Golden Turtle and Rainbow Bridge". In the Ming period, the lake south of Taiye Lake was expended which was referred to as "Nanhai"(south lake), and, as the lake north and south of the Golden Turtle and Rainbow Bridge were called "Zhonghai" (middle lake) and "Beihai" (norht lake), they got the generic name "Sanhai" (three lakes). On the north bank of Taiye Lake are Jiale Hall, Five-Dragon Pavilion, Lietinghongying Hall, etc; on the west bank are Yuxi Palace, Ziguang Pavilion, Shouming Hall; on the east bank are Chongzhi Hall, and the Dock; and on the south bank are Zhaohe Hall and Taibo, etc. As the banks of Taiye Lake were planted with beautiful trees and flowers, and dotted with palaces, pavilions and piled stones, the lake looks picturesque and graceful like a roll of paintings, making West Park the a beautiful royal park in the Imperial City of the Ming Dynasty.

West of the park and in the southwestern part of the Imperial City are Wanshou Palace, Daguangming Hall, Tu'er Hall, Xuanbotai, etc., all built on the site of Longfu Palace of the Yuan Dynasty. In the northwestern part of the Imperial City and west of the park are Qingfu Hall and warehouses of the government constructed on the site of Xingsheng Palace of the Yuan Dynasty.

The East Park lies on the east of Palace City. On the southeastern side of the provisional imperial palace is the little southern town, or simply called "South Town" composed of Chonghuan Palace, Zhongzhi Hall, Yuzhi Palace and other main structures. The Imperial Archives is on the southwestern part of the South Town where ancient records and books were stored.

Various government agencies and departments directly providing services to the royal court were scattered along east and west sides of the palace and in the northeastern and northwestern parts of the Imperial City. These were also areas for officially-run handicraft industries.

Road network was mainly aimed at linking various architecture and government departments, so they were unique because they were not square-shaped as the courtyards were. The framework for the main streets in the Qing Dynasty has already been formed including Donghuamen Street, Xi'anmennei Street, Di'anmennei Street, Dong and Xi Jingshan Street, Jingshanhou Street, Beichizi, Beichang Street, etc.

(2) Imperial City of the Qing Dynasty

The Qing Dynasty (1644-1911) designated Beijing as its capital, which was basically kept as it were during the Ming Dynasty. Only a few renovations and expensions were made such as the large-scale renewal project for the imperiral city in the middle of the dynasty. The Qing rulers, after entering Beijing, were mainly concerned about stability and how put various nationalities under their control in order to achieve long-term peace. So the urban construction and settlement of inhabitants all reflected this policy.

In 1645, after renovating the Huangji, Zhongji and Jianji Hall in the Forbidden City, they changed their names to the Hall of Supreme Harmony, Hall of Central Harmony and Hall of Preserving Harmony, highlighting the word of "harmony". As in the Mind period, the Imperial City of the Qing Dynasty also boated 4 gates, and in September 1651, The Chengtianmen was changed, after renovation, into "Tian'anmen" (heavenly peace). In July next year, the rear gate was changed to "Di'anmen", thus the word "peace" was accentuated in the 4 gates. The square, Chang'anzuo Gate and Chang'anyou Gate outside Tian'anmen remained unchanged, although Damingmen was changed to Daqingmen. The Duuanmen and Meridian Gate as well as the Imperial Ancestral Temple and Altar to the God of the Land and Grain inside Tian'anmen also remained as there were.

The inscribed boards on the gates of the Forbidden City were in Manchu and Han languages, so were the languages used in prayers at Ancestral Temple and Temple of Heaven when praying for good harvest. All this indicated the state policy of seeking stability and peace in the early years of the Qing Dynasty.

The temples in the Imperial City of the Qing period also demonstrated that Beijing was a political center of a unified country with multiple ethnic groups. For example, the Manchus believed in shamanism, so upon arriving in Beijing, a site was constructed east of the imperial bridge to worship their god. As Mongols believed in Lamaism, the Chonghua Palace in the Little Town built by the Ming Dynasty lying in the southeast of the Imperial City was changed into a lama temple during the reign of emperor Kangxi (the Imperial Archives of the Ming period remained unchanged) to appease the Mongol nobility. Also, a lama temple was built on the bank of Dongzi River east of Beichang Street, which is still the mansion of Panchen Lama today. It was said in the 23rd year of emperor Qianlong, a mansion called Baoyulou was built for Concubine Xiang of the Xinjiang Uygur nationality (today's Xinhuamen of Zhongnanhai) with a camp for Hui nationality and mosque in the front as a place for looking at hometown. And with the introduction of Catholicism, a catholic church was erected at Chanchikou.

The Longevity Hill south of Di'anmeng was changed to Jingshan, or Coal Hill. In the 15th year of emperor Qianlong (some say 16th year), pavilions were built on each

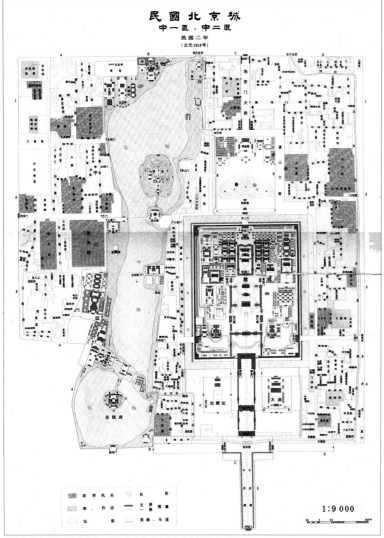

民国皇城平面图（资料来源：侯仁之主编《北京历史地图集（一）》）
Plane figure of the Imperial City of the Republic of China (from *A Collection of Historical Atlas of Beijing(Vol.1)* ed. by Hou Renzhi)

南池子（1961年摄）
Nanchizi (shot in 1961)

的障碍，在随后的城市建设中逐步得到了改造与清除。

首先是拆除皇城墙。民国初年，开通了位于紫禁城东西的南北池子和南北长街。嗣后在皇城东北角开一门，为北箭亭；皇城西北角开一门，为厂桥；皇城西南角开一门，为府右街。1917年，拆了东安门南段皇城城垣。同年又拆了西皇城根灵清宫一带以迄灰厂豁口的皇城墙，用它的砖石移修大明壕。1923年以后，又拆除了东安门以北转而向西至地安门南北城墙。从此，有500年悠久历史的皇城城墙在北京的地图上消失大半，"今所存者只天安门左右数十丈中华门内左右各百余丈耳"。

其次是天安门前宫廷广场的改造。民国初年，长安左右门首先拆去，仅剩门阙；民国四年（公元1915年），拆除了"均为禁地，不准车马行人来往"的大清门（中华门）内东、西千步廊以及东、西三座门两侧的宫墙，打通了天安门大街，并修砌了中华门至天安门右道，"惟杂植花木备市民游览休憩，过其地者心境为之一爽"。此外，1931年，故宫博物院为拓宽北门马路、便利交通，拆除了神武门外的北上东门和北上西门，筑大道于景山门与北上门之间，使原来的紫禁城和景山之间的禁地也变成为一条东西交通要道。

此外，清皇城内部分大建置的占地在民国时期逐渐被国家行政单位和学校占用，如国务院、参谋部、禁卫军、京师大学堂、医养院等，皇城的用地性质开始发生变化。

虽然皇城自民国初期以后发生了极大变化，但是，作为皇城原有的主体格局和中心内容仍然存在。居皇城之中的紫禁城、太庙、社稷坛、皇史宬、三海等古建筑群都基本保存完好，皇城区内的街道、胡同、

1943年皇城航拍
Aerophotography of the Imperial City in 1943

沙滩地区今昔对比
Contrast of the past and the present Shatan area

四合院平房区大部分还存在，特别是紫禁城周围的南北池子、南北长街和景山周围街区的四合院基本保持原貌，增加了少量仿西洋风格的建筑形式，胡同结构基本定型，奠定了今日皇城居住区建筑形态和空间形态的基本特色。

3.1.4 中华人民共和国成立以后

1949年9月，在北京中南海怀仁堂召开了中国人民政治协商会议第一届全体会议，会上一致通过决议：中华人民共和国的国都定于北平，自即日起，改北平为北京。从此，北京成为中华人民共和国的首都。

从1949到1959年中华人民共和国成立10年的时间内，完成了天安门广场的扩建，人民英雄纪念碑、人民大会堂、中国革命历史博物馆等大型建筑的建设，建成了世界上最大的城市中心广场——天安门广场，使其成为旧皇城前最受人民敬仰和喜爱的场所。

文化大革命后，随着社会的发展和人口的急增，旧皇城内的用地功能发生了相当大的变化。行政办公、医疗、教育、中小学校、工业等用地增加较快，挤占了原先的居住类用地，使得原本相对安静、以居住为主的环境和功能被部分侵蚀。同时，部分多层建筑的建造，民居四合院的拆毁等等具有破坏性的开发建设活动，导致皇城内的部分街道、胡同和四合院平房区的风貌受到不同程度的侵害。

道路系统的改造包括继续疏通民国时期开辟的通过紫禁城和景山之间的东西交通要道；改造北海大桥等。此外，由于大型事业单位的侵入，包括部分居住小区的修建，历史上服务于传统院落的细密的、小尺度的街巷已渐渐不适于新的交通需求，有的被取消或局部扩大，失去原来的尺度关系，有的甚至在用地性质改变的过程中几近荒废消失。

近年来，由于名城保护意识的加强，皇城保护的工作力度也不断加大。自1984年起，根据《中华人民共和国文物保护法》和《国务院批转国家建委等部门关于保护我国历史文化名城的请示的通知》，北京市先后在市域范围内划定了180项文物保护建控地带，公布了6批文物文物保护单位，进行了3次文物普查，明确了旧皇城内的文物保护对象。1999年初，完成了北京旧城区范围内25片历史文化保护区的重新认定工作，划定了保护范围和控制范围，并进行了保护规划的编制，其中包括旧皇城内14片历史文化保护区。2000年，为恢复皇城的完整性，东城区修建了以展示皇城墙为主题的遗址公园，掀开了新时期北京整体保护皇城传统风貌的历史篇章。

3.2 结论

北京皇城在历史上是一个专为封建君主服务的功能区，兼有国家最高权力中心的职能；现在转化为展示封建末期中国传统城市规划艺术、中国北方地区皇家建筑文化、造园文化、宫廷艺术，展示老北京传统居住文化（北京四合院），具备丰富历史文化内涵和浓郁古都风貌的最重要的历史文化保护区。北京皇城记录了北京城市的发展历史，

of the five peaks on the hill with the one in the middle named Wanchun. Wanchun is right on the urban axis, large in scale, and symmetrical with Guanmao and Zhouchang in the west and Jifang and Fulan in the east. The architecture in the West Park in the Imperial City such as Yuxi Palace, Qingfu Hall, Tu'ershan and Xuantaipo were either in disuse or extended.

Unlike the period of Ming, the areas within Dong'an, Xi'an and Di'an in the Imperial City of the Qing period were open to inhabitants belonging to "Eight Banners" (Manchu military-administrative system). And the areas for government departments in the Ming period were gradually transformed into residential hutongs with many eunuchs from the Office of the Imperial Household and their families moved in. These hutongs included Guanjian, Zhiranju, Jiusuju, Xixinsi. The use of land became more detailed and hutongs developed in a natural way.

The changed mentioned above had profound historical and social reasons. During the middle and later periods of the Ming Dynasty, the officially-run handicraft industry had started to decline. According to the statistics of the year 1485 under the reign of emperor Xianzong, The 32 kinds of trades owned by the Imperial Weaving and Dyeing Board in the early years of the Ming were reduced to only 11 artisans. Most of the artisans had quit their services for the imperial court. In the Qing period, the officially-run handicraft industry was under the Office of Imperial Household and the Ministry of Public Works and its scale had no match for that of the previous dynasty. The Imperial Weaving and Dyeing Board under the Office of Imperial Household supervised the following workshops: silver, copper, dyeing, clothing, embroidery, flower patterns, leather, iron, lacquer, and cannon-making. As the folk commerce prospered at that time, many needs of the imperial family turned to market. The expansion of private workshops was the inevitable outcome of development of market economy which constituted the basic reason for the sharp decline of state-run services. This also in turn led to the change of the nature of land used in the Imperial City. Another reason for the disappearance of the officially-run industry was that the Qing Dynasty drew a lesson from the corruption of the Ming officials, and reduced the number of the imperial institutions.

Although the Qing Dynasty did not renovate Beijing in large-scale way, it invested a lot of money and manpower in building imperial parks, employing hundreds of thousands of laborers. On the basis of the West Park of the Ming period, the "three lakes" areas were undergone large scale construction, its surroundings were rendered more graceful and picturesque, making it a genuine imperial park with unmatched scenery. The Yingtai, the main structure between Baoyulou and Wusong Bridge in Nanhai (south lake) and its name deprived from a legendary fairy land in the sea, was expanded in 1645. It was the place for emperors to keep heat in summer and appreciate snow in winter. It was also the place where Emperor Guangxu was house-arrested by Empress Dowager Cixi after the the reform movement of 1898 was supressed.

The Zhonghai (middle lake) was situated between Wusong Bridge and Golden Turtel and Rainbow Bridge (today's Beihai Bridge) which boated Qinzheng Hall, Fengzeyuan, Haiyanlou, Huairen Hall, Ziguang Pavilion, Water and Clouds Pavilion, etc. Looking across Yingtai at a distance, Qinzheng Hall was the place where emperors attended to state affairs while staying in the park. The Fengzeyuan west of Qingzheng Hall was built by Emperor Kangxi which boated the Hall of Prayer for Good Harvest. The Chrysanthemum Study and Chenghuai Hall were on the east and west of the Hall of Prayer for Good Harvest. There were 10 mu of paddy-field in Fengzeyuan where every spring emperors would go to give lectures on diligence and exhort people to cultivate land. The Haiyan Hall in Zhonghai (later changed to Juren Hall) was built on the ruins of Yiluan Hall by Empress Dowager Cixi to curry favor with foreigners, so it was a architectural compound combing Chinese and western styles.

Beihai (north lake) was located north of the Golden Turtle and Raibow Bridge, where Liao, Jin and Yuan Dynasties built Yaoyu Palace, Jade Islet and Taiye Lake respectively. The Dadu Palace was constructed around this area. In the Ming Dynasty, the place was called West Park where Yitian Hall was erected on its Round City. Other architecture such as Taisu, Ninghe and Yingcui Halls were built on the northern, eastern and western banks of Beihai. The Qing period saw further renovation of Beihai. In 1651, the White Pogada Temple or Yong'an Temple was erected on the original Guanghan Hall. During the reign of Emeperor Qianlong, many architecture that we see today were built, renovated and extended on the northwestern, eastern banks as well as on Jade Islet. From 1885 to 1888 during the reign of Empeor Guangxu, Cixi appropriated naval fund to rebuild Beijhai. She ordered the construction of a small railway called Ziguangge Railway along the western and northern banks, and added more structures for her enjoyment. Therefore, up to the Qing period, the "three lakes" not only became a famous imperial park, but also an important venue for political and entertaining activities, thus turning into another political center of the Qing Dynasty.

3.1.3 Period of the Republic of China

With the passage of time, the city wall, temples and palaces, altars, streets and parks also underwent tremendous changes.

After the 1911 Revolution, although Emperor Puyi was still allowed to live in the rear court, the outer court was changed into the Palace Museum in 1913 with its Supreme Harmony, Central Harmony and Preserving Harmon Halls open to the public. The other parts of the court were also open for visitors when Puyi was driven out in 1925. Other imperial parks, once sacred temples and altars also followed suit in opening to the public or transformed into other use. In 1914, the Altar to the God of the Land and Grain that lies to the west of Tian'anmen became Central Park (today's Zhongshan Park; The Imperial Ancestral Temple was changed into Peace Park. In 1926, Beihai was open to the public with the new name of Beihai Park; In 1929, Zhongnanhai was open and in 1928, Jiangshan became a park as well.

With the development of the city, the Imperial City, inner and outer cities, as well as overlapping walls and defending citadels and gates constituted many obstacles to traffic, so they were gradually either pulled down or revamped in the urban construction.

The first structure that was pulled down was the wall of the Imperial City. At the beginning of the Republic era, the Nanchiai and Beichizi and Nanchang street and Beichang Street on the east and west of the Forbidden City were connected. Then, a gate was opened in the northeastern corner of the Imperial City called Beijianting; another in the northwestern corner called Changqiao; and still another in southwestern corner called Fuyou Street. In 1917, the imperial wall at the south of Dong'anmen was torn down. In the same year, the imperial wall at Hukou around Lingqing Palace at West Huangchenggen was pulled down, the bricks of which were used to update the Daminghao. After 1923, the section of wall from Dong'anmen to Di'anmen was pulled down. Thus, the half of the wall of the Imperial City of 500 years disappeared from the map of Beijing except a mall portion of it around Tian'anmen and Zhonghuamen.

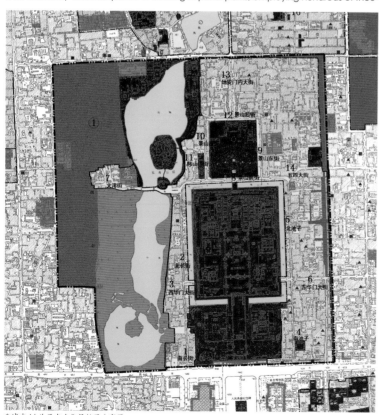

皇城内14片历史文化保护区分布图
Distribution map of the 14 conservation districts of historic sites in Imperial City

北京历史文化名城北京皇城保护规划
Conservation plan for the Historic City of Beijing and Imperial City of Beijing
北京皇城

延续了城市传统文化内涵,并随着城市发展不断变化。

功能的转变也导致了传统格局的变化。具体表现在:

3.2.1 城廓平面的位置、形状及出入口位置

自元代规划确定皇城位置以来,皇城城廓平面的变化的主要表现在,为扩大皇家后勤服务区用地,明代皇城墙向东、南、北三向拓展;为避让辽代双塔,形状也由原来方正平直的矩形平面转化为不规则形;东西出入口也因此由原来的对称布置转化为不对称安排。这种不强求礼制对称的变化进一步反映出皇城本身作为服务区的功能特点。

3.2.2 主要建筑物的位置、形制

元代皇城内主要的三组建筑群是宫城、隆福宫和兴圣宫,其中,宫城的位置在皇城中央偏南部。除宫城成为明清紫禁城(宫城)的建设基址外,其他两组宫殿已湮灭在后世的不断建设中,这种变化突出了宫城的中心地位。而且,宫城的建设较前世投入更大的人力和财力,形制日趋完备,突出礼制观念,创造出最恢弘壮丽,具有极高艺术价值的宫殿群体,成为今天弥足珍贵的"世界文化遗产"。

3.2.3 功能构成

元代皇城主要由宫殿和御苑构成;明代皇城主要由宫殿、御苑和内宫衙署构成;到了清代,除前述三种功能外,增加了寺庙和皇宫服务人员的居住功能;民国又增加了大学、教堂和行政办公用房;中华人民共和国成立后,增加了医疗、中小学教育配套和工厂。

3.2.4 街巷系统

元代皇城没有形成街巷系统,宫殿和园囿的过渡区域,空间辽阔舒展,与皇城外里坊式的居住建筑布局形成鲜明对比。明代由于内宫衙署的增设形成一些顺应衙署布局的大尺度的地块和街巷,形态自由;清代由于居住功能的增强,胡同肌理变得细密,这就是今天该区传统城市肌理的主要形成时期,也成为皇城独具的风貌特征。后世主要是在其上不断增补。中华人民共和国成立后,由于一些大型企事业单位的迁入和新增的部分功能,部分用地重新整合为大尺度的地块,导致部分传统城市肌理被破坏。

3.2.5 河湖水系

元代皇城内河湖水系曾经承担了漕运、宫廷生活用水和园林用水的供应。后代除对园林水系开挖和修葺整理外,漕运功能消失;到了民国以后,更有部分水系被添埋和转为暗沟。现存水系主要是以三海和筒子河等为代表的园林水系和景观水系。

3.2.6 开敞空间

由于皇城功能的转化,形成了一处崭新的超大尺度的广场空间——天安门广场。许多大型的庆祝活动和隆重的欢迎仪式都在这里举行,成为中外人士驻足游览的场所。随着朝阜干线(西安门内大街——五四大街)的开辟,原来封闭的皇家苑囿成为可供人游玩、近观和远眺的城市开放空间,如北海、景山、中南海。

3.2.7 空间形态

形成平缓开阔、起伏有致的城市天际线。在广大的平房四合院民居衬托下,形成以故宫为中心,以景山万春亭、钟鼓楼、正阳门等建筑为控制点的,并由旧城墙和各城楼拱卫的、起伏有致的城市天际线。传统街区中形成以寺庙、衙署、厂、库为节点,周边民居衬托的城市景观。

4. 现状情况

北京旧皇城地区历经元、明、清、民国约数百年的发展历史,传统城市中轴线上的主要建筑群、御用坛庙、皇家园林、寺庙、民居等较好地保存下来,成为旧皇城地区的保护核心和精华。同时,作为城市生活的有机体,该地区也深深刻上社会经济快速发展的烙印,具体体现在以下方面:

4.1 皇城边界

明清皇城拥有完整的皇城墙,皇城边界明确清晰,现状只留下沿长安街和府右街一侧的皇城墙,界定皇城南部和西南部边界;位于南北河沿大街和东黄(皇)城根的皇城根遗址公园界定出皇城东边界。皇城北边界和西边界城墙已拆,现状是平安大街和西黄(皇)城根南北街,边界模糊,缺乏辨别性。

4.2 用地现状

规划范围总用地683hm²,用地内文物古迹众多,人文风光和自然景观优美独特。主要用地除故宫、景山公园、中山公园、劳动人民文化宫、北海、中南海六大文物保护单位用地外,以居住、行政办公、中小学、医疗为主,还有少量商业、工业和其他性质用地。

4.2.1 主要单位

皇城内故宫、景山公园、中山公园、劳动人民文化宫、北海、中南海是六个国家级文物保护单位,它们占地面积339hm²,占皇城面积的49.7%,占地面积大,文物级别和开放程度高;其他有一定规模的单位大约有177个,总占地面积124.5hm²,占皇城面积的18%;二者合计单位总计有183个,总占地面积约463.5hm²,占皇城面积的67.7%。

皇城内的主要单位大致可分三类:第一类是故宫、景山公园、中山公园、劳动人民文化宫、北海、中南海,这六个单位是皇城中规模最大的,它们是皇城的主要景观和最具历史文化价值的组成部分;第二类是中央、军队所属单位和医院、学校,如文化部、民政部、国务

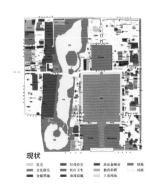

皇城历史演变图
Maps that show the evolution of the Imperial City's history

Another renovation involved the imperial square in front of Tian'anmen. In the early years of the Republic period, the left and right gates of Chang'an were pulled down. The Thousand-Step Corridors on the east and west sides within the Daqing Gate that used to be off limits to commoners and the walls on either side of the three doors on the east and west were all pulled down. As a result, the Tian'amen Street was opened up and the road between Zhonghuamen and Tian'amen was built where trees were planted to provide place for inhabitants to relax themselves. In 1931, in order to widen the road north of the Palace Museum, the west and east gates outside Shenwu Gate were pulled down and a road was constructed between the north gate of the Museum and Jiangshan Park, so that the forbidden piece of land between the Forbidden City and Jiangshan Park became an east-west traffic street.

Moreover, large tracks of land within the Imperial City were gradually occupied by state organs and schools, such as State Council, General Staff, Imperial Guard, Metropolitan University, and Hospitals, thus the nature of land use in the Imperial City has changed.

Even though dramatic changes took place in the Imperial City in the early days of the Republic period, the primary layout and central content remained unaffected. The Forbidden City, Ancestral Temple, Altar to the God of the Land and Grain, Imperial Archives and the three lakes at the heart of the Imperial City were still well preserved. So was the case with large numbers of streets, hutongs, quadrangle courtyards within the Imperial City, especially those courtyards in Nanchizi, Beichizi, Nanchang street and Beichang Street around the Forbidden City, and those around the Jiangshan Park. Their features were preserved combined with a small number of houses of quasi-western style. The layout of hutongs were fixed, laying down the basic characteristics of architectural and spatial patterns of today's residential quarters in the Imperial City.

3.1.4 After the Founding of New China

In September 1949, a proposal was unanimously adopted at the first plenary session of the National People's Political Consultative Conference at Huairen Hall of the Zhongnanhai in Beijing. The proposal stipulated that Beiping would be the capital of the People's Republic of China, and as of the same day, Beiping was changed to Beijing which became the capital of new China.

In the ten years from 1948 to 1959, large-scale construction involving the renovation of the Tian'anmen Square, building of the Monument to the People's Heroes, the Great Hall of the People, the Museum of Chinese History was undertaken, resulting in the completion of Tian'anmen Square, the largest city central square in the world and the most revered and loved place in front of the ancient Imperial City.

After the Cultural Revolution, the social development and dramatic increase of population caused drastic changes in the land functions within the Imperial City. The areas for office buildings, hospitals, education, primary and secondary schools and industry increased rapidly, Squeezing in the original residential quarters, undermining the once quiet environment of mainly residential function. Also, destructive construction of multi-story buildings, and demolition of courtyard housing resulted in various degree of damage of the features of part of the streets, hutongs and courtyards in the Imperial City.

Road revamping included the continuation of opening up the east-west street between the Forbidden City and Jiangshan Park initiated during the Republic period, and the renovation of Beihai Bridge. Owing to the intrusion of large office buildings including construction of some residential quarters, the small and narrow streets and alleys serving the traditional courtyards were not longer able to meet the needs, so some of them were got rid of, some expanded, with the result that the original scale relation was lost. Some even completely disappeared in the transformation of the nature of the land use.

In recent years, as the awareness of protecting the Imperial City has been heightened, the conservation work in this regard has been intensified. Since 1984, in line with the Cultural legacy Conservation Code of the People's Republic of China, and the Notice of Replying the Report of Preserving China's Historic Cities Submitted by the Ministry of Construction and Relevant Departments Approved by the State Council, the Beijing municipal government, within the whole city, has designated 180 construction control areas around relic sites, declared 6 groups of cultural relic units under conservation, undertaken three surveys on cultural relics, and clarified the targets of cultural conservation within the Imperial City. In the early 1999, 25 conservation districts of historical sites in the old city of Beijing were determined, the extent of conservation and control for them clarified, and compilation of conservation plans were drafted, including 14 conservation districts in the Imperial City. In 2000, to protect the entirety of the Imperial City, the Dongcheng District built a ruins park with the theme of illustrating the walls of Imperial City, heralding in a new historical chapter of Beijing's effort to protect the traditional features of the Imperial City in an overall manner.

3.2 Conclusion

Beijing's Imperial City in history was a functional district that primarily served the feudal monarch, also playing the role of the highest power of the State. But now it has become the most important conservation district of historic site in terms of demonstrating the traditional Chinese urban planning art of the late feudal society, the imperial architectural culture, gardening art and court art of north China, the traditional residential culture of old Beijing (quadrangles), as well as boating a rich historical and cultural content and strong features of ancient capital. The Imperial City in Beijing is a witness of the history of development of Beijing, a continuation of the historical urban culture, and will change with urban evolution in the future.

沿府右街一侧的皇城墙
Imperial wall along one side of Fuyou Street

西黄城根北街（皇城西边界现状）
North Street of west Huangchenggen (present Imperial City at west boundary)

The functional change also resulted in the change in traditional layout, which is manifested as follows:

3.2.1 The location, shape of the city outline plane as well as the location of entry and exit.

Since the Yuan Dynasty established the location of the Imperial City, the changes of the contour plane of the Imperial City is mainly shown as follows: in order to enlarge the servicing land for the royal family, the wall of the Imperial City of the Ming Dynasty expanded eastward, southward and northward; to make room for the twin towers of the Liao Dynasty, the original rectangular shape was changed into irregular shape; therefore, the east and west points of entry also changed from symmetrical into asymmetrical structures. This change of not strictly adhering to rules reflected the service function of the Imperial City.

3.2.2 Location and shape of primary architecture

The three main groups of architecture of the Yuan Dynasty were the Palace City, Longfu Palace and Xingsheng Palace, out of which the Palace City is situated a little bit south of the center of the Imperial City. The Palace City was used as the site for the Forbidden City of the Ming and Qing Dynasties, but the other two compounds have disappeared amid the construction of later generations, which illustrates the central position of the Palace City. Furthermore, more manpower and funding were invested in the construction of the Palace City which helped the completion of its shape and highlight the concept of the norms of etiquette. It also helped create a grand palace compound of extremely high artistic value, making it an invaluable "world cultural heritage."

3.2.3 Functional Composition

The Imperial City of the Yuan period is composed of palaces and imperial parks; that of the Ming period is composed of palaces, parks and government agencies in the inner court; and in the Qing Dynasty, apart from the above-mentioned three were added temples and residential quarters for attendants at court; universities, churches and office buildings were added during the Republic period, and, after the founding of new China, medical facilities, elementary and secondary schools and factories were built there.

3.2.4 Street system

No street and allay system was formed in the Imperial City of the Yuan Dyasty, so

院中央事物管理局、国防部、总参、北京口腔医院、北京第四中学、二十七中等，它们占地较大，建筑形态多为多、高层建筑，对皇城的传统风貌影响较大，多数分布在东西黄（皇）城根道路沿线，由于现状发展空间有限，呈现出向保护区中心渗透的趋势；第三类是一般规模的单位，它们占地较小，建筑规模不大，对皇城的传统风貌影响不明显。此外，部分单位占用文保单位用地，利用不尽合理。

4.2.2 土地使用现状

皇城的土地使用情况可分为15类，包括行政办公、商业金融、文化娱乐、医疗卫生、教育科研、宗教福利、特殊用地、普通住宅、一类住宅、中小学托幼、市政设施、仓储、工业、公共绿地、道路广场。依照占地规模大小，主要用地有：公共绿地约占皇城面积的33％；居住用地约占20％；文化娱乐约占17％；道路广场约占10％；行政办公约占10％，其他占10％。由此可见，绿地、道路用地占皇城用地的43％，近一半用地为景观类用地，为皇城保护打下坚实基础。居住、文化、行政用地占47％，突出了皇城的特色和性质。

（1）绿化用地

保护区内的绿化用地主要指中山公园、劳动人民文化宫、景山、三海、皇城遗址公园等公园绿地，占总用地的32％，是皇城保护区内自然景观的主体。

（2）居住用地

居住用地在保护区中占的比例较大，占总用地的20％。它又分为普通住宅和特殊住宅两类用地。普通住宅基本上是四合式建筑，大部分是一层，少数有局部二层。建造年代多数为清末至民国期间，其中建筑质量较好的一般为过去的衙署或官宦人家，院子宽敞，建筑高大，选材上乘，保存较好。多数建筑年久失修，院内搭建严重（搭建建筑大部分未反映在地形图中，原因是违章建筑，不予承认）。

特殊住宅用地在街区中占2％，一般建筑质量较好，建筑风格大多与传统风貌协调。但由于不注意保护原有四合院建筑和街巷肌理，随意拆建现象严重，缺乏有效控制，是实施规划管理中的一大难点。

（3）行政办公用地

行政办公用地主要为中央机关、军委、外省市驻京办事处等用地，建筑多为建国后建的低、多层仿古建筑，坡屋顶，但体量和色彩上与保护区整体风貌不太协调，需要改造，如国防部大院。也有部分近年新建项目，对传统风貌关注不足，在建筑造型、体量和色彩等方面明显不协调。这类用地的办公建筑对规模、交通有较高要求。

（4）中小学校用地

保护区内有中学6个、小学10个，中等专业技术学校两个。部分学校占用现状文物用地，对文物保护不利，如北长街小学、北池子小学等；部分学校的空间布局和形态与传统建筑差距较大，影响皇城的整体景观，如北京市二十七中、三十九中学。

（5）医疗卫生用地

医疗用地主要集中在皇城保护区西北部地区，占总用地的2％。主要有北京口腔医院、北大口腔医院、北京医科大学第一临床医学院、北大医院妇儿门诊部、305医院、妇产医院、公安医院等。建筑以多层建筑为主，用地较为局促，多有加建和扩建现象，新建筑层数较多，反映医院建筑的专业设计要求，但较少考虑与传统风貌的协调。

（6）工业用地

工业用地主要集中有京华印刷厂、北京微电子厂、北京低压电器厂、北京市证章厂等工业企业，占总用地的1％。目前这些工厂均有待改造。

（7）道路用地

指街区内除建设用地、绿地以外的街道、胡同所占的用地面积，约是总用地的10％。保护区内府右街地区、景山八片地区、南北长街地区、南北池子地区的街道、胡同基本保持了明清及民国时期的格局，胡同走向自由、曲折，宽度最窄处1.5m左右，最宽处也不过10m左右，供机动车通行的胡同较少，且不成体系。空间肌理反映出早期生活的特征，对此应在总体上予以保护。

4.2.3 用地平衡表

现状用地平衡表

用地代码	用地性质	用地面积(hm²)	百分比(%)
C1	行政办公	70.24	10.29%
C2	商业金融	20.3	2.97%
C3	文化娱乐	115.33	16.89%
C5	医疗卫生	13.28	1.95%
C6	教育科研	4.35	0.64%
C9	宗教福利	3.7	0.54%
R	住宅	124.45	18.23%
R1	一类住宅	12.07	1.77%
R5	中小学、托幼	17.92	2.62%
U	市政设施	0.46	0.07%
W	仓储	0.3	0.04%
M	工业	4.67	0.68%
G	公共绿地	218	31.93%
Z	在建	1.78	0.26%
S	道路广场	75.91	11.12%
合计	规划范围	682.76	100.00%

（注：用地构成将保护区内北海、中南海用地内的水体归入绿化用地，护城河归入故宫博物院文化娱乐用地）。

4.2.4 现状院落分布

皇城内有单位大院约60余个（如故宫等），居住性院落约2800余个。

皇城内的居住性院落主要分布在南北池子、景山东西街、地安门内大街、府右街、南北长街一带。除故宫等6大单位处于皇城中央，单位大院多分布在皇城的西北部沿西什库大街两侧和东部沿南北河沿西侧，形成了较为集中的多层建筑群落。目前，单位院落的扩展对居住性院落的冲击很大，应予以高度重视。

4.2.5 已审批在建和待建项目

皇城内目前已审批的在建项目有9项，占地面积约25hm²；待建项

建设中的牡丹园改造项目
Renovation project of Mudanyuan in construction

the open and vast areas between palaces and parks made a striking contrast with the layout of lanes and allays of the residential quarters outside of the Imperial City. In the Ming Dynasty, as some government offices appeared in the inner court, large scale of plots and streets came into being which were free in structure. As the residential function was strengthened in the Qing Dynasty, the system of hutongs was elaborately developed which presented the unique style of the Imperial City. Later generations only made additions to what was already available. The new China saw the moving in of man large enterprises and government offices which resulted in the appearance of large scale plots in some areas and the damage of parts of the traditional urban fabric.

3.2.5 River system

The river system of the Imperial City of the Yuan Dyasnty played the role of transporting grain to capital by water, water supply for the court and parks. Lager generations only renovated and further developed park river system, so the grain transportation function was discarded. After the Republic period, some river systems were filled up or turned into covered sewerage. The extant river systems are mainly park and landscape systems represented by three lakes and Dongzi River.

3.2.6 Open space

The change of function of the Imperial City gives way to an extra large space — the Tian'anmen Square. It has become a place where many Chinese and foreign people come for sightseeing because many large-scale celebrations and ceremonies are held here. With the opening of the Chao-Fu Road (Xi'anmennei Street-Wusi Street), the former closed royal parks became open urban spaces of sightseeing and relaxation, such as Beihai Park, Jiangshan Park, and Zhongnanhai.

3.2.7 Spatial Shape

To form a open, flowing and undulation urban skyline. This can be achieved by the vast tracks of quadrangles as the backdrop, with the Palace Museum, the Wanchun Pavilion, the Drum and Bell Towers and Zhengyangmen as the controlling nodes, and the old city and its defending towers. Urban landscape can also be formed by using temples, ancient government offices, warehouses in traditional streets as nodes which are set off by surrounding residential housing.

4. Current Situation

After experiencing hundreds of historical development through the Yuan, Ming and Qing Dynasties, the main building compounds, imperial temples, parks and residential housing of the Imperial City of Beijing on the old axis have been fairly preserved and become the core of conservation in the Imperial City. Meanwhile, as an organic part of the urban life, this area has also been affected by the rapid social and economic development, which can be shown in the following aspects.

4.1 Boundary of the Imperial City

The Imperial City of the Ming and Qing Dynasties had complete walls and the boundary of the Imperial City was clear-cut. Now only one section of the wall along the Chang'an and Fuyou Streets still exists which can be used to ascertain the southern and southwestern boundaries of Imperial City. The Huangchenggen Ruins Park at Nanheyan, Beiheyan and East Huangchenggen can be used to identify the eastern boundary; and, as the walls at northern and northwestern boundaries have been pulled down (today's Ping'an Street, Nan and Bei Streets at West Huangchenggen), the boundary is very unclear, and hard to ascertain.

4.2 Current Situation of Land Use

The total land use in the planned area covers 683 hectares, where cultural relics abound and human and natural landscape are beautiful and unique. Apart from the land taken by the six cultural relic units under conservation, namely, Palace Museum, Jiangshan Park, Zhongshan Park, Laboring People's Palace of Culture, Beihai Park and Zhongnanhai, the land is mainly occupied by residential housing, office buildings, elementary and secondary schools, hospitals as well as small number of business and industrial facilities.

4.2.1 Primary units

The Palace Museum, Jiangshan Park, Zhongshan Park, Laboring People's Palace of Culture, Beihai Park and Zhongnanhai in the Imperial City are 6 cultural relics under the State protection, covering about 399 hectares, which account for 49.7% of the Imperial City. The land they occupy is large and they enjoy a very high protection and open level. Other relics of considerable scale stand at 177, covering 124.5 hectares, which account for 18% of the Imperial City. The two categories, put together, stand at 183, covering a total area of 463.5 hectares and accounting for 67.7 of the Imperial City.

The buildings inside the Imperial City can be divided into three categories: the first include Palace Museum, Jiangshan Park, Zhongshan Park, Laboring People's Palace of Culture, Beihai Park and Zhongnanhai. Largest in scale, they constitute the primary landscape and boast the highest historic and cultural values in the Imperial City. The second group encompasses units of central government and military as well as schools and hospitals such as Ministry of Culture, Ministry of Civil Affairs, Government Offices Administration of the State Council, Ministry of Defense, Headquarters of the General Staff, Beijing Stomatological Hospital, Beijing No. 4 Middle School, Beijing No. 27 Middle School, etc. As they cover large areas, and are of multi-storied and high-rise buildings, they exert significant damage on the ancient features of the Imperial City. Scattered along the East and West Huangchenggen Roads, and the limitations of further space for development, they have the tendency of making forays into the center of the conservation district. The last category refers to units of ordinary scale. As the land they occupy is not large and they do not have large-scale construction, the damage they inflicted on the Imperial City is not obvious. Besides, some units that use the land of relics are not using the land in a reasonable way.

4.2.2 Current Situation of Land Use

The land use of Imperial City can be divided into 15 categories: office buildings, business and finance, cultural and entertainment, medical facilities, education and research, religion and welfare, ordinary housing, first-grade housing, elementary, secondary schools and day care centers, municipal installations, storehouses, industry, public green belt and road and square. According to the size of land used, public green belt accounts for 33% of the Imperial City; residential housing 20%; entertainment facilities 17%; road and squares about 10%; office buildings 10%; others account for 10%. It is obvious that green belt and road account for 43%; in other words, half of the land used is landscape, which paves a solid way for the conservation of the Imperial City. Residential housing, entertainment and office facilities account for 47%, highlighting the features and nature of the Imperial City.

(1) Land for greening

The land for greening in the conservation district mainly refers to the green belt parks such as Zhongshan Park, Laboring People's Palace of Culture, Jiangshan Park, Three Lakes and Imperial City Relics Park. They account for 32% of the total land used, constituting the main body of landscape within the conservation district of the Imperial City.

(2) Land for residential housing

Land for residential housing in the conservation district constitutes a large portion, accounting for 20% of the total land used. It is further divided for ordinary and special housing. The ordinary category refers to single-storied (occasionally two-storied) siheyuan (quadrangles) structures. They were mainly built in at end of the Qing Dynasty and early years of the Republic period. Most of high quality housing of this kind used to be government offices and belonged to wealthy families, which are high and constructed with quality materials, have wide courtyards and well preserved. However, most siheyuan structures are old and in disrepair, and there is the serious problem of added structures (they are not indicated on the map because additions are unauthorized architecture which cannot be approved).

Special housing account for 20%, most of which are of high quality and correspond to the traditional style. But as the protection of siheyuan and the fabric of alleys and streets were not taken good care of, random dismantling is rampant and effective control is lacking, which pose a serious problem in carrying out the planning management.

(3) Land use for office buildings

Land for office buildings mainly refers to those used by central government organizations, Central Military Commission, and provincial and municipal offices in Beijing. They are primarily low- and multi-storied, quasi-classical structures with reclining roofs, but incongruous in scale and color with the overall style of the conservation district, thus needing renovation. The courtyard of the Ministry of Defense is a case in point. Some newly built projects in the past years are also not in agreement with in terms of shape, scale and color as they do not attach importance to the classical style. These kind of office buildings demands high requirements for scale and traffic.

目10项，占地面积约8hm²；总计有19个项目，单位占地约33hm²。依据改造的规模和风格，这些项目的改造方式可分为三类。

建筑高度以1～2层为主，建筑形态采用四合院传统风格的改造项目有5项，如东城区南池子危改、菖蒲河公园改造；西城区中央警卫局某项目、兆泰公司四合院、赵朴初图书馆。除赵朴初图书馆以外的其他4项改造，均存在拆除平房四合院规模过大的问题。

建筑高度在4层以下，采用现代建筑风格的或尚未建设的项目有6项，如东城区中央政法委南报告厅、银闸B区、外交学会、某待建项目；西城区地安门加油站、中央警卫局某待建项目。这类项目的建设由于规模、尺度较小，对皇城的整体空间结构的影响不大。

建筑高度在4层以上，新建建筑尺度较大的项目有7项，如东城区东风电视机厂（牡丹园改造项目）、高检办公楼扩建、卫戍区北河沿住宅；西城区四中教学用房、北京口腔医院病房楼、三十九中教学楼、邮电局办公楼。这类项目对皇城传统风貌破坏较大。

4.3 人口状况

现状核心规划区内总人口77911人，其中，西城区行政区划内43346人，东城区34565人。

现状居住人口情况

行政区划	居委会名称	总户数(户)	总户籍人口数(人)
东城			
	骑河楼居委会		1941
	智德居委会		2323
	北池子居委会		1242
	东华门居委会		1159
	南池子居委会		2827
	缎库居委会		1733
	飞龙桥居委会		1609
	南河沿居委会		2168
	东吉祥居委会	940	1883
	北河居委会	830	1696
	织染局居委会	827	1820
	黄化门居委会	758	1919
	钟鼓居委会	1231	3064
	吉安所居委会	1064	2424
	大学夹道居委会	757	3064
	中老居委会	1173	2059
	五四大街居委会	659	1634
	小计		34565
西城			
	景山西街居委会	1200	2959
	西安门居委会	0	0
	恭俭居委会	878	1997
	米粮库居委会	1256	3037
	大拐棒居委会	998	2889
	爱民街居委会	878	2010
	爱民里居委会	1323	3463
	西什库居委会	1267	3250
	惜薪司居委会	959	2288
	后达里居委会	2022	5774
	光明居委会	784	2162
	西黄(皇)城根南街一区	774	2261
	互助巷	1353	3580
	北长街居委会	1152	3063
	南长街居委会	1775	4613
	小计		43346
	总计		77911

皇城保护区行政区划辖东城、西城两区，由32个居委会管辖，分别隶属于西城区的厂桥街道办、西安门街道和东城区景山街道办、东华门街道办。四个街道办事处共有32个居委会，其中西城区14个，东城区18个。

皇城内居住用地约有136hm²，现状人口密度（户籍人口数/100m²居住用地）可分为五类：

第一类（小于2人/100m²）占地15.0hm²，占11.0%；
第二类（2～4人/100m²）占地25.8hm²，占18.9%；
第三类（4～7人/100m²）占地54.3hm²，占39.8%；
第四类（7～10人/100m²）占地21.2hm²，占15.5%；
第五类（大于10人/100m²）占地20.2hm²，占14.8%。

其中，居住条件较好的一、二类住宅占地约30%，居住条件一般的三类住宅占地40%，居住条件较差的四、五类住宅占地约30%。可见，必须采取措施向皇城外疏导人口。

从现状调查的情况来看，居住人口分布的一般规律是私人院落、单位产权的传统院落和私人商住结合的用地居住密度较低，而房管局所属的公房居住密度较高，存在"苦乐不均"的现象。

4.4 现状街道、胡同分布

清朝时皇城内的道路主要为土路和石子路，尤以土路为多。到了清末越来越糟，"天晴时则沙深埋足，尘细扑面，阴雨则污泥满道，臭气熏天"。交通工具主要是骡车、骑驴和人力车。1931年修筑景山前街，使原来的禁区变成东西交通要道。

景山前街（资料来源：盛锡珊著《老北京市井风情画》）
Jingshanqian Street (from *Customs and Ways of Old Peking* by Shengxishan)

金鳌玉蝀桥旧貌（资料来源：盛锡珊著《老北京市井风情画》）
Old golden tortoise jade rainbow bridge (from *Customs and Ways of Old Peking* by Shengxishan)

(4) Land use for elementary and secondary schools

There are 6 middle schools, 10 elementary schools and 2 vocational schools in the conservation district. Some schools pose threat to relics as they are using the land for relics such as Beichang Street Elementary School and Beichizi Elementary School. Others, such as Being No. 27 and No. 39 Middle Schools, are damaging the overall landscape of the Imperial City due to their big difference with traditional architecture in with regard to spatial layout and shape.

(5) Land use for medical facilities

Medical facilities are mainly concentrated in the northwestern section of the Imperial City, accounting for 2% of the total land area. They include Beijing Stomatological Hospital, Stomatological Hospital Attached to Peking University, First Clinical Hospital Attached to Peking University, Women and Children Hospital Attached to Peking University, 305 Hospital, Beijing Hospital for Gynecology and Obstetrics, and Public Security Hospital. Multi-storied and using very limited land, they tend to add new structures and extend themselves. The newly built ones have more stories reflecting the requirements of medical design, but little attention was paid to be harmonious with classical style.

(6) Land use for industry

The land for industry mainly accommodates Jinghua Printing House, Beijing Microelectronic Plant, Beijing Low-Voltage Electric Device Plant, Beijing Badge Plant and others, accounting for 1% of the total land used. At present, these plants all need revamping.

(7) Land use of road

Referring to land area for streets and hutongs with the exception of land for construction and green belt. It accounts for about 10% of the total land used. The streets and hutongs in Fuyou Street, 8 areas around Jiangshan, Nanchang Street and Beichang Streets, Nanchizi and Beichizi Streets have all maintained their layout of the Ming, Qing and the Republic periods. The directions of hutongs are free and winding with the narrowest place being 1.5 meters and the widest place not exceeding 10 meters, thus few are able to accommodate motor vehicles. Lacking in system, their spatial fabric reflect the features of life in the past, so they should be conserved as a whole.

4.2.3 Balance Sheet of Land Use

Balance Sheet of Current Land Use

Code of Land Use	Nature of Land Use	Area of Land Use (hectare)	Percentage (%)
C1	Office Buildings	70.24	10.29%
C2	Business and Finance	20.3	2.97%
C3	Entertainment	115.33	16.89%
C5	Medical Care	13.28	1.95%
C6	Education and Research	4.35	0.64%
C9	Religion and welfare	3.7	0.54%
R	Housing	124.45	18.23%
R1	First-Grade Housing	12.07	1.77%
R5	Schools and Nursery	17.92	2.62%
U	Municipal Installations	0.46	0.07%
W	Storehouses	0.3	0.04%
M	Industry	4.67	0.68%
G	Public Green Belt	218	31.93%
Z	Under Construction	1.78	0.26%
S	Road and Squares	75.91	11.12%
Total	Extent of Plan	682.76	100.00%

(Note: the water body inside Beihai and Zhongnanhai is included to the land use for green belt, while the city moat goes under the land use for entertainment of the Palace Museum.)

4.2.4 Present Courtyard Distribution

There are 60 unit courtyards (such as Palace Museum) and 2,800 residential courtyards in the Imperial City.

The residential courtyards in the Imperial City are mainly located in Nanchizi and Beichizi, Dong and Xi Jiangshan Streets, Di'anmennei Street, Fuyou Street, Nanchang Street and Beichang Street. Except the 6 large sites such as the Palace Museum that are centered in the heart of the Imperial City, other unit courtyards, which are high-rise compounds, are concentrated on both sides of Xishiku Street in the northwestern part and the west side of Nanheyan and Beiheyan in the eastern part of the Imperial City. At the moment, the expansion of unit compounds is having a big negative impact on residential courtyards, which should be taken notice of.

4.2.5 Approved Projects under Construction and Yet To Be Constructed

There are 9 approved projects that are under construction in the Imprial City, covering 25 hectares. Those that are yet to be constructed are 10, covering about 8 hectares, bringing the total projects to 19, covering 33 hectares. In line with the scale and style, the methods of renovation for these projects are divided into 3 categories.

There are 5 renovation projects that involve single- or two-storied, siheyuan-styled housing, such as the ramshackle housing redevelopment at Nanchizi of the Dongcheng District, Changpu River Park Renovation, a project of the Guards Bureau of the Party Central Committee in Xicheng District, siheyuan of Zhaotai Company, and Zhaopuchu Library. Apart from Zhaopuchu Library, the other 4 all have the problem of dismantling too many courtyard housing.

There are 6 projects that are under three stories, in the form of modern architectural style, or yet to be constructed, such as the South Conference Hall of the Political and legislative Affairs Committee of the Party Central Committee, District B of Yinzha, Society of Foreign Affairs, a protect to be constructed, Di'anmen Service Station in Xicheng District, a yet-to-be constructed project of the Guards Bureau of the Party Central Committee. As these projects are small in scale, they have insignificant impact on the overall spatial structure of the Imperial City.

There are 7 large-scale projects involving 4-storied or above buildings, such as Dongfeng TV Manufacturing Plant in Dongcheng District (Mudanyuan Renovation Project), extension of the office building of the Supreme People's Procuratorate, Beihuyan Residential Quarters of Beijing Garrison Command, Teaching Building of the No. 4 Middle School, In-Patient Building of Beijing Stomatological Hospital, Teaching Building of No. 39 Middle School, and Office Building of the Post Office. These projects will have strong negative impact on the traditional features of the Imperial City.

4.3 Population situation

The population in the core planned district is 77,911, with 43,346 in the administrative district of Xicheng and 34,565 in the Dongcheng District.

The Imperial City administrative district is under the jurisdiction of Dongcheng and Xicheng Districts, directly under the supervision of 32 neighborhood committees. These committees are under the control of four sub-district offices: at Changqiao and Xi'anmen in Xicheng District, and at Jiangshan and Donghuamen in Dongcheng District. There are 14 sub-district offices in Xicheng District, and 18 in Dongcheng District.

Present Situation of Inhabitants

Administive division	Names of neighb-orhood committee	Total number of hou-seholds (household)	Total number of registered population (person)
Dongcheng District			
	Qihelou		1941
	Zhide		2323
	Beichizi		1242
	Donghuamen		1159
	Nanchizi		2827
	Duanku		1733
	Feilongqiao		1609
	Nanheyan		2168
	Dongjixiang	940	1883
	Beiheju	830	1696
	Zhiranju	827	1820
	Huanghuamen	758	1919
	Zhongguju	1231	3064
	Ji'ansuo	1064	2424
	Daxuejiadao	757	3064
	Zhonglao	1173	2059
	Wusidajie	659	1634

三眼井胡同(宽5～7m)
Sanyanjing Hutong (5-7 meters'wide)

中华人民共和国成立后，1950年，平整压实了胡同土路，对景山东西街等道路进行了绿化。1954年，结合拆除有轨电车轨道，拓宽了地安门东西街。1955年，将东华门大街以北的御河改为暗沟以后形成北河沿大街。1954~1957年打通修建了东西长安街。

为了疏通东西交通和开辟北京第一条无轨电车线路，1956年对朝阜干线进行了较大规模的改建，拓宽了路面达到14至20m，改建了北海大桥（旧称金鳌玉蝀桥），打通了五四大街，使文津街——景山前街——五四大街成为北京内城一条东西交通干线。近几年内陆续打通、扩建了部分城市主要道路，如府右街、平安大街（经地安门东、西大街）、大红罗厂街、西什库大街等。

通过以上改造，皇城基本形成了现有的道路交通体系，即北以平安大街、东以南北河沿大街、南以长安街、西以德内大街和西单大街形成外围交通环线；中间以西安门内大街——文津街——景山前街——五四大街为东西交通干道，以地安门内大街、景山后街、景山东、西街、南北池子、南北长街为南北干道。

随着北京皇城的开放和旅游的迅速发展，皇城内的主要城市道路，如沿中轴线分布的地安门内大街、景山后街、景山东、西街、南北池子、南北长街、沿朝阜干线分布的西安门内大街、文津街、景山前街、五四大街等逐渐成为以旅游观光为主的城市道路。

4.4.1 现状道路网结构

皇城内现状道路系统由城市主要道路和胡同组成。

皇城内现状主要城市道路有15条。贯穿全区的主要城市道路东西向为西安门内大街——文津街——景山前街——五四大街，南北向为地安门内大街——景山后街——景山东、西街——南北长街、南北池子。联系区域对外交通的主要有东华门大街、府右街、大红罗厂街、西什库大街。

4.4.2 现状胡同

保护区现状有各类胡同137条，总长度近35000m。

皇城内胡同依宽度不同可分为5类：

宽度大于9m的胡同，总长度约3300m，占10%；

宽度在7～9m的胡同，总长度约3600m，占10%；

宽度在5～7m的胡同，总长度约8700m，占25%；

宽度在3～5m的胡同，总长度约12600m，占36%；

宽度小于3m的胡同，总长度约6600m，占19%。

皇城内3m以下的胡同，以组织步行交通为主；3～7m的胡同，可组织机动车单向行驶；7m以上的胡同，可组织机动车双向行驶。

4.5 建筑现状

保护区内的区级及区级以上的文物建筑由于国家相关文物保护法规的规定大多保存尚可，风貌依旧。普查文物由于发现较晚，且多被居民占用，维护情况较差。

皇城内的单位建筑大多是近年新建，建筑质量较好，层数较高，与整体传统风貌协调较差。

占地面积较大的居住建筑多数为四合院。由于皇城内的胡同是自由形成的，导致用地形状很不规整，四合院大小不一，进深差别较大；建造时期多为清末至民国年间。20世纪60年代末期，在北京旧城见缝插针的建设和在市区盲目发展工业的做法，对皇城地区造成了进一步的破坏。20世纪70年代后期推广"推、接、扩"的方针，即允许在四合院住宅内推出一点，接长一点，扩大一点，以解决缺房的燃眉之急。其结果使四合院变成了大杂院，许多密度低、建筑较好的四合院遭到破坏，环境更加恶化。插建高楼，使不少地方原有的景观遭受破坏。

产权属单位或个人的院落维护尚可，产权属房管所的院落由于房租低廉导致缺乏维护。

现状四合院内私搭乱建现象严重，由于属于违章建设不予承认，故在新测绘的地形图中没有表示。调查按照图纸上的现有建筑进行，主要调查建筑的高度、质量、风貌等内容。

4.5.1 建筑高度

文物保护单位应遵循原貌保护的原则，因此暂不考虑其建筑高度对周边环境的影响，除文保单位外，皇城内现状建筑高度可分为：1层、2层、3～4层、5～6层、7层以上五类，五类建筑总占地面积约为244hm²。其中：

1层建筑的用地面积为146.4hm²，占59.9%；

2层建筑的用地面积为22.6hm²，占9.3%；

3～4层建筑的用地面积为30.7hm²，占12.6%；

5～6层建筑的用地面积为36.4hm²，占14.9%；

7层及以上建筑的用地面积为8.2hm²，占3.4%。

由此看出，1～2层建筑占地169hm²，占69.2%，这类建筑是皇城保护的基础，它们维持了皇城平缓的基本空间格局。5层及以上建筑占地44.6hm²，占18.3%，这类建筑属于与皇城整体风貌不协调的元素，应加以控制。

4.5.2 建筑质量评价

根据房屋质量状况，参照北京市房管局《房屋完损等级分类标准》，皇城内现状建筑质量分为三类：质量良好、质量一般、

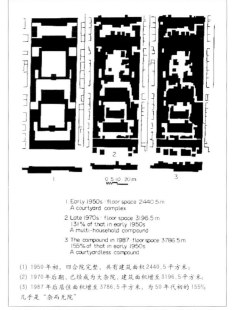

(1) 1950年初，四合院完整，共有建筑面积2440.5平方米；
(2) 1970年后期，已演变为大杂院，建筑面积增为3196.5平方米；
(3) 1987年后居住面积增为3786.5平方米，为50年代的155%，几乎"杂而无院"

北京某四合院的历史变迁（资料来源：朱嘉广、傅之京图）
Historical changes of a quadrangle in Beijing (paintings offered by Zhu jiaguang and Fu zhijing)

规划说明

Xicheng District			
	Subtota		34565
	Jingshanxijie	1200	2959
	Xi'anmen	0	0
	Gongjian	878	1997
	Miliangku	1256	3037
	Daguaibang	998	2889
	Aiminjie	878	2010
	Aiminli	1323	3463
	Xishiku	1267	3250
	Xixinsi	959	2288
	Houdali	2022	5774
	Guangming	784	2162
	Nanjieyiqu, West Huang chenggen	774	2261
	Huzhuxiang	1353	3580
	Beichangjie	1152	3063
	Nanchangjie	1775	4613
	Subtota		43346
	Total		77911

The land for residential quarters in the Imperial City is 136 hectares, and the present population density (number of registered population/100 square meter residential land) can be divided into 5 categories:

The first category (smaller than 2 persons/100 sm.) covers 15.0 hectares, accounting for 11.0%.

The second category (2-4 persons/100 sm.) covers 25.8 hectares, accounting for 18.95.

The third category (4-7 persons/100sm.) covers 54.3 hectares, accounting for 39.8%.

The fourth category (7-10 persons/100sm.) covers 21.2 hectares, accounting for 15.5%.

The fifth category (larger than 10 person/100sm.) covers 20.2 hectares, accounting for 14.8%.

Among these, the first- and second-grade housing of good conditions covers 30% of land; the third-grade housing of ordinary condition covers 40% of land; and the fourth- and fifth-grade housing of poor condition covers 30% of land. So measures must be taken to evacuate people out of the Imperial City.

According to the recent survey, the private courtyards, the old courtyards whose property right belonging to units and the land belonging to both private and business units have a low living density, whereas the public housing belonging to the Housing Management Bureau have a high living density, hence the phenomenon of "hapiness gap".

4.4 Current distribution of roads and hutongs

The roads of the Imperial City in the Qing Dynasty were usually made of earth or cobbles, especially earth. Things god worse towards the end of the period because "when the weather was fine, the dust can bury your feet and blow on the your face, and when it was raining, the earth became very muddy and smelly" Mule-driven carts, monkey-riding and rickshaws were the only means of transportation. In 1931, the Jiangshanqiang Street was built so that the land off limits to the commoners became an east-west road.

In 1953 after the new China was founded, the roads of hutongs were leveled and rammed, and trees were planted on the sides of East and West Jiangshan Streets. In 1954, while the streetcar rails were dismantled, the East and West Di'anmen Streets were widened. In 1955, Beiheyan Street was developed after the imperial river north the Honghuamen Street was turned into a covered sewarage. From 1954-1957, The East and West Chang'an Streets were constructed and connected.

In order to develop east-west traffic and the first trolleybus line in Beijing, a large-scale renovation was undertaken on the Chao-Fu Street by widening the road to 14 to 20 meters, revamping the Beihai Bridge (former Golden Turtle and Rainbow Bridge) and opening up Wusi Street, thus changing the Wenjin-Jiangshanqian-Wusi Streets into an east-west road in the inner city of Beijing. In the past few years, some of main roads in the city have been opened up and extended such as Fuyou Street, Ping'an Street (through East and West Di'anmen Street), Dahongluochang Street and Xishiku Street.

The present road system of the Imperial City became into being after the above-mentioned renovation: an outer traffic ring has been formed with Ping'an Street in the north, Nanheyan Street in the east, Chang'an Street in the south, and Denei and Xidan Streets in the west. In the middle, the east-west trunk line is the Xi'anmennei-Wenjin-Jingshanqian-Wusi Streets, and the north-south trunk line is made up of Di'anmennei, Jinghsanhoujie, Dong and Xi Jianghsan, Nanchizi and Beichizhi, Nanchang and Beichang Streets.

Damage done to traditional features by multi-storied buildings around Ranzhiju Hutong

With the opening up of the Imperial City and the rapid development of tourism, the main roads there such as Di'anmennei Street, Jingshanqianjie Street, Dong and Xi Jingshan Streets, Nanchizi and Beichizi Streets, Nanchang and Beichang Streets on both sides of the axis, as well as the Xi'anmennei Street, Wenjin Street, Jingshanqianjie Street and Wusi Street on the Chao-Fu trunk line have gradually become urban roads mainly for sightseeing.

4.4.1 Present road network structure

The present road system in the Imperial City is primarily composed of main roads and hutongs

There are 15 main urban roads in the Imperial City. The east-west roads that run through the whole district are Xi'anmennei-Wenjin-Jingshanqian-Wusi Streets, and the north-south trunk line is made up of Di'anmennei, Jianghsanhou, East and West Jingshan, Nanchizi and Beichizhi, Nanchang and Beichang Streets. The roads that are connected with places outside the district are Donghuamen Street, Fuyou Street, Dahongluochang Street, and Xishiku Street.

4.4.2 Current situation of hutongs

There are 137 hutongs of various types in the Imperial City with a total length of 35,000 meters.

Five categories can be divided of these hutongs based on their width:
Those wider than 9 meters cover 10%, with a total length of 3,300 meters.
Those of 7-9 meter wide cover 10%, with a total length of 3,600 meters.
Those of 5-7 meter wide cover 25%, with a total length of 8,700 meters.
Those of 3-5 meter wide cover 36%, with a total length of 12,600 meters.
Those less than 3 meter wide cover 19%, with a total length of 6,600 meters.

Hutongs less than 3 meter wide in the Imperial City are mainly for pedestrians; those of 3-7 meter wide can allow one-way motor vehicle; and those wider than 7 meters can have two-way motor vehicle traffic.

4.5 Present situation of architecture

Relics in the Imperial City under the conservation of the district level or above are well preserved in term of their original features due to relevant relic conservation regulations of the State. Relics discovered in the survey are not maintained in good shape as they were occupied by inhabitants and discovered too late.

The architecture built by units in the Imperial City were mainly erected recently, which are of good quality and of many stories. But they are incongruous with the general classical style.

The residential buildings that occupy a large area are mainly siheyuan. As the hutongs in the Imperial City were formed in a natural way, the terrain lacks regularity resulting in various sizes and depths of sihuyuan. Most of these quadrangles were built in the end of the Qing and early years of the Republic periods. Towards the end of the 1960s, the practice of squeezing in construction projects in the old city and developing industry in a blind way brought about further damage to the Imperial City. At the end of the 1970s, in order to solve the problem of housing shortage, extension of siheyuan by adding structures became prevalent, which resulted in many households squeezing in a compound, so that many low-density siheyuans of high quality were

质量较差。质量良好的建筑指近20~30年建设的建筑;质量一般的建筑指建筑质量维护较好的传统建筑或建设时期较早的近现代建筑,质量较差的建筑指建设时期较早、建筑质量维护较差的建筑,特别是平房式的民居。

由于文物保护单位的特殊性,对其暂不做建筑质量评价。

三类建筑的总建筑面积约232万m²。其中:

质量良好的建筑约136万m²,占58%;

质量一般的建筑约74万m²,占32%;

质量较差的建筑约22万m²,占10%。

4.5.3 建筑风貌评价

根据现状建筑的传统历史文化背景、建筑空间布局与建筑形态,对其传统风貌和历史文化价值进行评价,并分为5类。从表中可以看出,不协调的建筑面积占47.8%,且主要集中在皇城保护区的西北部地区和东部东黄(皇)城根南北街道路沿线,大多为多高层建筑。保护区内以众多文保单位、普查文物为主的地区和大部分居住用地内的建筑基本保持着清末民国初年的历史风貌。

院落绿化
Greenery in courtyard

保护区内建筑风貌评价

序号	分类名称	建筑面积(万m²)	所占比例(%)
1	国家、市、区级文物保护单位	31.9	12.1%
2	具有一定历史文化价值的传统建筑和近现代建筑	8.9	3.4%
3	与传统风貌比较协调的一般传统建筑	63.4	24.0%
4	与传统风貌比较协调的现代建筑	33.4	12.7%
5	与传统风貌不协调的建筑	125.8	47.8%
	总计	263.4	100%

4.6 绿化水系

保护区内的绿化分布比例大,现状保存较好,主要包括文保单位绿化、街道绿化和保护古树三个层次。文保单位绿化主要包括景山公园、北海公园等市级公园,养护较好。街道绿化指主要街道两侧的行道树,树木茂盛,绿化覆盖率高,特别是景山前街、后街、东街、西街、地安门内大街、府右街、文津街、五四大街、南、北池子大街、南北长街、黄化门街等绿化情况较好。

皇城内挂牌保护树木的分布可以划分为两部分。一部分分布在故宫、景山公园、中山公园、劳动人民文化宫、北海公园、中南海六大单位内;另一部分分布在六大单位外部的街区中。

目前,六大单位内的保护树木未做统计。六大单位外的保护树木包括:挂A牌的保护树木有10棵,挂B牌的保护树木有186棵,未挂牌的保护树木15棵,总计211棵。

此外,在传统居住院落中,也都较好地保留了许多长年生的高大乔木,与低矮的庭院绿化高低结合,形成了独具旧城风貌的绿化景观。对于其中树径大于30cm的大树,应按照市园林局的有关规定进行保护。

保护区内的水系除中华人民共和国成立后部分次要河道成为盖板河外,主要水体如三海和筒子河、内外金水河等水系维护较好。

4.7 市政设施现状

4.7.1 现状

(1) 排水设施

皇城范围内的排水体制为雨污分流制与雨污合流制并存,主要大街下基本上建有雨污水管道,但旧街坊内由于市政设施建设滞后,排水仍旧为雨污合流。随着城市排水设施完善、河湖水系截污工程的建设,皇城内沿河均已设置污水截流井,将合流管道的污水在入河前截入污水管道。

皇城雨水排除分为两大系统。文津街、景山后街、五四大街以北地区的雨水排除属于四海下水道(地安门大街下)的流域范围,排水尾闾为东护城河暗沟;文津街、景山后街、五四大街以南地区的雨水排除分别属于府右街雨水方沟、御河下水道(南、北沿大街)的流域范围,排水尾闾为前三门暗沟。

皇城污水排除属于高碑店污水处理厂的流域范围。皇城内现状主要污水干线有地安门大街污水管、西黄(皇)城根北街污水管、府右街污水管、南北长街污水管、景山前街——五四大街污水管、南北河沿污水管,污水出路分别为东二环路污水管和前三门大街污水管。

(2) 供水设施

皇城范围内供水基本上由市区自来水管网供给,主要大街和大部分街坊内已经建有自来水管道,部分单位内有少量自备井,基本不作为生活饮用水。

皇城内现状主要供水管道有地安门大街供水管、西黄(皇)城根供水管、府右街供水管、南北长街供水管、南北池子供水管、南北河沿供水管、文津街——景山前街——五四大街供水管、长安街供水管。

(3) 供气设施

皇城范围内供气气源基本上都是天然气。主要大街下均建有供气管道,在街坊内安排有中低压调压站。

皇城内现状主要供气管道有地安门大街供气管、府右街供气管、南北河沿供气管、文津街——景山前街——五四大街供气管、长安街供气管。

(4) 供热设施

皇城范围内部分主要大街下建有供热管道,为一些单位和多层住宅供热,街坊内供热以烧煤和电采暖为主。

(5) 电信设施

皇城范围内主要大街下均建有电信管道,在街坊内安排有交换间。

(6) 供电设施

皇城范围内有现状西单110kv变电站,主要大街下均建有10kv电缆线路,局部地区为架空线。

4.7.2 存在问题

皇城内部分地区为老的平房区,近些年没有进行大的改造和修缮,

undermined, and environment became worse. A lot landscapes were also affected by insertion of many skyscrapers.

The courtyards belonging to private people or units were in fairly good shape, but those belonging to Housing Management Bureau were not due to their low rent.

It is now very popular to add new rooms in siheyuan in a disorderly way, so they were not acknowledged as they are unauthorized buildings, and not indicated in the new typographic map. The survey is conducted according to the architecture on the map, involving their height, quality and style.

4.5.1 Architectural Height

The principle of conserving the original features is applied to relic sites, so the impact of their height on surroundings will not be considered for the time being. The present architecture in the Imperial City can be put into 5 categories: single-storied, two-storied, three-four-storied, five-six-storied, seven-storied or above. They cover an areas of 244 hectares, out of which

single-storied covers 146.4 hectares, accounting for 59.9%;
two-storied covers 22.6 hectares, accounting for 9.3%;
three-four-storied covers 30.7 hectares, accounting for 12.6%;
five-six-storied covers 36.4 hectares, accounting for 14.9%; and
seven-stories and above covers 8.2 hectares, accounting for 3.4%.

It is clear that the single- or two-storied housing covers 169 hectares, accounting for 69.2%. which, as they maintain the flat spatial layout of the Imperial City, paves the way for the conservation of that area. Buildings of or above five stories cover 44.6 hectares, accounting for 18.3%. As they are elements not in agreement with the style of the Imperial City, they should be controlled.

4.5.2 Evaluation of Architectural Quality

Based on the quality of house, and in reference with the Classification Criteria of Quality of Houses issued by the Beijing Housing Management Bureau, the quality of the present architecture in the Imperial City can be divided into 3 categories: fairly good, average, and poor. The first category refers to buildings constructed in the past 20 to 30 years; the second category refers to traditional or modern architecture that have been in existence for a long time but well preserved; the last category refers to architecture that were built in the past and not well maintained, especially single-storied residential houses.

Quality evaluation will not be done for relic sites due to their uniqueness.

The total area of the three categories of architecture is 2.32 million square meters, out of which

The first category covers 1.36 million square meters, accounting for 58%;
The second category covers 740,000 square meters, accounting for 32%; and
The last category covers 220,000 square meters, accounting for 10%.

4.5.3 Evaluation of Architectural Style

The evaluation of traditional feature and historic and cultural values of architecture is conducted in terms of traditional and cultural background, architectural space and shape, thus 5 categories can be divided. From the following diagram, it is obvious that the area of incongruous architecture accounts for 47.8%, and they are mainly concentrated in the northwestern part and on both sides of the Nan and Bei Streets of East Huangchenggen in the eastern side of the Imperial City. Most of them are multi-storied buildings or high-rises. The historical features of the late Qing and early Republic periods in the conservation district have been preserved by the many relic places and residential architecture.

Architecture Style Evaluation in the Conservation District

Number	Classification	Floor space (10,000 sqm.)	Ratio (%)
1	Relic sites under State, municipal and district protection	31.9	12.1%
2	Traditional, modern and contemporary architecture of certain historic and cultural values	8.9	3.4%
3	Average traditional architecture fairly corresponding to traditional style	63.4	24.0%
4	Contemporary architecture fairly corresponding to traditional style	33.4	12.7%
5	Architectural incongruous with traditional style	125.8	47.8%
	Total	263.4	100%

4.6 Afforestation of River System

The afforestation of the conservation district is widely spread and well preserved encompassing cultural relics, streets and ancient trees. The afforestation of cultural relics is well maintained including Jingshan and Beihai and other municipal-level parks. The street landscaping refers to the sidewalk trees which have a high coverage rate with exuberant foliage. Some of the trees are well maintained such as those on the following streets: Jingshaniqian, Jingshanhou, Jingshandong, Jiangshanxi, Di'anmennei, Fuyou, Winjin, Wusi, Nanchizi, Beichizi, Nanchang, Beichang, and Huanghuamen, etc.

The trees in the Imperial City that are on the protection list are divided into two parts: one part is in the 6 relics sites: Palace Musuem, Jingshan Park, Zhongshan Park, Laboring People's Palace of Culture, Beihai Park and Zhongnanhai; the other part is scattered in the roads outside the 6 relics sites.

At present, the trees under protection within the 6 relic sites have not been counted, but the trees under protection outside the 6 sites include: 10 with A-grade label; 186 with B-grade label; 15 without labels, bringing the total number to 211.

Besides, in old courtyards, many tall arbors are also well kept. Contrasting pleasantly with low courtyards trees around, they constitute the unique landscape in the old city. Trees exceeding 30 centimeters in diameter should be protected in accordance with the regulations of the Municipal Gardening Bureau.

The primary river systems in the conservation district are all well preserved such as Three lakes, Dongzi River, Inner and Outer Golden Water Rivers, except for some secondary river-courses that have been turned into covered rivers after the founding of new China.

4.7 Current Situation of Municipal Installations

4.7.1 Current Situation

(1) Drainage Facilities

Separate system and combined system are the ways of drainage in the Imperial City. Rainwater and drainage pipes are installed under the main streets but due to the fact the construction work lags behind in old streets, the combined drainage system is still used there. With the perfection of urban drainage facility and the construction of sewage interception projects in rivers and lakes, interceptors have been installed along the rivers in the Imperial City, intercepting the sewage into foul water pipes before it flows into rivers.

There are two big systems for rainwater drainage in the Imperial City. The rainwater drainage north of Wenjin, Jingshanhoujie and Wusi Streets belongs to the Sihai sewerage (under Di'anmen Street) area, with its mouth at the covered sewerage of east city moat. The rainwater drainage south of Wenjin, Jingshanhoujie and Wusi Streets belongs to the Fuyou Street and imperial river (Nanheyan and Beiheyan Street) sewerage areas with the mouth at Qiansanmen covered seweage.

由北海望中海、南海（2002 摄）
Zhonghai and Nanhai viewed from Beihai (shot in 2002)

因此在这些地区内各种市政设施没有得到及时的更新,各种市政管网系统没有深入到街坊中,各种市政站点配置不足。例如供水管道没有进户,排水系统为雨污合流制,供热靠自家的煤炉,炊事靠煤气罐或煤炉,电线以架空线为主,等等。

4.8 市民生活

保护区内居住用地的空间结构大致分为两类,一类是依托传统四合院建筑的以"街—胡同—院落—户"为主的结构。院落和胡同成为居民交往和活动的主要空间。由于旧城基础设施较差,再加上机械人口的增加,居住的生活质量在下降;且居住人口情况复杂,社会地位和经济收入差距较大,住房条件和生活方式也有很大不同,其中,中低收入家庭占有多数,急需改善住房条件。由于位居市中心,紧靠大公园,生活比较方便,不少人靠小生意或出租房屋为生。加上部分老住户居住时间较长,地域和邻里感情上难于割舍,且搬迁住房多位于外城地区,位置较远,缺乏相应的服务设施,所以尽管住房紧张,愿意搬走的人不多。另一类是以新建多层居住小区为主的"小区—邻里单位—住宅楼—户"结构,居住条件多数中等偏上,如果可以进一步改造社区的绿化条件,整治周边商业秩序,则要求改造的呼声不高。

总体说来,中华人民共和国成立后近20年,是对旧皇城改造最为剧烈的时期。具体表现为:

土地使用性质——性质变更。行政办公用地、医疗卫生用地、教育科研、中小学等用地明显增加,大多建设标准高,建筑质量较好,尤其是部分新建建筑;

风貌和质量——城市传统肌理遭到不同程度的破坏。机关大院等单位的圈地和建设导致原有街巷空间缺失,新建多层居民小区完全破坏了肌理,虽无风貌,但建筑质量和居住条件好;一些具有传统风貌的居住院落由于使用和维护不当,已逐步破败,而且缺乏市政配套设施,居民生活不便,尤其是住在公房里的条件更差;已更新的部分新四合院尚能满足风貌要求,有关各方评价不一;

建筑高度——拓宽城市道路后,皇城周边的新建建筑突破了"控规"高度限制,影响历史保护区的传统空间形态;

交通——机动车交通量增加后挤占行人步道,步行环境差;缺乏机动车、非机动车停放设施;

文物保护——部分文物得不到合理利用,周边建控地带实施不易。

5. 皇城的性质与历史文化价值

5.1 性质

北京明清皇城是以紫禁城为核心,以皇家宫殿、衙署、坛庙建筑群、皇家园林为主体,以满足皇家工作、生活、娱乐之需为主要功能的、具有浓厚传统文化特色的历史文化保护街区。

明清皇城在元代皇城基础上发展起来,虽历经五百余年的沧桑岁月,但仍较完整地保留了历史上象征一国之都——皇城的布局规模和建筑特色,这是我国惟一保存较完整的代表封建国都的皇城建筑群,是中国几千年封建社会国都建筑的代表,也是中华民族优秀的历史文化遗产。

这就是北京,古老而又鲜活,博大而又精深,高远而又亲切,迷人而又难解。它是单纯的,单纯得你一眼就能认出那是北京。它又是多彩的,丰富得你永远无法一言以蔽之。而无论久远深厚的历史也好,生机勃发的现实也好,豪雄浩荡的王气也好,醇厚平和的民风也好,当你一进北京,它们都会向你扑面而来,让你目不暇接,不知何读起。
——易中天《读城记之北京城》

5.2 历史文化价值

北京皇城始建于元代,主要发展于明清时期,是一座拱卫紫禁城安全和为帝王统治服务的皇宫外围城。皇城以其独特的规划布局,杰出的艺术成就和成熟的建造技术,成为中国几千年封建王朝统治的象征,体现了历史上皇权的至高无上,具有极高的历史文化价值。

皇城的惟一性:明清皇城是我国现存惟一保存较好的封建皇城,它拥有我国现存惟一的、规模最大、最完整的皇家宫殿建筑群,是北京旧城传统中轴线的精华组成部分。

皇城的完整性:皇城以紫禁城为核心,以明晰的中轴线为纽带,城内有序分布着皇家宫殿园囿、御用坛庙、衙署库坊等服务机构和服务人员居所,体现出为封建帝王服务的鲜明的规划理念和完整的功能布局。

皇城的真实性:皇城中的紫禁城、筒子河、三海、太庙、社稷坛和部分御用坛庙、衙署库坊、四合院等传统建筑群至今保存较好,真实地反映了古代皇家生活、工作、娱乐的历史信息。

皇城的艺术性:皇城在规划理念、规划手段、建筑布局、建造技术、色彩运用等方面具有很高的艺术性,展现了历史上皇权的至高无上的等级观念。

城市是人类最伟大的创造之一,它以一种简单然而不断变化的方式表现着人类丰富多彩和永无止境的创造活动。而北京皇城的产生与发展正是这一伟大活动的真实记录。

5.2.1 皇城的惟一性

中国古代大规模的建筑组群如城市、宫殿等其建筑外部空间的布局处理,鲜明显现了中国传统建筑艺术的一大突出特色和卓越成就。北京皇城的整体布局,突显故宫的功能与作用,其作为我国现存惟一保存

皇城鸟瞰
Bird's eye view of the Imperial City

The sewer drainage in the Imperial City falls in the range of the Gaobeidian Sewerage Treatment Plant. The sewage lines in the Imperial City include Di'anme Street pipeline, West Huangchenggenbei Street pipeline, Fuyou Street pipeline, Nanchang and Beichang Streets pipelines, Jinghsanqian-Wusi streets pipelines, and Nanheyan and Beiheyan Streets pipelines. The mouths of sewage are East Second Ring Road conduit and Qiansanmen conduit.

(2) Water Supply Facilities

The water supply in the Imperial City is provided by the tap water network of the city, and tap water pipelines have been installed in main streets and most part of the neighborhoods. Some units have stand-by wells, which basically are not used as drinking water.

The water pipelines in the Imperial City are at the following places: West Huangchenggen, Fuyou Street, Nanchiang and Beichang Streets, Nanchang and Beichang Street, Nanchizi and Beichizi, Wenjin-Jianghsanqian-Wusi streets, Nanyeyan and Beihuyan and Chang'an Street.

(3) Gas Supply Facilities

The gas in the Imperial City is mainly natural gas, with main streets equipped with gas pipelines underneath. Medium and low pressure adjusting facilities are installed in neighborhoods.

The primary gas supply conduits are at the following places: Di'anmen Street, Fuou Street, Nanheyan and Beheyan, Wenjin-Jiangshanqian-Wusi Streets and Chang'an Street.

(4) Heating Supply Facilities

Some Streets in the Imperial City are equipped with heating conduits underneath providing heating for some units and multi-storied housing. The heating supply in neighborhoods mainly comes from coal and electricity.

(5) Telecommunication Facilities

Telecommunication pipelines are installed under the major streets in the Imperial City with switchboards equipped in the neighborhoods.

(6) Electricity Supply Facilities

Within the Imperial City, there is the Xidan 110 KV transformer substation, and 10 kilovolt cable lines are installed under main roads. Aerial lines are installed in some areas.

4.7.2 Existing Problems

As some parts of the Imperial City are single-storied housing areas which have not received large-scale renovation, their municipal installations were not updated promptly. The neighborhoods did not have access to citywide piping systems and various supplies are inadequate. For example, water pipes have not reached the households, the drainage system is still the combined system, and heating is provided only with home-made coal stove, cooking is done by gas containers and aerial line is the major means of providing electricity.

4.8 Life of Inhabitants

The spatial structures in the conservation district is mainly divided into two categories. One is the structure of the "road-hutong-courtyard-household" type based on the ancient siheyuan. The courtyards and hutongs are the major activity space for inhabitants. As the infrastructure in the old city is poor and with the increase of population, the living standards have declined. The population composition is also complex as indicated in social status and big gap in incomes. Most of people there are medium- and low-income families so their housing conditions need prompt improvement. Due to its location next to the heart of the city and big parks, life is easy so many inhabitants there are involved in small business or live by renting houses. Many old households have known each other for a long time so they do not want to part company by moving to the outer city where service facilities are not adequate enough. Therefore, even though the housing is very cramped, few of them are willing to move away. The second category is multi-storied housing estate type, with a medium or better housing conditions. If the green belt of the community can be further improved, the business order in surroundings rectified, they would not be likely to call for renovations.

Generally speaking, the past 20 years witnessed the most drastic renovation of the Imperial City, which was particularly illustrated as follows:

Nature of land use—the nature has changed. The land for office buildings, hospitals, education and research and primary and middle schools have increased dramatically. These buildings are of good quality and have high standards, especially some of the newly completed ones;

Features and quality—the traditional urban fabric has been damaged in various degrees. Big compounds of organizations and their constructions have resulted in the disappearance of the original space of streets and allays, and newly built housing estates have undermined the fabric. Although they all lack styles, their quality and living conditions are good. Some courtyards that boast traditional fea-

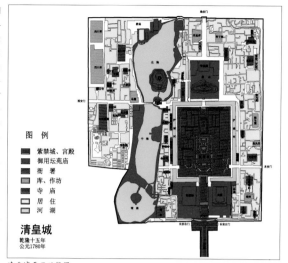

清皇城平面功能图
Functional plane figure of the Imperial City of the Qing Dynasty

tures have disintegrated due to unreasonable use and lack of maintenance. They are also inconveniencing the inhabitants, especially those living in poorly appointed public houses because of the absence of municipal installations. The newly updated siheyuans are in agreement with ancient style but are drawing different comments and opinions.

Architectural height—after the roads have been widened, the new buildings around the Imperial City have exceeded the limits of controlled height, affecting the traditional spatial shape of the conservation district.

Transportation—the increase of vehicles are have resulted in the squeezing in of sidewalks, and there is not enough room for parking motor and non-motor vehicles.

Cultural relic conservation—some of relics are being used in a sensible way, and the controlled areas around them are not well regulated.

5. Nature and Historic and Cultural Values of Imperial City

5.1 Nature

The Imperial City of the Ming and Qing Dynasty is centered around the Forbidden City, surrounded by imperial palaces, yamens, temples, altars and imperial gardens, satisfying the needs of work, life and entertainment of the imperial family. It is a conservation district of rich traditional and historical culture.

The Imperial City of the Ming and Qing Dynasties was constructed on the basis on the Yuan Dynasty imperial city. Although it has a history of 500 years, it has well preserved the layout scale and architectural features of the Imperial City, the symbol of a capital of a country. It is the only imperial architectural compound representing the feudal capital, the modal of thousands of years of feudal capital architecture and also the excellent historical and cultural heritage of the Chinese nation.

This is Beijing, old and energetic, profound and expansive, far-away yet intimate, charming yet inscrutable. It is so simple that you know what it is at a glance. Yet it is also colorful, too rich to be summed up in a word. Whatever it is, profound history, energetic reality, domineering air of the kings, or the naive customs of the folkways, once you put your foot in Beijing, you will encounter everything, without any moment to ascertain and read it from the very beginning. From Yi Zhongtian's City of Beijing—A Series of Cities with Walls.

5.2 Historic and Cultural Values.

The Imperial City of Beijing, first constructed in the Ming and Qing periods, is a outer city defending the Forbidden City and serving the royal rulers. It is the token of thousands of years of feudal society in China with its unique planning layout, outstanding artistic achievement and mature construction technology. It represents the unchallenged power of imperial rulers in history, so has a very high historic and cultural values.

Uniqueness of the Imperial City: Being the only well preserved Imperial City of the Ming and Qing Dynasties, the Imperial City is the sole largest and most complete imperial architectural complex, forming the cream of the traditional axis of the old city of Beijing.

较好的封建皇城和世界现存古代皇城的杰出代表，在建筑群外部空间设计方面表现出了高超造诣，产生了震撼人心的气势与魄力。

皇城内拥有我国现存惟一的、规模最大、最完整的皇家宫殿建筑群——故宫，其单体建筑通过院落组合，运用赋有哲理的中国独特空间观念，以轴线控制形成群体，群体在功能要求下按照一定的序列逐渐展开，形成相应的秩序空间，从而达到空间恢弘的气势以及惟我独尊的目的。

5.2.2 皇城的完整性

（1）体现封建统治的需要，按《周礼·考工记》建造起来的封建王城。

皇城按照《周礼·考工记》上所记的"前朝后市，左祖右社"的王城规划制度，在宫城南侧沿中轴线对称的布置了太庙和社稷坛这两座重要的礼制建筑。

（2）城市中轴线及重要礼制建筑。

由中轴线组织各个部分的布局形式，一方面承袭了传统礼仪制度，另一方面，又是所尊崇的制度与自己的特殊条件相结合产生的灵活运用。皇城布局紧密联系实际用途，又经过艺术的处理达到了高度成功。

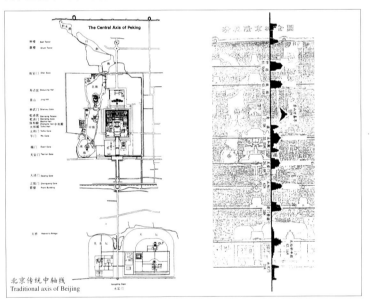

北京传统中轴线
Traditional axis of Beijing

以宫城为内城核心，布局皇城乃至整个北京城的是一根直线。一根长达8km，全世界最长，也最伟大的城市中轴线穿过了全城。北京独有的壮美秩序就是由这条中轴线的建立而产生。前后起伏左右对称的布局以及空间分配都是以这条轴线为依据的。气魄之雄伟就在其南北引伸，一贯到底的规模。

在皇城内，由天安门起，是一系列轻重不一的宫门和广庭，金光闪耀的琉璃瓦顶，一层又一层的起伏峋峙，一直引导到太和殿顶，便到达中轴线前半的极点，然后向北、重点逐渐退削，以神武门为尾声。再往北，又有"奇峰突起"的景山作为宫城的衬托。由此向北一波又一波进入广阔无垠的大平原。

（3）御用坛庙、衙署库坊、寺庙、民居四合院等共同组成的功能群体。

皇城内分布着的坛庙、衙署库坊、寺庙和民居四合院，无一不为皇城的核心——宫城服务，皇城内皇家御用的祭祀风、云、雷神的庙宇，皇家道观，为多民族宗教领袖进京面圣、讲经说法设置的行宫和寺院，为皇宫提供衣食住行的各种后勤供应部门，为皇宫服务人员建造的就近居所……所有这些建筑实体，构成了能正常运行的完整的皇城。

（4）街巷、胡同、绿地、河湖水系等系统。

皇城内自由分布的街巷和胡同系统，深刻地反映出历史的变迁，与四合院建筑一起，成为北京皇城独具的风貌特征。河湖水系和绿地分布一方面起到调节城市生态的作用，另一方面，也成为城市历史的生动见证。

（5）空间形态。

形成平缓开阔、起伏有致的城市天际线。在广大的平房四合院民居衬托下，突显以故宫为中心，沿景山万春亭、钟鼓楼、正阳门等建筑展开的城市中轴线。皇城内园林三海里，用高的亭、塔和低的小丘或桥梁与郊区的西山遥相联系，扩大空间的深度，并和不同形体的建筑、旧城墙、城楼一起，构成了参差起伏的城市轮廓。其城市空间的特色在于主体建筑及其构筑的空间形态雍容大度、平缓开阔，从容不迫；在于其从属建筑收放有致的生活情趣和悠闲的市井情态，"处处有空儿，可以使人自由地喘气"。

（6）居住建筑形式。

产生了北方居住建筑的典型代表——北京四合院。北京四合院院落宽绰舒朗，四面房屋各自独立，起居十分方便，具有很强的私密性。院内植树栽花，饲鸟养鱼，居住者不仅享有舒适的住房，还可分享大自然的美好天地。它是中国传统城市艺术的画面结构，展示着立体轮廓、外观形式、色彩、装饰、绿化等城市艺术个性。此外，四合院蕴涵的深刻的文化内涵，四合院包含的轴心统驭、端庄中正、秩序等级、九宫八卦等功能象征因素，使其成为中华传统文化的载体。

5.3 保护皇城的意义

保护皇城，就是延续城市的历史。

皇城的价值不仅是个别建筑类型和个别艺术杰作，最重要的还在于各个建筑物的配合，在于它们与北京旧城的全盘计划、整个布局的关系，在于这些建筑的位置和街道系统的相辅相成，在于全部部署的严肃秩序，在于形成了宏伟而美丽的整体环境。

中轴景观(景山——钟鼓楼)
Landscape along the axis (from Jingshan to Bell and Drum Towers)

Completeness of the Imperial City: Imperial city is centered around the Forbidden City and situated on the axis. Imperial palaces, gardens, temples, yamens and warehouses are all scattered in the city, demonstrating the complete idea and functional layout of serving the feudal emperors.

Authenticity of the Imperial City: Traditional architectural complex of the imperial city such as the Forbidden City, Dongzi River, Three Lakes, Imperial Ancestral Temple, Altar to the God of the Land and Grain, government agencies, warehouses and courtyard houses have been well preserved, providing solid evidences of the historical information on the life, work and entertainment of the royal families.

Artistic Quality of the Imperial City: The Imperial City has attained a high artistic quality in planning concept, architectural layout, construction technique, and color application. So it boasts a high historical and cultural value symbolizing the absolute authority of the royal monarch.

City is the greatest creation of mankind, presenting the endless and colorful creative activities of man with in a simple yet ever-changing way. And the creation and development of the Imperial City in Beijing is exactly the real record of this great activity.

5.2.1 Uniqueness of the Imperial City

The large ancient architectural compounds in China such as cities and palaces and the layout treatment of their outer space represent one of the outstanding features and great achievement of traditional Chinese architectural art. The overall layout of the Imperial City highlights the functions and role of the Palace Museum. As the only extant and well preserved feudal city and an outstanding ancient imperial city in the world, the Imperial City has manifested excellent accomplishment in outer spatial design, producing an effect that is both grandiose and exciting.

The imperial City boasts the only extant, biggest and most complete imperial palace compound in China—the Forbidden City. Each of its individual buildings is arranged through courtyards and, by employing the unique and philosophical Chinese spatial concept and forming a compound around the axis, is unfolded in accordance with certain order and functions, thus producing a spatial grandeur and unparalleled greatness.

5.2.2 Completeness of the Imperial City

(1) A feudal city presenting the needs of feudal rule and constructed according to the Book of Diverse Crafts.

In line with the rule of the capital of kings stipulated in the Book of Diverse Crafts that "the court at the front and market in the rear; and ancestral temple on the left and altar to the god of the land and grain on the right", the two important buildings of rites—the Ancestral Temple and the Altar to the Land and Grain God—were built symmetrically on both sides of the axis at the southern side of the palace.

(2) Urban Axis and Important Architecture of Rites

Various layout forms were organized around axis, which not only inherits the traditional system of rites, but also through the combination and flexible application of revered system and particular conditions. Great achievement was made by the layout of the Imperial City which is closely related with practical purpose plus artistic treatment.

A long line runs through the Imperial City and even the city of Beijing with the palace as the core. Eight kilometers long, this greatest urban axis is the longest in the world, by which the unique beautiful order of Beijing was generated. The undulating and symmetrical layout and spatial distribution is also produced around this axis, whose grand charm lies in its scale that runs from north to south.

Starting from Tian'an man in the imperial City, a stretch of palace gates and courtyards of various magnitudes with shining glazed tile roofing and undulating shapes goes all the way to the Hall of the Supreme Harmony, the apex of the first half of the axis. Then the axis tails off northwards until the Gated of Prowess. Further north, however, rises the Jingshan Park, another high peak, as the foil to the palace, thus wave after wave rolled northward into an unfolding plain.

(3) Functional Compound Made up of Imperial Temples, Storehouses and Government Offices, Altars and Single-Storied Residential Houses

The temples, government offices, altars and residential siheyans dotted in the Imperial City all but serve the palace, the center of the Imperial City. The temples used to worship gods of rain, thunder and wind, the Taoist temples, the temporary dwelling places for leaders of other nationalities to have an audience with emperors and deliver sermons, all the service departments providing food and clothing and other services to the court, the living quarters built for the court attendants all made it possible for the Imperial City to function in a normal way.

(4) Street and Allays, hutongs, Green Belt and River Systems

The streets and hutong network in the Imperial City, coupled with siyeyuan, typically reflect the evolution of history, thus constituting a particular style of Beijing. The river system and green belt not only regulate the urban ecology, but also become the living monument to the urban history.

(5) Spatial Pattern

To form a open, level, and undulating urban skyline. Set off by the vast stretch of quadrangles, the urban axis starts from the conspicuous Forbidden City and runs through the Wanchun Pavilion of the Jiangshan Park, Drum and Bell Towers, Zhengyangmen and other architecture. In the three lakes of the parks of the Imperial City, the tall pavilions, pagodas and low hills or bridges echo the far-away West Mountain in the suburbs, thus expending the depth of space and creating a rolling city horizon along with the architecture, old city walls and towers of various sizes. The characteristics of its urban space is that its principal architecture and spatial patterns are graceful and poised, level and open, natural and peaceful, and that its secondary architecture have good taste and a leisurely life attitude, so that "there is space everywhere for people to breath freely."

(6) Forms of Residential Architecture

A typical residential architectural of the north was born—siheyuan (quadrangles). The siheyuan in Beijing is roomy an comfortable, each room inside being self-contained, thus both convenient and having privacy. The inhabitants through planting trees and flowers, breeding fish and raising chickens, would feel satisfied with their life, but can share the joys with the nature. This urban art is like a picture with artistic characteristics in terms of outer pattern, color, decoration and green plants. Also, the profound cultural implication, the symbolic elements of hierarchy and order have made it the carrier of the Chinese tradition.

5.3 Significance of Protecting Imperial City

Protecting the Imperial City amounts to the continuation of the urban history.

The value of the Imperial City lies not only in individual architectural type or individual artistic accomplishment, it also lies in the coordination of different buildings, in the relation of overall planning and layout of Beijing, in the fact that the locations of these buildings and the road system complement each other, in the serious order of generic arrangement, and in the formation of a grand and beautiful environment.

Although the land layout inside the Imperial City, apart from cultural relics and water bodies with green belt, does not have an obvious planning purpose as that outside the Imperial City, its freely developed layout and road network mostly reflect the functional features of the place, including the historical information of the urban evolution. It has the cooperative relation between individual architecture and the

街巷胡同及绿地水系
Streets, hutongs, green belt and river system

皇城内除文物古迹和绿化水体之外的用地布局虽然不象皇城外的区域显现明显的规划意图，但其自由发展的布局和路网最能反映旧皇城的功能特点，包含着城市变迁的历史信息。它们有着值得保存的建筑个体和城市整体的配合关系，有值得保护的传统空间形态，有值得保护的文物环境。对于这些，只保护单个的文物建筑和历史街区是不够的，必须对整个皇城进行整体保护。

6.规划目标、原则及依据

6.1 规划目标

明确皇城保护区性质；正确处理皇城历史文化保护区的保护和城市现代化的关系；制定保护和整治的原则和方法，严格控制皇城内的建设；提出保护和整治的相关政策和实施方式的建议；保证皇城传统整体风貌和空间格局的延续。

6.2 规划原则

坚持整体保护与分类保护相结合，最大限度保存真实历史信息的原则。

坚持有重点、分层次、分阶段逐步整治、改善和更新的原则。

坚持逐步适度疏导人口的原则。

坚持严格控制皇城内建设规模的原则。

坚持文物保护单位的保护与合理利用相结合的原则。

6.3 指导思想

把皇城作为一个整体加以保护，特别是搞好故宫等重点文物的保护。

正确处理皇城的保护与现代化建设的关系，皇城内新的建设要服从保护的要求，保证皇城整体风貌与空间格局的延续。

要用辩证和历史的观点，正确认识和处理皇城的发展和变化，区别和慎重对待已有新建筑。

贯彻"以人为本"和"可持续发展"的思想，努力改善皇城中的居住、工作条件和环境质量。

6.4 规划依据

《中华人民共和国城市规划法》（1989年12月）
《中华人民共和国文物保护法》（2002年10月）
《北京市文物保护管理条例》（1987年6月）
《北京城市总体规划（1991~2010年）》
国务院关于《北京城市总体规划》的批复（1993年10月）
《北京旧城历史文化保护区保护和控制范围规划》（1999年4月）
《北京市区中心地区控制性详细规划》（1999年9月）
《北京市文物保护单位范围及建设控制地带管理规定》（1994年）
《北京25片历史文化保护区保护规划》（2001年3月）
《北京历史文化名城保护规划》（2002年4月）

7.保护规划

历史是动态的发展过程，皇城保护规划应本着充分尊重城市历史、体现皇城文化价值的态度，一方面严格保护传统城市空间和建筑的精华部分，另一方面不影响居民的正常生活，不妨碍该区的发展和社会进步。通过展示北京皇城的建设发展过程，主次分明地勾勒出一个历史剖面。

7.1 分区与地块编号

根据皇城内土地使用特征及街道的自然分布，按照由南至北、由东至西的原则，将皇城用地划分为10个区域。

01区：皇城内故宫与劳动人民文化宫以东，五四大街以南的区域；

02区：皇城内地与劳动人民文化宫以东，五四大街以南的区域；

03区：由故宫、中山公园和劳动人民文化宫三部分组成的区域；

10区基本为多层建筑区
10 Zones are mainly composed of multi-storie buildings

04区：景山公园所在的区域；

05区：中南海以东，文津街以南，故宫与中山公园以西的区域；

06区：北海以东，景山前街以北，地安门内大街、景山西街以西的区域；

07区：中南海所在区域；

08区：北海所在区域；

09区：文津街以南，中南海以西的区域；

10区：文津街以北，北海以西的区域。

10个区各类地块总计有572个，在今后的改造中应按照地块编号进行管理。

各分区的特征：

01、02、05、06、09五个区为传统平房四合院建筑相对集中的地区，胡同和四合院构成街区的主体。03、04、07、08四个区为文物保护单位，皇家建筑群传统风貌的精华，它们对皇城的保护具有决定性作用。10区基本上为多层建筑区，传统四合院建筑仅剩两小片，是新的建设对皇城冲击最大的区域。

7.2 土地使用功能规划

用地原则：作为记录城市发展历史的传统风貌保护区和完整的传统城市核心区，规划区的用地原则应以保护城市历史信息、完善特定的城市功能为基本原则，故其土地使用功能规划应以调整功能为基础，并应符合以下原则：

7.2.1 尽可能保持现状用地中具有历史延续性、符合保护区性质、功能和形式的城市功能用地；

7.2.2 扩大保护用地，为在严格控制下的皇城保护区的未来发展创造条件；

7.2.3 尽可能调整与保护区的性质、功能、形式和环境不协调或造成破坏的城市功能用地。

皇城的土地使用功能规划可分为12类，包括行政办公、商业金融、文化娱乐、医疗卫生、教育科研、宗教福利、普通住宅、一类住宅、中小学和托幼、市政设施、绿地、道路广场。

依照占地规模大小，主要用地有：绿地（226hm²），约占皇城面积的33%；居住用地（123hm²），约占18%；文化娱乐用地（124hm²），约占18%；道路广场（84hm²），约占12%；行政办公（67hm²），约占10%，其他占9%。

规划以调整为主，现有公园绿地、行政办公用地、医疗卫生等用

whole city, the traditional spatial patterns and cultural relic environment, all of which are worth conserving. Because of this, it is not enough to protect individual architecture or an ancient road block; rather, the whole Imperial City should be protected.

6. Objective, Principles and Basis of the Plan

6.1 Objective of the Plan

Clarify the nature of the conservation district of the Imperial City; correctly address the relation between conservation of the conservation district of the Imperial City and urban modernization; formulate principles and approaches of conservation and rectification; strictly control the construction in the Imperial City; put forward relevant policies, means and advice of the implementation of the conservation and rectification; and guarantee the continuation of the overall features and spatial layout of the Imperial City.

6.2 Principles of the Planning

The principle of combining overall and classified protections, and protecting the authentic historical information to the full.

The principle of rectifying, renovating and updating in a gradual way, and in accordance with order of importance.

The principle of gradually relocating inhabitants in a modest way.

The principle of strictly restricting the construction scale within the imperial city.

The principle of combining the conservation and sensible utilization of cultural relic places.

6.3 Guidelines

Protecting the imperial city as a whole, especially the key cultural relics such as the Forbidden City.

Correctly address the relationship between the protection of the Imperial city and construction of modernization, with the new constructions within the Imperial City meeting the requirements of protecting the Imperial City, its overall features and spatial layout.

From the dialectical and historical perspectives, correctly recognize and deal with the development and changes of the Imperial City, and treat new constructions in a discriminate and cautious way.

Following the idea of "caring for man" and "sustainable development", try to improve the living, working and environmental conditions within the Imperial City.

6.4 Basis of the Plan

"Urban Planning Code of the People's Republic of China" (December, 1989)

"Cultural Legacy Conservation Code of the People's Republic of China" (November 2002)

" Ordnance of Conservation of Cultural Relics in Beijing" (June, 1987)

"Master Plan of Beijing Construction (1991-2010)"

"The Written Reply by the State Council to the 'Master Plan of Beijing'" (October, 1993)

"Conservation and Control Scope Plan for the Conservation Districts of Historic Sites in Old City of Beijing" (April, 1999)

"Detailed Plan of Central Beijing" (September, 1999)

"Scope and Construction Control Regulations for Cultural Relic Sites in Beijing" (1994)

"Conservation Plan for the 25 Conservation Districts of Historic Sites in Beijing" (March, 2001)

"Conservation Plan for the Historic City of Beijing" (April, 2002)

7. Conservation Plan

As history is a dynamic process, the Imperial City plan should be based on fully respecting the urban history, and reflecting the cultural value of the Imperial City. On the one hand, it should strictly protect the traditional urban space and architectural cream; on the other hand, it should not affect the normal life of inhabitants and the development and social progress there. It should try to delineate a series of historical pictures by demonstrating the evolutionary process of the Imperial City.

7.1 Zoning and Numbering of Plots

According to the features of land use and the natural layout of streets within the imperial city, as well as the principle of going from south to north, and east to west, the land of Imperial City is divided into 10 districts.

01 District: area east of the Forbidden City and the Laboring People's Palace of Culture, and south of Wusi Street within the imperial city;

02 District: area east of Di'anmennei Street, Jingshandongjie Street and north of Wusi Street;

03 District: area composed of the Forbidden City, the Zhongshan Park and the Laboring People's Palace of Culture;

04 District: area where Coal Hill Park is located;

05 District: area east Zhongnanhai, south of Wenjin Street, and west of the Forbidden City and the Laboring People's Palace of Culture;

06 District: Area east of Beihai Park, north of Jingshanqian Street, west of Di'anmennei Street and Jingshanxijie Street;

07 District: area where Zhongnanhai is located;

08 District: area where Beihai Park is located;

09 District: area south of Wenjin Street, and west of Zhongnanhai;

10 District: area north of Winjin Street and west of Beihai.

There are 572 planned plots of various kinds in the 10 zones, which will be renovated according to their numbers.

Characteristics of Each Zone

01, 02, 05, 06, and 09 are areas with concentrated traditional courtyard and single-storied housing, and are dominated by allays as well. 03, 04, 07, and 08 are located key cultural relics sites which are the best of imperial architectural complex featuring traditional style. They play a decisive role in protecting the Imperial City. Number 10 is an area with many new multi-storied buildings, having the most negative impact on the Imperial City.

7.2 Plan for Land Use Functions

Principle of land use: as a conservation district of traditional style and a complete core district of ancient city that has witnessed the history of urban development, the planned district should stick to the principle of protecting the information of city history and perfecting the urban functions in drafting its land use principle. So its plan for land use functions must bases itself on adjusting functions and adhere to the following principles:

7.2.1 Try to conserve the urban functional land that can continue history, and is in tune with the nature, functions and forms of the conservation district.

7.2.2 Expand the conserved land, with a view of creating conditions for further development in the strictly-controlled conservation district of Imperial City.

7.2.3 Try to adjust the urban land that is incongruous with the nature, functions, forms and environment of ,or has done damage to the conservation district.

Land use functions in the Imperial City is divided into 12 categories, including office, business and finance, cultural and entertainment, medicare and health, education and research, religion and welfare, ordinary housing, first-grade housing, daycare centers, primary and secondary schools, municipal installations, greenbelt, and roads and squares.

Based on the land covered, they are: 226 hectares of greenbelt accounting for 33% of the area of the imperial city; 123 hectares of residential quarters accounting for 18%; 124 hectares of recreational land accounting for 18%; 85 hectares of roads and squares accounting for 12%; 67 hectares of office buildings accounting for 12%, and others accounting for 9%.

The plan gives priority to adjustment with areas of present park greenbelt, office buildings and land for medical facilities unchanged.

In principle, land for industrial use and warehouse that are not in tune with the nature of Imperial City will be moved. There are only two large pieces of land which are used by Microelectronic Plant, Low-Voltage Electric Plant and Jinghua Printing House. The plan will remove these pieces of industrial land by dismantling the plants and changing the nature of land use. The vacated land will be turned into residential house and entertainment area. At present, the change of land use nature in parts of these areas have taken place, but the scale and forms are in not congruous with the environment style of the conservation district, and pose direct disturbance to surrounding relics and street landscape. So they should be brought under strict control.

A 10-meter wide greenbelt along both sides of Di'anmennei Street (within 50 meters of the road red line) will be planned; another 5 to 10 meter greenbelt in the

地面积基本不变。

原则上外迁与皇城性质不符的工业用地和仓储用地。保护区内仅有两处规模较大的工业用地，分别为微电机厂、低压电器厂用地和京华印刷总厂用地。规划通过拆除厂房、改变用地性质的方法，取消工业用地，用地性质调整为兼容居住和文化娱乐的用地。现状该工业用地已出现部分土地使用性质变更现象，但建筑尺度和形式不符合历史文化保护区的环境风貌，对周边文物和街道景观有直接干扰，应严格控制。

将地安门内大街两侧（道路红线50m范围内）规划各10m宽的绿地；景山后街北侧规划约5～10m宽的绿地；拆除景山西街东侧、景山公园西墙外侧的房屋，恢复成绿地或河道。

在原御河古河道所在的位置，规划一条宽15～20m宽的绿带，严格控制相关建设，恢复古御河。

由于朝阜干线是一条西起阜城门东至朝阳门穿过北京旧城的交通干道，是独特的城市景观大道，汇集人文景观、自然风光和商业市肆。其经过皇城区域内沟通西什库教堂、北海、景山、北大红楼，以人文胜迹、自然景观为主。因此，保护规划应强化这种特质，调整区内住宅、公建混合用地。规划调整西安门内大街两侧用地为绿化用地，保留少量商业设施。

在皇城内不应设置大规模的商业设施，规划将为街区服务的小型商业设施分别安排在：01区东华门大街两侧；02区东板桥大街、纳福胡同两侧；05区西华门大街两侧；06区陟山门街两侧；09区文津街一侧；10区大红罗厂街一侧。

结合我院编制的全市教育资源整合规划，将本地区原有小学、中学和职高予以合并调整，取消了编制不健全和配套设施严重不足的学校。

对占用皇城内的文物保护单位且使用不合理的单位，应积极创造条件进行外迁，改变用地性质为文化娱乐等用地，从根本上改善文物的使用环境；如大高玄殿、智珠寺等。

规划用地平衡表

用地代码	用地性质	用地面积(hm²)	百分比(%)
C1	行政办公	66.89	9.80%
C2	商业金融	17.43	2.55%
C3	文化娱乐	123.55	18.10%
C5	医疗卫生	13.23	1.94%
C6	教育科研	2.54	0.37%
C9	宗教福利	3.26	0.48%
R	住宅	111.4	16.32%
R1	一类住宅	11.54	1.69%
R5	中小学、托幼	21.29	3.12%
U	市政设施	1.3	0.19%
G	公共绿地	226	33.10%
S	道路广场	84.33	12.35%
合计	规划范围	682.7	6100.00%

7.3 人口疏解

保护区内现状人口密度已达570人/hm²，应严格调控。迁出具有历史、文化和艺术价值的居住院落的人口；保持原居住人口密度低于4人/100m²以下的院落的居住密度；控制原居住人口密度在4～7人/100m²的现状院落居住人口的发展；降低现状居住人口大于7人/100m²的院落到7人/100m²以下；并停止建设3层及以上的居住建筑。力争达到居住用地内居住人口的密度在4人/100m²以下，使皇城历史文化保护区内人口控制在4万以下。

7.4 道路和胡同体系规划

根据《北京历史文化名城保护规划》中旧城道路调整原则，在风貌保护核心区以及与文物建筑有冲突的地段，各级道路的走向和空间尺度必须在服从保护要求的前提下兼顾交通功能，其建设标准应有别于其他地区，但应为发展公共交通提供支持条件。

旧城道路网调整中涉及到的皇城内的主要城市道路有：朝阜干道

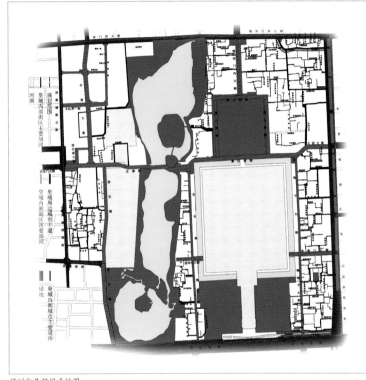

规划街巷胡同系统图
Systematic map of streets and hutongs planning

（西安门内大街——五四大街段）、旧城中轴路（地安门内大街、景山后街、景山东西街、南北长街、南北池子大街）

7.4.1 朝阜干道

朝阜干道原规划红线宽度70m，其中的文津街、景山前街、五四大街等历史文化保护区内的主要街道维持现状道路宽度，不再拓宽。

7.4.2 旧城中轴路

旧城中轴路的道路性质为城市主干路，规划红线宽度从35m到80m不等。穿越皇城的景山后街、景山东街、景山西街、南北长街、南北池子，现状道路宽度基本保持不变。考虑到特勤的要求，力争布置3条机动车道，其中两条车道供社会车双向行驶，另一条做为公交专用道。上述道路自行车高峰小时流量为700辆左右，可以考虑适当拓宽人行道，将自行车布置在人行道上。

与皇城直接关联的道路和胡同体系可分为四类：

（1）皇城周边的城市干道，包括长安街、平安大街、南北河沿大街、西黄（皇）城根南北街、灵境胡同和府右街6条。

（2）皇城内部的城市主要道路，包括地安门内大街、景山前街、后街、东街、西街、南北长街、西华门大街、南北池子、东华门大街、五四大街、文津街、府右街、西什库大街、大红罗厂街、爱民街，共计15条。

area north of Jianghshanhou Street will also be planned; the houses east of West Jiangshan Street and outside the west wall of the Jiangshan Park will be demolished and replaced by a greenbelt, paving the way for recover the historical river course.

A 15 to 20-meter-wide greenbelt will be planned at the original imperial river, where construction shall be strictly restricted with a view of recovering the imperial river.

The Chao-Fu traffic line runs through old Beijing from Fuchengmen in the west and to Chaoyangmen in the east. As a unique urban line, it encompasses many cultural, natural scenes as well as commercial centers. When it reaches the Imperial City, it will pass Xishiku Church, Beihai and Jiangshan Parks, the red building of former Peking University and many other cultural and natural spots. So this characteristic should be accentuated by adjusting the land where residential and public buildings are located. The plan will turn the land on both sides of Xi'anmennei Street into green belt, with a few commercial facilities untouched.

Large commercial business should not be place in the Imperial City according to the plan, so small businesses serving the neighborhood community will be arranged as follows: both sides of Donghuamen Street in 01 District; both sides of Dongbanqiao Street and Funan hutong in 02 District; Both sides of Xihuamen Street in 05 District; both sides of Zhishanmen Street in 05 District; one side of Wenjin Street in 09 District; and one side of DaHongluochang Street in 10 District.

Referring to the resources combination plan of the city education drawn up by our commission, we will merge and adjust the primary, secondary and vocational schools in this areas in order to get rid of schools that are inadequate in size and facilities.

For units that are occupying and abusing the cultural relics in the Imperial City, conditions must be created to resettle them elsewhere so that the nature of land use will be turned into that for cultural and entertainment activities, and the environment for relics such as Dagaoxuan and Zhizhu Temples will be fundamentally improved.

Balance Diagram of Planned Land

Code of	Land Use Nature	Land usearea (hectare)	Percentage (%)
C1	Office building	66.89	9.80%
C2	Business and finance	17.43	2.55%
C3	Cultural entertainment	123.55	18.10%
C5	Medical care	13.23	1.94%
C6	Education and research	2.54	0.37%
C9	Religion and welfare	3.26	0.48%
R	Residential housing	111.4	16.32%
R1	First-grade residential housing	11.54	1.69%
R5	Primary, middle schools and kindergarten	21.29	3.12%
U	Municipal installations	1.3	0.19%
G	Public greenbelt	226	33.10%
S	Roads and squares	84.33	12.35%
Total	Extent of Plan	682.7	100.00%

7.3 Resettlement of Residents.

The current population density in the conservation district has reached 570 per hectare, so strict control must be imposed. People living in courtyards that have historic, cultural and artistic values should be evacuated. Measures should be taken to maintain the 4 person per meter density, control the development of courtyards that have the density of 4-7 person per meter, reduce the courtyard of more than 7 person per meter to 7 persons per meter, and stop constructing residential buildings of three or over three-stories. Try to limit the density to 4 persons per meter, and the population within the conservation district of Imperial City to 40,000.

7.4 Plan for Road and huong Systems

According to the adjusting principle of the roads in old city prescribed in Conservation Plan of Historic City of Beijing, in the conservation district and sections that are in conflict with cultural relic architecture, the direction and spatial scale of roads of various levels must have traffic functions in compliance with the requirements of conservation. Their construction should be differentiated from other areas, but provide supporting conditions for developing public transportation.

The main roads in the Imperial City involved in the adjustment of road network in the old city include Chao-Fu Road (Xi'anmennei Street-Wusi Street section), axis road of the old city(Di'anmennei Street, Jingshanhoujie Street, Dong and Xi Jingshan Streets, Nanchang Street, Beichang Street, Nanchizi and Beichizi Streets).

7.4.1 Chao-Fu Road

The original width of the planned red line for Chao-Fu Road is 70 meters, out of which the principal streets in the conservation district such as Wenjin, Jinghshanqian and Wusi Streets will not be expanded but keep their current width.

7.4.2 The Axis in the Old City

The nature of the axis line in the old city is an urban principal road, whose planned width of red line varies from 35 to 80 meters. Running through Jingshanhoujie, Dong and Xi Jingshan streets, Nanchang and Beichang Streets, Nanchizi and Beichizi Streets, it will keep its current width unchanged. Three lanes of motor vehicle will be installed on this line, two for two-way vehicles, one for public transportation exclusively. The flow of traffic for bicycles during rush hour on the above-mentioned road is about 700 units per hour, so sidewalks could be widened to accommodate bicycles.

The road and allay systems closely related with the imperial city can be divided into 4 categories:

(1) the 6 main roads around the Imperial City: Chang'an Avenue, Ping'an Street, Nanheyan and Beiheyan Streets, Nanbei Street at Xihuangchenggen, Lingjing Allay and Fuyou Street.

(2) The 15 main roads within the Imperial City: Di'anmenei Street, Jingshanqian Street, Jingshanhou Street, Jingshandong Street, Jingshanxi Street, Nanchang and Beichang Streets, Xihuamen Street, Nanchizi and Beichizi Streets, Donghuamen Street, Wusi Street, Wenjin Street, Fuyou Street, Xishiku Street, Dahongluochang Street, and Aimin Street.

(3) Main hutongs located in the streets within the Imperial City include Zhishanmen Street, Shatanbei Street and Caolanxi hutong, Qinlao hutong, the total number of which is 60 and are wider than 7 meters.

(4) Minor hutongs within the Imperial City stand at 105, such as Ji'ansuoyouxiang, Qiangzhai, etc., which are narrower than 7 meters.

To keep the overall spatial scale and old boundary of the Imperial City, the urban roads around the Imperial City should pay attention to the extant imperial walls (under or above ground), strictly control the damage of walls brough about by the expansion of roads, and make room for future exhibit of the walls. The width of the roads within the conservation district should remain unchanged. Ancient style cannot be undermined and valuable architecture cannot be torn down because of the widening of roads.

Traffic construction in the conservation area should make use of the hutongs of various widths, respect the fabric of street blocks, and try to sustain the current road system status quo based on the hutong framework. Priority should be given to fully making use of hutongs exceeding 7 meters in width. Some of the dead-end and narrow hutongs can be linked together to make room for traffic in the conservation district. And the specific planning requirements are as follows:

Hutongs less than 3 meters in width are mainly for pedestrians and non-motor vehicles. Those that are 3 to 5 meters in width are also mainly for walking and non-motor vehicles, but can also serve as one-way road for occasional motor vehicles that can provide service for local people. Hutongs that are 3 to 7 meters in width can be used as one-way road for occasional motor vehicles for the benefits of the local residents. Seven to 9 meter wide hutongs can become two-way traffic road for motor vehicles, not only providing service for the local residents, but also shouldering part of the traffic load of the city. Hutongs as wide as or wider than 9 meters can shoulder the local traffic load.

In the process of renovation of crumbling houses, the location, layout, direction, and names of current hutongs should be protected and maintained.

To satisfy the convenience of inhabitants' activities, and to meet the requirements of the improvement of municipal installation, it is allowed to open up and connect some hutongs, but in general, conservation requirements must be met.

7.5 Afforestation Plan

The afforesting system within the Imperial City can be divided into 4 levels:

First is the afforestation of large parks, referring to the Jiangshan Park, Zhongshan Park, Laboring People's Palace of Culture, Beihai Park, Zhongnanhai Park (not open

(3) 皇城内部街区主要胡同，如陟山门街、沙滩北街、草岚子胡同、勤劳胡同等，总计有60条，胡同宽度一般大于7m。

(4) 皇城内部街区次要胡同，如吉安所右巷、前宅胡同等，总计约有105条，胡同宽度一般小于7m。

为保证皇城的整体空间尺度和旧有边界，皇城周边的城市道路应注意保护皇城现存城墙（地上和地下部分），严格控制道路拓宽对城墙的破坏，并为未来城墙展示留有空间。区内的城市主要道路应维持现状道路宽度不变，不应因拓宽道路而破坏风貌或拆除有价值的建筑。

历史文化保护区内的交通组织应充分利用现有不同宽度的胡同系统，尊重街区的街巷肌理，以胡同框架为基础，基本维持道路系统现状；特别是7m以上的胡同，更应在规划中充分利用。可适当打通一些尽端胡同，拓宽一些"瓶颈"胡同以改善区内的交通组织状况。其具体规划要求如下：宽度小于3m的胡同为步行和非机动车道路；宽度为3~5m的胡同主要是步行和非机动车道路，同时可以是非穿行性机动车单向道路，为就地的居民服务；宽度为5~7m的胡同为非穿行性机动车单向道路，为就地居民服务；宽度为7~9m的胡同可以组织为机动车双向道路，除了为就地居民服务外，适当承担局部地区的穿行性交通；宽度大于9m的胡同可适当承担局部地区的城市交通。

在危房改造中，应当尽量保护现有胡同的位置和格局，维持胡同的基本走向，保护并延续传统的胡同名称。

根据居民出行或改善街区市政设施条件的要求，允许适当打通或拓宽个别胡同，但总体上要满足保护的要求。

7.5 绿化规划

皇城内的绿化系统可分为四个层次：

一是大型公园绿化，指景山公园、中山公园、劳动人民文化宫、北海公园、中南海（暂不开放）。这类绿化中，有四个是对外开放的公园，一个是不对外开放的中央所在地（中南海）。历史上它们曾是皇家祭祀和娱乐之所，公园内部有大量的古树、大树和绿地。五个公园总占地面积约207hm²，占皇城面积的30%。

二是小型公园，指新近建设的东皇城根遗址公园和菖蒲河公园，占地面积约7.7hm²，占皇城面积的1.1%；

三是小型集中绿地，指规划的景观绿地和社区绿地。景观绿地包括沿中轴线（地安门内大街、景山前街、后街、东街、西街）的带状绿地，沿古御河河道绿地、沿平安大街、西黄（皇）城根南北街、灵境胡同的带状绿地和筒子河的西北、东北、东南角绿地，象征古古安门、地安门所在地的绿地、普渡寺绿地；社区绿化指分布在街区中的小型绿地。小型集中绿地总占地面积约11.3hm²，占皇城面积的1.7%。四是沿街行道树和分布在街区、

四合院中的古树和大小树木。皇城中成百上千的树木构成了独特的自然景观，是营造皇城生态环境的重要组成部分。应积极保护树木，杜绝砍伐行为。

以上四个层次的绿化，构成了皇城的绿地系统，应加以保护。

7.6 保护体系

在皇城的发展历史中，不难找寻到城市发展的印记。虽然，有些印记已经淡化或消失，但依然能整理出发展的脉络。在现代城市建设中着经营、加强深化发展脉络，就能延续和继承城市历史。保护体系在明确四个层次的保护对象的基础上，有机地组合各构成皇城整体城市意象的组成元素，使人体会到皇城地区城市文脉的延续与跳跃。

保护元素：

● 气势恢弘的帝王之城（故宫、社稷坛、太庙、景山）
● 风景优美的皇家园林（北海、中海、南海）
● 御用坛庙汇集之所（昭显庙－雷神（道教）、宣仁庙－风神（道教）、凝和庙－云神（道教）、永佑庙－皇城城隍）
● 多民族宗教活动场所
（大高玄殿－皇家道观，嵩祝寺－明代的汉经厂和番经厂遗址，为蒙古活佛章嘉胡图克图在京修建的宗教场所（藏传佛教）、普度寺－原为睿亲王多尔衮府，康熙三十三年改建为玛哈噶喇庙（藏传佛教）、万寿兴隆寺－明为兵仗局佛堂，康熙年间改为万寿兴隆寺、福佑寺－班禅驻京办事处（藏传佛教）、大马关帝庙－又名汉寿亭侯庙（道教）、三官庙（道教）、华严寺遗址、西什库教堂（天主教）、天主教共进行会、协和教堂旧址）
● 衙署库坊遗存之地（吉安所－明代为司礼监，清代为公眷停灵处、明番经厂、汉经厂遗址、清御史衙门、雪池冰窖、帘子库遗址、会计司旧址、内务府铁库、升平署、饽饽房（传））
● 皇室服务人员之居所（李莲英宅（传说）、太监宅若干处）
● 自发形成之街巷胡同（各历史时期的延展演变）
● 教书育人之各类学府（京师大学堂、北大红楼、北大地质馆旧址、盛新中学与佑贞女中旧址、西什库小学、北京四中）
● 民主革命运动之地（五四广场、草岚子监狱旧址）
● 古今中外名人故居（毛泽东故居、陈独秀旧居、张自忠故居、庄士敦故居、邓小平故居（传）、胡适故居（传）、伊斯雷尔·爱泼斯坦故居（传）、汉斯·米勒故居（传））
● 其他碑刻、地下埋藏、历史河道遗迹

皇城经过几百年的沧桑岁月，保存至今，仍然具有独特的魅力和无

皇城的各保护要素
Conservation components for the Imperial City

at the moment), which cover altogether an area of 207 hectares, accounting for 30% of the area of the Imperial City. These park, 4 of them open to the public, one is not as it is the seat of the Central Government (Zhongnanhai), used to be venues for emperors to enjoy themselves and worship gods. Inside parks are many tall and old trees as well as greenbelt.

The level is small Parks which refer to the Ruins Park at sections close to the east wall of the imperial palace and Changpuhe Park. The two cover an area of 7.7 hectares, account for 1.1% of the areas of the Imperial City.

The third level is small and concentrated greenbelts including planned landscape greenbelts and green areas scattered in the neighborhoods, covering an area of 11.3 hectares, accounting for 11.3% of the area of the Imperial City. The landscape greenbelts are in the following places: axis line (Di'anmennei Street, Jingshangqian Street, Jiangshanhou Street, East and West Jiangshan Streets); along the ancient imperial river; along the Ping'an Street, Nan and Bei Streets at West Huangchenggen and Lingjing hutong; at the northeastern, northwestern and southeastern corners of Dongzi River; at places symbolizing ancient Xian'an men and Di'anmen; and at Pudusi Temple. Neighborhood greenbelts refer to those scattered in the communities.

The fourth level are sidewalk old trees and other kinds of trees in the neighborhoods, and courtyards, which must be protected because these trees, hundreds of thousands in number, constitute a vital part in creating an ecological environment unique natural landscape in the Imperial City.

These levels of afforestation constitute the greening system in the Imperial City, so must be protected.

7.6 Conservation System

It is not difficult to find out the traces of urban development in the evolution of the Imperial City. Although some of them have been faded, or disappeared, their clues of development can still be ascertained. So long as these traces or clues are worked upon and enhanced in the process of urban modernization, the continuation and progress of cities will go on.

On the basis of clarifying the 4 protection targets, one can sense the vitality and continuation of the Imperial City by integrating various elements that constitute the generic urban image of the Imperial City.

Conservation elements:
● Magnificent city of kings (Forbidden City, Altar to the God of the Land and Grain, Imperial Ancestral Temple, Jiangshan Park)
● Imperial Parks with beautiful scenery (Baihai, Zhonghai and Nanhai)
● Concentrated place of imperial temples (Zhaoxian Temple[God of thunder, Taoism], Xuanren Temple [God of Wind, Taoism], Ninghe Temple [God of Cloud, Taoism], Yongyou Temple [town-god of the Imperial City].
● Places of religion of multi-nationalities

Dagaoxuan Hall--imperial Taoist temple, Songzhu Temple-ruins of the Hanjingchang and Fanjingchang of the Ming Dynasty, also a religious venue built in Beijing by Mongolian living Buddha Zhangjiahutuketu (Tibetan Buddhism), Pudu Temple-former mansion of Prince Duo'erguan, turned into a Lama temple in the 33nd year of emperor Kangxi (Tibetan Buddhism), Wanshouxinglong Temple-former Buddhist temple of the Weaponry Board of the Ming Dynasty, turned into Wanshouxinglong Temple in the reign of emperor Kangxi, Fuyou Temple-Panchen Lama's office in Beijing (Tibetan Buddhistm), Damaguandi Temple-otherwise called Hanshoutinghou Temple (Taoism), Sangguan Temple (Taoism), Relics of Huayan Temple, Xishiku Church (Catholicism), and Relics of Xihe Church.

● Offices and Storehouses: Ji'ansuo-former department of rites of the Ming period, turned into mortuary for imperial families in the Qing Dynasty, Relics of Haningchang and Fanjingchang of the Ming period, Imerial Censor Office of the Qing period, Xuechibing Storehouse, Relics of Lianziku, Sited of former Accounting Department, Blacksmith Shop of the Office of Interior Household, Shengping Department, and Pastry Shop.
● Residences of imperial attendants (Lianlianying's mansion [rumored], several mansions of eunuchs).
● Naturally formed streets and hutongs (evolution and changes of various periods)
● Education institutions (Metropolitan University, Red Building of Peking University, Former Site of Geology Hall of Peking University, Former Sites of Shengxin Middle School and Youzhen Women's Middle School, Xishiku Elementary School, Bejing No. Middle School).
● Places for democratic revolution (Wusi [May Fourth Movement] Square, Former Site of Caolianzi Prision).
● Residences of famous people (former residences of Mao Zedong, Chen Duxiu, Zhang Zizhong, Zhuang Shidun, Deng Xiaoping [alleged], Hu Shi [alleged], Isbel Ainstain [alleged], Hans Miller [alleged]).
● Others such as monuments, buried artifacts and historic rivercourses.

The Imperial City still boasts unrivalled charm and historic value even after hundreds of yeas of vicissitudes. Four categories can be divided from the perspective of conservation system.

7.6.1 Cultural Relic Places under the Protection of Various Levels

Cultural relic sites of various levels within the Imperial City total 63, covering an area of 369 hectares, accounting for 54.1% of the area of the Imperial City. Nine sites are under the State protection, covering an area of 234 hectares and accounting for 63.4% of the total area of cultural relic places including Palace Museum, Imperial Ancestral Temple, and Jiangshan Park; 18 are under the municipal protection with an area of 127 hectares, accounting for 34.4% of the total area of cultural relic sites including Zhongnanhai, Great Hall of Pudu Temple, and Xuanren Temple; 6 are under the protection of districts with an area of 3 hectares, accounting for 1% of the total area of cultural relic places including the Society of Returned Overseas Scholars, Imperial Wall and Wanshouxinglong Temple; and 30 are registered in the general survey with an area of 5 hectares, accounting for 1.3% of the total area of cultural relic sites such as Xishiku Elementary School, Relics of Accounting Department and Jingmo Temple.

It is obvious that the cultural relic places under the protection of various levels account for half of the total area of the Imperial City, so they constitute the core content of conservation. At the moment, the urgent task for conserving the relics within the Imperial City is renovation, maintenance, vacation and sensible utilization. All kinds of construction related with relics must be managed in line with the requirements of conservation range and control belt of relic places at various levels (including relics embraced in general surveys).

7.6.2 Architecture or Courtyards of Historic and Cultural Values

The number of courtyards of various sizes in the Imperial City stands at 3,264, out of which 204 have certain degree of historical and cultural values (non-relics), accounting for 6.3% of the total courtyards of the Imperial City. They cover an area of 21 hectares, which equals 3.1% of the area of the Imperial City.

These kind of courtyards refers to traditional architecture that have not been

Former sites of Shengxin Middle School and Youzhen Girls' Middle School

与伦比的历史文化价值。从保护体系的角度，可以分为如下四个类别。

7.6.1 各级文物保护单位

皇城内拥有各级文物保护单位63个，总占地面积约369hm²，占皇城面积的54.1%。其中，国家级文物保护单位9个，占地面积约234hm²，占文保单位总面积的63.4%，如故宫、太庙、景山等；市级文物保护单位有18个，占地面积约127hm²，占文保单位总面积的34.4%，如中南海、普渡寺大殿、宣仁庙等；区级文物保护单位有6个，占地面积约3hm²，占文保单位总面积的1%，如欧美同学会、皇城墙、万寿兴隆寺等；普查登记在册文物有30个，占地面积约5hm²，占文保单位总面积的1.3%，如西什库小学、会计司旧址、静默寺等。

可以看出，各级文物保护单位占皇城总面积的一半以上，是皇城保护的核心内容。目前，文物的修缮、维护、腾退、合理化使用是皇城内文物保护工作的紧迫任务。与文物相关的各类建设活动必须按照各级文保单位(包括普查项目)的保护范围和建设控制地带的要求进行管理。

7.6.2 有历史文化价值的建筑或院落

皇城内共有各类大小院落3264个，其中具有一定历史文化价值的建筑或院落(非文物)有204个，占院落总数的6.3%；这些建筑或院落占地约21hm²，占皇城面积的3.1%。

所谓有历史文化价值的建筑或院落，是指那些尚未列为文物保护单位，但建筑形态或院落空间反映了典型的明清四合院格局或近代建筑特征的，具有真实和相对完整的历史信息的传统建筑。它们有的曾是寺庙，有的曾是衙署库坊，有的曾是官宦府邸，有的是名人故居或大户居所，由于保存较好，成为历史某一阶段生活的真实写照。

有历史文化价值的建筑或院落和各级文物保护单位相加，占地约390.6hm²，占皇城面积的57.2%。

目前，这部分建筑或院落由于没有列入文保单位，正面临着被损坏或破坏的危险。规划要求按照文物的保护要求对其进行保护，并挂牌向公众明示。对其进行修缮、维护、腾退和合理化使用，同样也是皇城保护工作的一项紧迫任务。

7.6.3 城墙、坛墙和水系

尽管皇城的主要城墙、城楼在民国时期已被拆除，许多王府、坛庙的院墙也残缺不全，一些历史河道已改为暗沟，但从现在依然保留的一些墙体和水系，仍可窥见往日皇城的辉煌。

现状皇城城墙南部保存较好，东部随着皇城根遗址公园的建设已对部分城墙和城门遗址进行展示。沿平安大街南侧、西黄（皇）城根东侧、灵境胡同北侧结合环境整治，规划一条不小于8m的绿化带，作为西、北皇城边界的象征。

据初步统计，皇城内现保存的城墙、坛墙遗存约41处，其中一些依托于文物保护单位的墙体保存完好，如景山、故宫等；有一些已被埋没在密集的平房四合院建筑群落中，缺乏良好的展示环境，伴随着危房改造和环境整治，应对这些历史墙体加以保护、修缮和展示，改善其周边环境。

历史上皇城内的主要水系北海、中南海、筒子河、金水河至今保存完好；菖蒲河正在恢复中；御河、织女河、连接筒子河和菖蒲河的古河道、连接北海和筒子河的古河道，有的已被填埋，有的已改为暗沟。对现有的水面及周边环境必须严格加以保护，对一些有恢复价值和可能的古河道，如连接北海和筒子河的古河道、连接筒子河和菖蒲河的古河道和御河，对古河道用地上的建设应加以严格控制，为将来恢复古河道创造条件。

7.6.4 传统胡同

北京皇城源于元代，是为了满足皇帝的生活、起居、娱乐和工作之需而修建的，皇城内还有宫城，宫城之内是皇帝及其家室成员的居所，宫城之外是服务于皇室的各类机构、园囿。皇城空间辽阔、宏大，随着封建社会的衰落，皇城内的居住功能增强，才形成了现今曲折、自由的胡同体系，与皇城外整齐、规则的胡同、街巷体系形成鲜明的对比。

根据北京大学侯仁之先生对北京历史地理的研究成果，可以对皇城内胡同的发展演变得出初步认识：明代形成的胡同有16条，如刘兰塑胡同、陟山门街、光明胡同等；清代形成的胡同109条，如会计司胡同、沙滩北街等；民国时期形成的胡同有11条，如草岚子胡同、南红门胡同等。除现有的西什库大街、文津街等15条城市主要道路外，皇城内各个时期形成的有名称可考的胡同共有137条。

胡同和四合院是北京旧城的基本组成单位，缺一不可，因此必须进一步提高保护传统胡同的意识，危房改造不能无视传统胡同的存在。

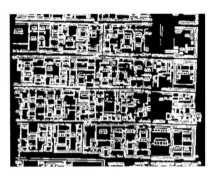

皇城内胡同与元大都胡同遗存对比
Comparison of hutongs within the Imperial City with those of the capital of the Yuan Dynasty

7.7 建筑高度控制

皇城是北京旧城保护的核心，严格保护其传统的平缓、开阔的空间形态是皇城保护的中心任务之一。由于历史原因，在皇城内建设了许多多层和高层建筑，严重地破坏了皇城优美的空间格局和形态。1999年编制的《北京市区中心地区控制性详细规划》由于没有将皇城整体作为历史文化保护区，因此将皇城的部分地区划定为18m建筑高度控制区。为了加强对皇城的整体保护，进一步认识建筑高度控制对皇城保护的重要性，必须严格皇城内的建筑高度控制，重新规定建筑高度控制标准。

在皇城内，对现状为1~2层的传统平房四合院建筑，在改造新建时，建筑高度应按照原貌保护的要求进行，禁止超过原有建筑的高度；对现状为3层以上的建筑，在改造更新时，新的建筑高度必须低于9m。

总体来说，除了严重破坏皇城空间景观的少数多、高层建筑必须拆除外，一般情况下对现状多层建筑可采取暂时承认其存在，通过将平屋顶改造成坡屋顶的整饰手段，使其与传统风貌相协调。当它们重建时再按照新的建筑高度控制标准执行。对现状为平房四合院建筑的地区，必须停止审批建设3层及3层以上的房屋，以确保新的建设对皇城空间不造成新的破坏。

7.8 保护与更新方式

综合现状建筑质量和建筑风貌的评估，皇城内建筑的保护与更新方式可分为6类：

7.8.1 文物类建筑

指国家级、市级、区县级以及普查登记在册文物，占地约369hm²，占皇城面积的54%。这类建筑必须严格按照国家或北京市文物保护的

Portions of poorly preserved wall

listed as cultural relic sites under conservation, but their shapes and courtyard spaces reflect typical courtyard layout of the Ming and Qing Dynasties and modern architectural styles, as well boast authentic and comparatively complete historical information. They used to be offices and storehouse, high-ranking officials' mansions, and residences of famous or wealthy people that have become authentic depictions of life of certain periods due to their well preserved statues.

The combined number of architecture and courtyards of historic and cultural values and the relics under the conservation of various levels covers 290.6 hectares, accounting for 57.2% of area in the Imperial City.

As they have not been filed as places under conservation, they are facing the risk of being damaged or destroyed. The plan required that they should be protected in line with the conservation requirements of cultural relics, and be marked as such. It also a part of the urgent task of conserving the Imperial City to conduct renovation, and maintenance of these courtyards. They should also be vacated of inhabitants and utilized in a sensible way.

7.6.3 City and Altar Walls As Well As Water System

Although the principal city walls and watch towers of the Imperical City have were torn down during the Republic period, many princes' mansions and altars are broken and some rivers have become covered sewerage, the former glory of the Imperial City can still be glimpsed through the extant wall and water systems.

The southern wall of the Imperial City is in good shape, the eastern wall and some gates there become more conspicuous owing to the construction of the Huangchenggen Relic Park. A no less than 8-meter-wide greenbelt will be planned south of Ping'an Street, east of sections close to the west wall of the imperial palace, and north of Lingjing hutong to make it the symbol of the western and northern borders of the Imperial City.

Initial statistics show that there are 41 broken walls in the Imperial City. Among them those under the State and municipal protections are well preserved such as the Forbidden City, the Jingshan Park, etc. Other temple, altar and residential walls were lost and scattered during the renewal of risky housing and environmental facelift amidst single-storied housing without any chance of being seen. They should be protected, updated and on display along the rectification of environment.

The main ancient rivers systems within the Imperial City such as Beihai, Zhonganhai, Dongzi River, and Jinshui River are well preserved, and Changpu River is being recovered. Some ancient river-courses have either been filled up or turned into covered sewerage, such as the Imperial River, Zhinu River, the ancient river linking Dongzi with Changpu Rivers, the ancient river linking Beihai with Dongzi Rivers, etc. Strict control must be imposed on these bodies of water and their surroundings Construction on those river courses that have the possibility of being salvaged such as the ancient river linking the Beihai with Dongzi River and that linking Dongzi with Changpu Rivers must be strictly restricted to make room for recovering the ancient rivers in the future.

7.6.4 Traditional hutongs

The Imperial City in Beijing, built in the Yuan Dynasty, was aimed at meeting the needs of life, entertainment and work of the emperors. Inside the City is the palace where emperors and their royal families lived, and around which are various institutions and parks that provide services to the royal family. The space of the Imperial City is huge and spacious and, with the decline of the feudal society, the living function of the Imperial City shrank, giving way to the current winding and free-floating hutong system, which makes a striking contrast with the neat and regular hutong and street systems outside City.

The following can be concluded on the evolution of Beijing's hutong based on the research on the history and geography of Beijing by Mr. Hou Renzhi of Peking University: the hutongs developed in the Ming period stand at 16 such as Liulansu, Zhishanmen, Guangming; those formed in the Qing period are 109 such as Kuaijisi, Shatanbeiji; and the Republic period produced 11 hutongs such as Caolanzi, and Nanhongmen. Except the 15 principal urban roads such as Xishiku and Wenjin Streets, the hutongs formed in various periods and bear names in the Imperial City total 137.

As an integral part of the old city of Beijing, hutongs and siheyuans are indispensable, so the awareness of protecting them must be further raised, and their existence cannot be ignored in the process of revamping of crumbling houses.

7.7 Control of Building Height

As the Imperial City is the key component of protecting the old city of Beijing, protecting its traditional flat and open spatial layout is the central task in the conservation of the Imperial City. Owing to historical reasons, many multi-storied and high-rises were erected in the Imperial City which severely damaged the beautiful spatial pattern and layout there. As the Detailed Plan of the Control of the Center of Being drafted in 1999 failed to designate the Imperial City as a conservation district as a whole, buildings of 18 meters were allowed to be built in some places inside the Imperial City. In order to strengthen the protection of the area as a whole, and recognize the importance of height control of buildings on the conservation of the Imperial City, imposition of height control must be exercised and standard of height control of buildings re-formulated.

During renovation in the Imperial City, the height of traditional single- and two-storied courtyard houses must be kept according to their original shapes. Height exceeding the original one will be prohibited. The height of buildings of or above three stories shall not exceed 9 meters.

Generally speaking, apart from those multi-storied and high-rises that severely undermine the landscape space of the Imperial City, and that must be torn down, most buildings of this height will be left as they are for the time being, but they must be transformed to be in agreement with the traditional style by turning their flat roofs into reclining ones. When they are reconstructed, new height control must be followed. In areas full of single-storied courtyards, buildings of or above three stories will not be approved in order to make sure that new constructions will not bring about damage on the traditional features of the Imperial City.

7.8 Conservation and Updating Method

In accordance with the quality and style evaluations of present architecture, the means of protecting and updating the buildings within the Imperial City can be divided into 6 categories.

7.8.1 Architecture of cultural relics

These refer to cultural relics under the State, municipal and district protections and are registered in the general survey. Covering 369 hectares and accounting for 54% of the area of the Imperial City, they must be protected and managed strictly according to the relevant regulations of the State and Beijing municipal government.

7.8.2 Architecture that should be protected.

They refer to buildings or courtyards that have certain degree of historical and cultural values, which must be protected and managed according to the relevant regulations of the State and Beijing municipal government. They cover 21 hectares and account for 3.1% of the area of Imperial City.

7.8.3 Architecture that should be renovated.

They refer to ordinary single-storied and courtyard structures in the neighborhood,

文物类建筑
Architecture of cultural heritage

保护类建筑
Architecture under preservation

改善类建筑
Architecture that should be rehibisitated

保留类建筑
Architecture that needs maintaining

相关法规进行保护和管理。

7.8.2 保护类建筑

指有一定历史文化价值的建筑或院落，占地约21hm²，占皇城面积的3.1%。这类建筑应参照国家或北京市文物保护的相关法规进行保护和管理。

7.8.3 改善类建筑

指街坊中一般的平房四合院建筑，占地约28hm²，占皇城面积的4.1%。这类建筑应以修缮、维护为主，对具有传统四合院空间形态的建筑，应尽量保持四合院的空间肌理。

7.8.4 保留类建筑

指街坊中和传统风貌较协调的新建筑，占地约38hm²，占皇城面积的5.6%。这类建筑大部分为新建的仿四合院建筑，与传统风貌较协调，应予以保留。

7.8.5 更新类建筑

指建筑质量很差，又没有传统风貌特征的破旧平房或对皇城传统空间具有破坏作用的多层或高层建筑，占地约21hm²，占皇城面积的3.8%。对破旧平房可采用改造的方式，恢复传统四合院的形态；对破坏性的多、高层建筑有条件时应予以拆除。

更新类中多、高层建筑的确定可遵循几个原则：
(1) 破坏文物周边环境的建筑；
(2) 破坏皇城内重要景观视廊的建筑；
(3) 破坏皇城中轴线景观，距中轴线较近的建筑；
(4) 破坏北海、中南海和故宫、景山联系区域的建筑。

7.8.6 整饰类建筑

指一时无条件拆除的多、高层建筑或沿主要街道、胡同的一些被与传统风貌不协调的广告牌、墙面砖、色彩所充斥的店铺，占地约66hm²，占皇城面积的9.6%。对这类建筑只有通过统一的设计规则，对屋顶、墙面材料、色彩等要素统筹考虑，进行整饰，以期达到与皇城风貌协调的目的。

7.9 市政设施规划

7.9.1 规划原则

由于皇城内部的城市主要道路应维持现状道路宽度不变，尽量保护现有胡同格局和肌理，倡导以院落为单位小规模地逐步改造的方式，强调危旧房改造的微循环，因此对于市政基础设施建设提出了新的要求。对于皇城内的市政设施的建设，不同于一般区域的建设，在以下四个规划原则中分别论述。

(1) 排水体制

皇城占地面积约6.8km²，其中包含第一批25片历史文化保护区中的14片，第一批25片历史文化保护区中的其他11片占地面积约为3.9km²，皇城和历史文化保护区总占地面积约为10.7km²，占市区面积的1%，占高碑店污水处理厂流域面积的10%。

现状皇城和历史文化保护区里的街坊内排水基本上以合流制为主，为保护胡同格局和历史风貌，现有胡同格局和宽度基本保持不变，因此会有部分胡同无法安排下雨污分流的管道。如果皇城和历史文化保护区街坊内的排水体制采用合流制，主要道路下安排雨污分流的排水管道，同时在外部市政雨污水管道和河道沿线设置截流设施。这样不仅可以截流污水和初期雨水，还清河道，还可以满足皇城和历史文化保护区保护的要求。因此，应因地制宜，在有条件的地区实行雨污分流的排水体制，不具备条件的地区实行雨污合流的排水体制。

(2) 能源结构

在皇城内应以清洁能源为主，杜绝燃煤小锅炉。现状皇城内热力管道覆盖的面积较大，主要供给单位和多、高层住宅。由于胡同宽度的限制，热力管道无法接入到一些平房区内，考虑到照明和炊事为百姓的生活基本需求，因此在供热管道无法覆盖的地区，可以采用电采暖或燃气采暖。有条件地区也可以采用太阳能。

(3) 胡同内市政管道布置

由于胡同宽度变化较大，有宽有窄，因此在布置市政管道时应统筹规划、全面考虑。局部路段市政管道如不能满足安全间距要求，应对管道采取特殊处理。胡同宽度较窄时，可以采用合流制排水管道。详见图24胡同内市政管道横断面图。

(4) 实施顺序

由于皇城保护强调原真性和逐步更新，不得采用大拆大建的方式，因此市政基础设施的建设上也要遵循皇城保护的要求。在外部市政道路上，能够结合城市基础设施建设计划进一步完善市政管网系统；在街坊内的主要胡同应结合皇城内的改建统一规划、设计、建设，一次性将市政管道实施到位。对于街坊内的次要胡同和院落里的市政设施，应会同院落的改建一并实施，这部分将是一个缓慢的、逐步的过程，不能操之过急。

7.9.2 市政设施规划

(1) 排水设施

按雨污分流制进一步完善皇城范围内的主要雨污水排除系统，将符合排水要求的合流管道用做雨水管道，解决现状管道能力不足、地势低洼、排水出路不畅等问题。在街坊内合流管道接入外部市政排水管道前修建污水截流设施。在皇城内的雨水管道入河前修建污水截流设施。在皇城内提倡设置"低绿地"(或降低绿地高度)，使雨水能够充分渗透，不仅可以涵养地下水，还可以减少城镇的洪涝损失。

(2) 供水设施

更新类建筑
Architecture that needs renovating

整饰类建筑
Architecture that needs decoration

which cover 28 hectares and account for 4.1% of the area of the Imperial City. They should be mainly renovated and rebuilt according to their original style. For those that boast spatial shapes of traditional courtyard, the courtyard fabric should be retained.

7.8.4 Architecture that should be kept.

It refers to new buildings that are in harmony with traditional features, most of which are pseudo courtyard housing. They cover 38 hectares and account for 5.6% of the area of the Imperial City. As they are in tune with traditional features, they should be retained.

7.8.5 Architecture that needs renovation.

It refers to dilapidated single-storied houses of poor quality which are not compatible with traditional character. It also includes multi-storied buildings or high-rises that have destructive effect on the traditional space of the Imperial City. For the former, renovation should be done in order to salvage the shapes of the traditional courtyard; for the latter, they could be demolished if needed. These buildings cover 21 hectares, accounting for 3.8% of the area of the Imperial City.

The following principles can be adhered to in the need to renovate multi-storied and high-rises:

(1) those that damage the environment of cultural relics;

(2) those that damage the important visual effect of landscape in the Imperial City;

(3) those that damage the landscape on the axis or too close to the axis; and

(4) those that damage the related areas of Beihai, Zhongnanhai, Palace Museum and Jingshan.

7.8.6 Architecture that needs rectifying.

Referring to the multi-storied buildings and high-rises that can no be torn down at the moment, as well as single-storied houses along the streets and allays which are covered with billboards, facing tiles and colors. They cover 66 hectares and account for 9.6% of the area of the Imperial City. These kinds of architecture must be revamped in terms roofing, wall materials and colors through unified design.

7.9 Plan for Municipal Installations

7.9.1 Plan Principles

New requirements for construction of municipal installation infrastructure have been set owing to the following considerations: the width of the main urban roads within the Imperial City will remain unchanged, the current fabric of hutongs will be retained as much as possible, the approach of gradual and small-scale renovation with courtyard as module is encouraged, and micro-circulation in revamping risky and old houses is highlighted..

(1) Drainage System

The area of the Imperial City is 6.8 square kilometers, including 14 districts of the first group of 25 conservation districts of historic sites. The area of the remaining 11 district of the 25 districts cover 3.9 square kilometers, thus, the total area of the Imperial City and the conservation districts is 10.7 square kilometers, accounting for 1% of the city, and 10% of the drainage area of Gaopeidian Sewage Treatment Plant.

The drainage in the communities of the Imperial City and conservation districts is still in the form of combined system. In order to protect the fabric and historical features of hutongs, and the current width and style of hutongs will basically remain unchanged, so some of the hutongs cannot use separate drainage system. If the neighborhoods of the Imperial City and conservation districts use combined drainage system, and pipelines of separate system are installed under principal streets and meanwhile, intercepting devices are installed in the combined system pipes outside the area and along the rivers, not only can the sewage and early rainwater be intercepted and river be cleared of dirt, protection requirements for the Imperial City and conservation districts can also be met. Therefore, separate system should be applied in areas where conditions allow, whereas combined system can be used in areas where conditions are not very good.

(2) Energy Structure

Clean energy should be used in the Imperial City, and coal-burning boiler must be banned. At present, the heating piping covers a large area in the Imperial City, providing heating primarily for units and multi-storied buildings and high-rises. As the hutongs are narrow, heating pipelines cannot reach some areas of quadrangles. As the needs of lighting and cooking of the inhabitants' life have to be met, heating generated by electricity or gas cannot applied in areas that have no access to heating pipes. Solar energy can also be use in areas where conditions are fit.

(3) Arrangement o Municipal Installations in hutongs

As hutongs vary in width, municipal installations should be conducted with a unified plan and overall consideration. If the installations in some parts of the roads cannot meet the requirements of safety distance, pipelines should be treated in a special way. When hutong are too narrow, combined system can be used. See cross-section diagram no. 24 for installations in hutongs.

(4) Order of Implementation

The construction of municipal installations must satisfy the requirements of the conservation of the Imperial City because original style and gradual renovation are stressed in the protection of the area in stead of dismantling and construction on a large-scale. On the outer roads, municipal piping network should be further perfected in coordination with the urban infrastructure construction, while in hutongs, installations should be in right positions once for all in a unified planning, design and construction in cooperation with the renovation of the Imperial City. For secondary hutongs in the neighborhoods and installations in courtyards, installation work must be carried out together with the renovation of the courtyards, which will be a slow and gradual process. Any undue haste would spoil the efforts.

7.9.2 Plan for Municipal Installations

(1) Drainage Facilities

The major rainwater and sewage draining system of the Imperial City shall be further perfected by the separate system, turning qualified combined pipes into rainwater draining pipes so as to solve the problem of insufficient pipes, low and wet terrain and blockage of sewer. Sewage interception devices will be installed before the combined pipes in the community are connected with the outside municipal piping. Sewage interception devices will also be installed before the rainwater piping of the Imperial City flow to the rivers. Imperial City is encouraged to cultivate "low greenbelts" (or lower the greenbelt) so that the rainwater can fully soak up the ground. In this way, not only underground water can be nourished, the losses of flooding in the city and can also be reduced.

(2) Water supply facilities

Gradually perfect the main and branch lines of water supply alongside the housing renovation and road construction in the Imperial City.

(3) Gas supply facilities

Gradually perfect the main and branch line system of gas supply and medium and low pressure adjusting facilities along with the housing renovation and road construction in the Imperial City. The performance capacity of heating by gas must be taken into account to meet the needs of the inhabitants.

(4) Heating facilities

Heating supply in the Imperial City should be dealt with differently in different circumstances. Areas that can make use of the urban heat supple network should fully utilize the central heating system. Areas that cannot be covered by heating pipes can use heating generated by gas or electricity. The load of heating should be taken into account in planning the electricity and gas supplies.

(5) Communication facilities

Gradually perfect the main and branch line systems of communication and the telecommunications bureaus along with the housing renovation and road construction in the Imperial City. Cable TV and Broadband networks should be planned and constructed with communication lines.

随皇城内的房屋改建和道路系统的建设，逐步完善供水干、支线系统。

（3）供气设施

随皇城内的房屋改建和道路系统的建设，逐步完善供气干、支线系统和中低压调压设施。将燃气采暖负荷考虑在内，以满足居民使用燃气采暖的需要。

（4）供热设施

皇城供热应结合具体情况分别对待。能够利用城市热力管网的地区应充分利用城市集中热力系统；热力管道无法覆盖的地区可以结合居民要求采用燃气采暖或电采暖。在供电设施和供气设施规划中均应考虑采暖负荷。

（5）电信设施

随皇城内的房屋改建和道路系统的建设，逐步完善电信干、支线系统和电信局所。有线广播电视网、宽带网络应与通信线路统一规划、建设。

（6）供电设施

随皇城内的房屋改建和道路系统的建设，逐步完善供电干、支线系统和开闭站、配电室；有条件地区应逐步将架空线入地，全部采用电缆线路。在沙滩规划建设一座220kv变电站。

8. 整体环境的整治

皇城曾是历史上具有明确实体（皇城墙）界定的区域范畴，发展到今天，要试图界定出完整的皇城保护区，必须建构起完整的区域意象，使人可明确感知到皇城的存在。

具体内容如下：

道路——构筑皇城区域内的主要道路、胡同系统。强化朝阜干线沿线的景观风貌特征，通过沿线用地功能调整给予明确界定，形成"传统居住——自然风光——宫殿群体——人文景观"的风景线。强化故宫东西两侧传统中轴路的文化旅游特征。

边缘——皇城的四至边界。通过城市设计手法明确界定皇城西侧和北侧边界。

区域——宫城、绿化和水体、传统街区和一般区域。在保护好以宫城为核心的区域景观外，强化传统街区的风貌特征，严格控制建筑风貌（高度、色彩、材料、造型等）保持原貌，整治街区内区别于民居的寺庙、衙署、厂库类历史建筑及其周边环境；对现状风貌不协调的一般区域进行整饰。

节点——天安门、地安门、东安门、西安门。重点处理除天安门、东安门遗址外的两处皇城入口空间。

标志点——中轴线上的重要建筑及皇城内其他具有地标性建筑及构筑物。结合文物保护传统和新增的各主、次要标志点，通过环境整治剔除现存不协调的建筑物和构筑物。

（注：原貌——指北京皇城发展历史上民国之前各个时期具有代表性和普遍意义，或标识历史事件的建筑物（群）、构筑物、构件、遗址等的突出特征如装饰风格、彩画建筑形制、体现本地区的传统与文化，以及其形成和发展的状况。）

8.1 皇城传统空间格局分析

皇城传统空间的特色集中体现下列要素：

封闭的皇城城墙；严谨的中轴线；雄伟的紫禁城；神圣的太庙和社稷坛；巍峨的景山；秀美的三海；以及起伏有序的宫殿、坛庙、塔寺和城楼。这些空间要素，保存到至今，成为皇城内的精华和核心。

一圈皇城城墙：高大、封闭而连续，明确界定了皇城和外部城市空间的关系，体现了天子的至高无上。

一条中轴线：由一系列城门、城楼、宫殿组成，经过天安门、端门、午门、太和门、太和殿、中和殿、保和殿、乾清门、乾清宫、神武门、景山万春亭、寿皇殿、地安门。

一座宫城：紫禁城，其总体规划和建筑形制体现了封建宗法礼制和象征帝王权威的精神感染作用。

两座重要的礼制建筑：太庙和社稷坛

一座风水山（景山）

一座园林（三海）

数个景观标志点（宫殿、坛庙、塔寺、城楼）

相关服务和居住区（皇家生活的服务功能区）

其中，前六项内容经历各时期的逐步建设，形制日趋完备，空间艺术效果更加突出，成为皇城内的精华和核心。而相关的服务和居住区随着时代变迁，产生了深刻的变化，是保护规划需要进行重点研究的区域。其功能简述如下：

在元代，该区域是宫殿和园囿的过渡区域。空间辽阔舒展，与皇城外里坊式的居住建筑布局形成鲜明对比。

在明代，该区域逐渐分布了专为皇室服务的管理机构、局、厂、作、库，经由内府管理，形成自给自足的小社会。

在清代，官办手工业衰落后，皇室物资需要转由民间供应，大量的内宫衙署逐渐萎缩，该区域转化为皇宫服务人员的居住地。这就是现今皇城城市肌理的主要形成时期。

在民国及以后时期，该区域性质不断变化调整。新增了部分功能，如行政办公、医疗、教育、工业等。这一时期的建设活动对皇城的传统城市风貌和肌理有一定程度的破坏。

皇城边界现状（北）
Present Imperial City boundary (north)

8.2 皇城现状空间格局分析

皇城现状空间格局的特征可从下列几个层次来表现：

8.2.1 皇城的边界

现状皇城的南边界、西南边界和东边界比较清晰；西、北边界由于皇城墙已不存在，成为皇城边界的模糊区域。现状皇城的边界特征：已由传统的封闭性、明确性转化为开敞性、模糊性。

皇城边界现状（东）
Present Imperial City boundary (east)

8.2.2 皇城的城门

传统皇城有天安门、地安门、东安门、西安门四座城门，作为地标联系皇城内外。现状只有天安门保存下来，

皇城边界现状（南）
Present Imperial City boundary (south)

(6) Electricity supply facilities

Gradually perfect the main and branch line systems of electricity supply and switching stations and distribution rooms along with the housing renovation and road construction in the Imperial City. Areas where conditions allow, aerial lines should gradually be buried underground and use cable lines. Construction of a 220 KV performer substation is planned in Shatan.

8. Rectification of Overall Environment

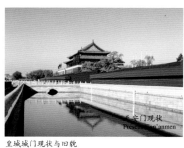

皇城城门现状与旧貌
Old and present gate of the Imperial City

The Imperial City used to have a clear-cut boundary (demarcated by the wall of the City). Today, in order to redefine the complete boundary of the conservation district, a complete image of the area must be established, so that people can sense the existence of the place.

The specifics are as follows:

Roads—principal roads, hutong system that constitute the Imperial City. Enhance the landscape features along the Chao-Fu Road. Establish a line of landscape characterized by "traditional residence-natural scenery-palace compound-spots of cultural interest" through the adjustment of land use functions along the Chao-Fu Road.

Boundary—four boundaries of the Imperial City. Clearly clarify the borders on the west and north via urban design.

Zoning—palace, greenbelt and body of waters, traditional and ordinary districts. Apart from conserving the landscape of the core area with the palace at the center, features of traditional districts should be enhanced, original style (height, color, material and shape) of architecture should be strictly retained, temples, offices and yamens and storehouses that are different from residences and their surroundings must be rectified, and ordinary areas that are incongruous with traditional features will be revamped.

Nodes—Tian'anmen, Di'anmen, Dong'anmen, and Xi'anmen. Two entrance spaces of the Imperial City excluding the relics of Tian'amen and Dong'anmen will be treated.

Landmark Points—main architecture on the axis and other landmark buildings and monuments in the Imperial City. Along with the tradition of conservation of cultural relics and the newly added primary and secondary landmark points, get rid of the incongruous architecture and other buildings by way of improvement of environment.

(Note: original features—refers to the salient characteristics of architecture, monuments, buildings, relics, compounds (decorating styles, color painting patterns, anything representing local and traditional cultures as well as the development of localities) that are representative, have universal significance and related with historic incidences in various historical periods before the Republic period in the development of the Imperial City.)

8.1 Analysis of Traditional Space Pattern of the Imperial City

The following elements are represented by the features of the traditional space of the Imperial City:

Closed walls; exact axis; magnificent Forbidden City; sacred Imperial Ancestral Temple and Altar to the God of the Land and Grain; towering Jingshan; graceful Three Lakes; and undulating palaces, temples pagodas and watchtowers. These extant spatial elements have become the quintessence and core of the Imperial City.

The round wall of the Imperial City is tall, closed and continuous, which define the spatial relationship between the Imperial City and outside city and the absolut power of the monarch.

An Axis is strung with a series of gates, watchtowers, and palaces, running through Tian'anmen, Duanmen, Meridian Gate, Taihemen, the Hall of Supreme Harmony, Hall of Middle Harmonly, Hall of Preserving Harmony, Gate of Heavenly Purity, Palace of Heavenly Purity, Gate of Prowess, Wanchun Pavilion of Jiangshan, Shouhuang Hall

and Di'anmen.

The palace—Forbidden City—represents the feudal patriarchal rites system and the symbol of supreme power of emperor through its master plan and architectural pattern.

Two architecture of rites system: Imperial Ancestral Temple and Altar to the God of the Land and Grain.

A hill of scenery: (Jingshan).

A park: (Three Lakes)

Several landmarks: (palaces, temples, pagodas and watchtowers).

Relevant services and living quarters (service functional district of royal families)

Among these, the first 6 have been constructed during various periods, so their patterns have almost been perfected, their artistic effect of spaces more conspicuous, thus becoming the essence and core of the Imperial City. However, great changes have taken place in the service and residential sectors with the passage of time, which are the key districts of research in the conservation plan. Their functions are as follows:

In the Yuan period, the service and residential district was transitional areas between palaces and parks. With open and vast space, it made a striking contrast with the neighborhood-type residential quarters outside the Imperial City.

In the Ming period, many management institutions, boards, storehouses, workshops serving the royal court were established in this district under the charge of the Office of Imperial Household, thus forming a self-contained small society of its own.

In the Qing period, after the decline of government-run handicraft industry, the royal necessities had to be supplied by society, which led to the withering of government institutions of the court. Thus, this district was turned to residential quarters for the attendants of the court. Theses are the main formative periods of the urban fabric of the Imperial City.

The nature of the district has been changing and adjusting ever since the Republic period by adding new functions such as office buildings, medical, educational institutions and industry. So the constructions of this period have caused a certain degree of damage to the traditional urban style and fabric of the Imperial City.

8.2 Analysis of Current Spatial Layout of the Imperial City

The spatial pattern of the Imperial City can be illustrated in the following aspects:

8.2.1 Boundary of the Imperial City

Currently, the southern, southwestern and eastern borders of the Imperical City are clear; western and northern borders are unclear due to the disappearance of the walls there. The features of the current borders: have changed from closeness and clearness of the past into openness and unclearness.

8.2.2 Gates of the Imperial City

The traditional Imperial City had four gates: Tian'an, Di'an, Dong'an and Xi'an,

东安门遗址已展示,西安门和地安门已不存在,地标的作用丧失。

8.2.3 皇城的中轴线

现状皇城中轴线南起天安门,经过端门、午门、故宫三大殿、神武门、景山万春亭,北至已消失的地安门,全长2.7km,是老北京城7.8km传统中轴线的最核心的部分。未来有条件时建议象征性地恢复地安门城楼。

8.2.4 皇城内的主要景观标志点

可分为三种类型:一、明清时代传统的主要标志点有天安门、午门、故宫三大殿、神武门、景山万春亭、北海白塔6处;二、明清时代传统的次要标志点有故宫四个角楼、东华门、西华门、北海团城、西什库教堂8处;三、民国以来建设的次要标志点有北京图书馆、北大红楼、总参宿舍楼、总政宿舍楼4处。总计18处景观标志点。

8.2.5 皇城内的区域性景观

皇城内除故宫等六大景观单位外,还有许多寺庙、衙署分布于街区中,形成了区域性景观。大致可分为三类:一、保存较好的区域性景观,有故宫、景山公园、中山公园、劳动人民文化宫、北海公园、中南海、西什库教堂、永佑庙、双吉寺、嵩祝寺、福佑寺、凝和庙、普渡寺、皇史宬、欧美同学会15处;二、新增的区域性景观,有佑贞女中旧址、北京图书馆、北大红楼、军调部、外交学会5处;三、由于维护不当处于消失边缘的区域性景观,有祉园寺、大马关帝庙、三官庙、大高玄殿、智珠寺、京师大学堂旧址、万寿兴隆寺、昭显庙、静默寺、关帝庙、升平署、宣仁庙12处。这12处景观应进行抢救性修缮。

8.2.6 皇城内与传统不协调的区域和建筑

皇城内有许多区域分布着多层或高层建筑,这些建筑群对皇城的传统空间结构有较大的破坏性,应逐步采取措施制止进一步的破坏行为。这部分区域的占地面积约有87hm²,占皇城面积的12.6%。现状中明显不协调的建筑有陟山门街北6层住宅楼、西什库教堂南侧京华印刷厂、南长街中央警卫局大楼、北池子大街北京证章厂大楼、欧美同学会北侧市房管局大楼5处,均位于关键的景观视廊内,对皇城的空间格局有较大的负面影响。

大高玄殿现状
Present Grand Gaoxuan Hall

智珠寺现状
Present Zhizhu Temple

万寿兴隆寺现状
Present Wanshouxinglong Temple

8.3 整体环境的整治

通过分析皇城的历史演变和现状情况,皇城整体环境的整治内容可分为如下几个方面:

8.3.1 严格保护6大文物保护单位,保护皇城内的中轴线及其现存的重要景观点。

景山公园、故宫、中山公园、劳动人民文化宫、北海、中南海这6大文物保护单位是皇城保护区内最精华的部分,是保护工作的重中之重。

8.3.2 重塑皇城边界,特别是皇城的西、北边界。

首先,准确勘测皇城墙的位置,沿城墙古迹的地面上做特殊铺装处理,向人们展示;其

北池子地区多高层建筑集中区
Concentrating area of Multistory buildings and high rises in Beichizi

次,沿皇城西、北的建筑边界规划一条不小于8m的绿化带,示意皇城的范围。在西安门、地安门所在位置两侧的用地中规划集中绿地,示意皇城城门的所在地。研究探讨西安门、地安门复建的必要性、可行性和合理性。

8.3.3 强化中轴路景观。

保持地安门内大街现有道路红线50m宽度不变,在红线范围内路板东西两侧规划10m宽的绿化带;沿景山后街、西街、东街的两侧,结合环境整治规划一条绿化带,拆除景山西墙外侧的破旧平房,恢复为绿地,为将来恢复连接北海和筒子河的古河道创造条件;保持景山后街、西街、东街、前街、南北长街、南北池子大街、东华门大街、西华门大街现状道路宽度不变,保护道路两侧的行道树木,形成沿中轴路的绿地系统,强化中轴路的城市景观。

8.3.4 进一步加强文物保护单位的保护、修缮、腾退和合理利用。

应使皇城内文保单位成为皇城文化的重要展示场所,首先改善大高玄殿、宣仁寺、京师大学堂、智珠寺、万寿兴隆寺等文物的环境,逐步采取措施改善社稷坛、太庙、北海蚕坛、故宫等文保单位内的环境。

8.3.5 加强对有历史文化价值的建筑或院落的保护和修缮。

先期选择2~6片有价值院落(真如镜胡同、张自忠故居、互助巷47号、会计司旧址、帘子库、北河等周边地区)比较集中的区域进行房屋修缮、腾退居民、拆除违章建筑、改善居住环境的试点,为普通民居的保护积累经验。

8.3.6 对在皇城环境整治过程中发现的,经考证为真实的历史建筑物或遗存,必须妥善加以保护,并加以标示。

8.3.7 采取坚决措施,分阶段拆除一些严重影响皇城整体空间景观的多层建筑。

近期需拆除的建筑有:陟山门街北侧的6层住宅楼、景山公园内搭建的圆形构筑物、京华印刷厂烟囱、欧美同学会北侧的市房管局办公楼、北池子大街北端的北京证章厂办公楼等5处;中期应重点整治文物保护单位内部及周边、主要道路两侧的多、高层建筑,如故宫内的第一历史档案馆以及京师大学堂旧址内、皇史宬周边、西什库教堂周边、北海北侧多高层住宅楼、南长街的警卫大楼等。经过努力,在未来逐步还给故宫一个良好的人文、历史和生态环境。

8.3.8 对于与皇城风貌不协调的建筑,应加强整饬。

对建筑质量较好且与皇城传统风貌不协调的多层住宅,应加强整饬工作。现状多层建筑特别是多层住宅大多为平屋顶,建筑形式、色彩与皇城传统风貌差异很大,应采取强制性措施将多层建筑的平屋顶改造成坡屋顶,坡屋顶颜色以青灰色调为主,不得滥用琉璃瓦。对新审批的建设项目,无论建筑高低,均应严格要求屋顶的形式。增加坡屋顶是减弱多层建筑对传统风貌破坏的最有效和可操作性的手段,应

serving as linking landmarks between inside and outside of the Imperical City. Only Tian'anmen still exists today; and the relic of Dong'anmen is on display, but the other two have lost, together with their role as landmarks.

8.2.3 Axis of the Imperial City

The current axis of the Imperial City is 2.7 kilometers long, being the core of the 7.8 kilometer-long traditional axis of the old city of Beijing. Starting from Tain'anmen in the south, it goes through Duanmen, Meridian Gate, the three halls of the Forbidden City, Gate of Prowess, Wangchun Pavilion of Jingshan, till Di'anmen which has ceased to exist. It is suggested that Di'anmen gate tower be rebuild in the future if conditions allow.

8.2.4 The Principal Landmark Landscapes in the Imperial City

Three types can be divided: 1. Of the primary ones in the Ming and Qing periods: 6 in number including Tian'anmen, Meridian Gate, the three halls of the Forbidden City, Gate of Prowess, Wangchun Pavilion of Jingshan and theWhite Pagoda of Beihai; 2. Of the minor ones in the Ming and Qing periods: 8 in number: four watchtowers of the Forbidden City, Donghuamen, Xihuamen, Round City of Beihai, Xishiku Church; 3. Of the minor ones in the Republic period: 18 in number: National Library, Red Building of Peking University, Residential Building of the General Staff Headquarters, and four residential buildings of the General Political Department.

8.2.5 Landscape Sites in the Imperial City

Apart from the 6 big landscapes such as the Forbidden City, there are also many temples, government offices scattered in communities, constituting local landscapes. They can be mainly divided into three groups: 1. Landscape sits in good shape including Forbidden City, Jiangshan Park, Zhongshan Park, Laboring People's Palace of Culture, Beihai Park, Zhongnanhai, Xishiku Church, Yongyou Temple, Shuangji Temple, Songzhu Temple, Fuyou Temple, Ninghe Temple, Pudu Temple, Imperial Archives, Union of Returned Overseas Students, altogether 15; 2. Newly added landscapes: former site of Youzhen Women's Middle School, National Library, Red Building of Peking University, Military Allocation Department, Foreign Affairs Society, altogether 5; 3. Landscapes that are in the brink of being extinct due to lack of protection including Zhiyuan Temple, Guandi Temple at Dama, Sanguan Temple, Dagaoxuan Temple, Zhizhu Temple, former sites of Metropolitan University, Wanshouxinglong Temple, Zhaoxian Temple, Jingmo Temple, Guandi Temple, Shengping Board, Xuanren Temple, altogether 12. The last 12 sites must be urgently recovered.

8.2.6 Districts and Architecture not in Agreement with the Tradition within the Imperial City

Measures must taken to deal with some of the multi-storied buildings or high-rises that are scattered in the Imperial City which have seriously impacted the overall spatial landscape of the imperial city. They cover 87 hectares, accounting for 12.6% of the area of the Imperial City. The ones that fall into this category include: the 6-storied apartment building north of Zhishanmen Street, the Jinghua Printing House south of Xishiku Church, the building of the Guards Bureau of the Party Central Committee at Nanchang Street, the office building of the municipal Housing Administration north of the Union of Returned Overseas Students, and the office building of the Beijing Badge Plant at Beichizi Street. These 5 sites are all located in key visual passage of landscape, causing serious negative impact on the spatial layout of the Imperial City.

8.3 Rectification of Overall Environment

Through the analysis of the historical evolution and current situation of the Imperial City, the rectification of the overall environment of the Imperial City are concerned with the following aspects:

8.3.1 Strictly protect the 6 big cultural relics, the axis of the Imperial City and the extant key landscape sites.

The 6 relics places of Jingshan, Beihai, Zhongshan Parks, Forbidden City, Laboring People's Palace of Culture and Zhongnanhai are the cream and core of the conservation district of the Imperial City, the key priority in the conservation campaign.

8.3.2 Redefine the Boundary of the Imperial City, Especially the Western and Northern Borders

Fist, carefully investigate the locations of the walls and pave the ground of the wall ruins in a special way to illustrated the places. Secondly, establish a greenbelt of no less than 8 meters wide along the northern and western borders of walls to mark the range of the wall. Plan concentrated greenbelts on both sides of the sites of Xi'an and Di'an Gates to indicated the places of the gates. Research the possibility, feasibility and the rationality of rebuilding Xi'an and Di'an Gates.

8.3.3 Improve the Landscape on the Axis

The keep unchanged the current 50-meter-wide red line at Di'anmen Street, and plan a 10-meter-wide greenbelt along the east and west sides of the red line. A greenbelt will be planed with the revamping of environment of both sides of the Jingshanhoujie, Dong and Xi Jingshan Streets. Dismantle the old and crumbling single-storied houses outside the western wall of Jingshan Park and turn the plot into a greenbelt, with the view of clearing the way for future recovery of the ancient river-course linking Beihai and Donzi River. The road width of Jingshanhoujie, Dong and Xi Jingshan and Jingshanqianjie Streets, Nanchang and Beichang Streets, Nanchizi and Beichizi Streets, Donghuamen Street, Xihuamen Street will remain unchanged. The sidewalk trees will be protected to form a greenbelt along the axis, highlighting the urban landscape on the axis.

8.3.4 Strengthen the Protection, Facelift, Vacating and Reasonable Utilization of the Cultural Relics

The cultural relics in the Imperial City should become important venues for exhibiting culture of the Imperial City. The first step should be to improve the environment of cultural relics such as Gaoxuan Hall, Xuanren Temple, Metropolitan University, Zhizhu Temple, Wangshoulong Temple, etc.. Measures should also be gradually adopted to improve the environment of the Altar to the God of the Land and Grain, Imperial Ancestral Temple, The Silkworm Altar in Beihai and Forbidden City.

8.3.5 Enhance the Conservation and renovation of architecture and courtyard houses that have historical and cultural values.

First select 2 to 6 areas with concentrated valuable courtyards (areas around Zhenrujing hutong, former residence of Zhang Zizhong, No 47 of Huzhuxiang, old site of Accounting Department, Lianziku, Beihe River) to conduct revamping, vacating inhabitants, demolishing unauthorized buildings, improving living environment, and accumulating experience in preverving ordinary residences.

8.3.6 With regard to architecture or relics that are discovered and found as a result of research to be authentic relics in the process of improving the environment of the Imperial City, measures should be taken to preserve and label them as such.

8.3.7 Take measures to tear down by stages some of the multi-storied buildings or high-rises that have seriously impacted the overall spatial landscape of the Imperial City.

The ones that must be demolished recently include 5 places: the 6-storied apartment building north of Zhishanmen Street, the round-shaped structure within the Jiangshan Park, the chimney of the Jinghua Printing House, the office building of the municipal Housing Administration north of the Union of Returned Overseas Students, and the office building of the Beijing Badge Plant north of the Beichizi Street. During the middle stage, priority should be given to rectify the multi-storied and high buildings along the sides of main roads and around cultural relics, such as those inside the First Historical Archive (located at the Forbidden City) and former Metropolitan University, and those next to the Imperial Archives, Xishiku Church, and north of Beihai. They also include the building of Guards of the Party Central Committee on Nanchang Street. Effort should be made to give back to the Forbidden City a nice ecological environment with sites of cultural and historical interest.

8.3.8 Renovating Buildings Not in Agreement with the Style of the Imperial City

With regard to buildings that are of good quality but not in agreement with the character of the Imperial City, revamping should be done. Measures must be taken to change the flat-roofs of high-rises, especially that of the residential architecture, which do not correspond with the ancient style of the Imperial City, into reclining roofs. The dominant color of the reclining roof should be greenish gray and no over-use of glazed tiles is allowed. Strictly control the newly approved constructions in terms of their roofs regardless of the height. Reclining roofing must be applied, which, as the most effective and feasible approach to alleviate the impact by the multi-storied buildings on the traditional style, should be actively promoted in the old city.

8.3.9 Gradually Reduce the Population Density of the Imperial City

With the help of property rights reform and land development of new areas outside the old city, the relocation of population from the Imperial City should be conducted in order to cut the population density by 40-60%, and keep the number of inhabitants in the old city at a reasonable level of under 40,000.

8.3.10 Strengthen the measure of demolishing buildings that are in defiance of rules and regulations, and prohibit construction of large public buildings for business,

北河沿地区四合院风貌
Courtyard housing in Beiheyan

北河沿地区四合院风貌
Courtyard housing in Beiheyan

在旧城范围内积极推广。

8.3.9 逐步降低皇城内的人口密度。

结合住房产权制度的改革以及旧城外新区的土地开发，逐步向外疏解皇城内的人口，降低人口密度，使现有户籍人口降低40～60%，达到约40000人以下的合理规模。

8.3.10 应加大拆除皇城内的违法建设的力度，不在皇城内新建大型的商业、办公、医疗卫生、学校等公共建筑。

8.3.11 在环境整治的基础上，应进一步宣传皇城作为历史文化保护区的重要意义。

提高全民对皇城整体保护的意识。同时，应加快立法工作，尽快制定《北京皇城保护规划》实施管理办法，加强对皇城的保护和建设。

9. 实施建议

虽然目前在皇城保护区没有大规模的开发改造，但作为城市的生活和工作空间，其内在的改造和整治的动力从来没有停止过。随着经济的多元化和部分居民收入的提高，在这些地区出现了小规模改造的实践，如仿古式四合院开发改造、用户自动改造、搭建等。但是，缺乏明确规划引导的自发建设和居住人口密度过高的现象，往往严重影响着这些地区历史环境的整体保存，对传统风貌造成不同程度的侵害。

为加强对皇城保护区的保护，应进一步广泛宣传保护皇城历史文化保护区的重要意义，提高全民对皇城整体保护的意识。结合旧城外的土地开发，与皇城保护区的保护和改造内外对应，采取切实措施，逐步降低保护区的居住人口密度。

加快对皇城内文物保护单位利用不合理情况的调整和改善。

皇城保护区内的道路改造要慎重研究，以保护为前提，逐步降低交通发生量。原则上皇城内的城市主要道路应维持现状道路宽度不变。

应积极制定皇城历史文化保护区保护管理条例，加强对皇城保护和建设的管理力度。

进一步深化皇城保护区的工作，制订详细的行动计划。

制定典型民宅保护、维修、使用条例，通过立法，使各有关住户认真执行。

进行改造的特殊区域，应设计多种改造和人口疏散方案，以便于政府操作及居民选择。

建立开放的交流平台，向专家学者及规划区内居民征询关于规划区历史文化价值及风貌保护的意见和建议。

皇城保护是政府行为，是由政府运作的一种城市公益活动，不应以获取经济效益为目的，只能在政府的政策引导下进行规划整治。

10. 结语

2002年10月15日和10月17日，市政府第142次市长办公会及首都规划建设委员会第22次全体会议审查并原则通过了《北京皇城保护规划》。2003年4月，北京市人民政府京政函[2003]27号正式批复《北京皇城保护规划》，我们期望在《北京皇城保护规划》的指导下，皇城的保护和规划管理能走上法制化的轨道，使皇城的历史风貌得以长久延续下去。

2003年4月

参考书目及资料：

- 陈宗蕃编著. 燕都丛考.
  北京古籍出版社，1991年
- 董光器著. 北京规划战略思考.
  中国建筑工业出版社，1998年
- 邓辉编著. 帝都风韵聚幽燕北京.
  中国地质大学出版社，1997年11月
- 傅公钺编. 北京旧影.
  人民美术出版社，1989年
- 方可著. 当代北京旧城更新：调查·研究·探索.
  中国建筑工业出版社，2000年
- 侯仁之主编. 北京历史地图集(一).
  北京出版社，1997年
- 梁思成著. 梁思成全集. 第四卷.
  中国建筑工业出版社，2001年
- 盛锡珊绘画. 老北京市井风情画.
  外文出版社，1999年
- 王景慧等编著. 历史文化名城保护理论与规划.
  同济大学出版社，1999年
- 吴建雍等著. 北京城市生活史.
  开明出版社，1997年
- 王其明著. 北京四合院.
  中国书店，1999年
- 王同祯著. 老北京城.
  北京燕山出版社，1997年
- (明)张爵著. 京师五城坊巷胡同集.
  北京古籍出版社，1982年
- (清)朱一新著. 京师坊巷志稿.
  北京古籍出版社，1982年
- 张宗平、吕永和译. 清末北京志资料.
  北京燕山出版社，1994年
- 北京旧城25片历史文化保护区保护规划
  ——景山八片：
  北京市规划委员会、中国城市规划设计研究院编制
  ——北池子地区：
  北京市规划委员会、北京工业大学建工学院建筑系编制
  ——南池子、东华门大街：
  北京市规划委员会、清华大学建筑学院编制
  ——南北长街、西华门大街：
  北京市规划委员会、北京市城市规划设计研究院编制

多层住宅楼
Multi-storied buildings in Zhishanmen Street

北海北侧多高层建筑
Multi- and high-storied buildings on the north of Beihai

office, medical and educational uses within the Imperial City.

8.3.11 On the basis of environmental rectification, further promote the important significance of regarding the imperial city as a conservation district of historic city

Raise the public awareness of protecting the Imperial City in an overall manner, quicken the legislative work, work out the methods of implementing and managing the Conservation Plan for the Imperial City of Beijing, and enhance the conservation and construction work of the Imperial City.

9. Advice for Implementation

Although there are no large-scale development and construction in the Imperial City, it has, as a life and working space in a city, never seen the cessation of the renovation and rectification in itself. With the pluralistic economic development and the increase of income of some inhabitants, small-scale redevelopment has occurred in some portions of the place, such as the development of quasi-classical siyeyuan, self-renovation and added structures. However, the self-initiated construction without the guidance of any plan and the high density of population seriously affect the overall conservation of historical environment of these places, bringing about damage of various degrees to the ancient style.

In order to heighten the conservation of the conservation district of the Imperial City, efforts should be made to further promote the important significance of preserving the this conservation district, and raise the awareness of protecting the Imperial City among citizens. The campaign of preserving and renovating the Imperial City should be coupled with the land development outside the old city, and measures be adopted to reduce the population density in the conservation district.

Step up efforts to improve and adjust the unreasonable utilization of the cultural relics in the Imperial City.

The renovation of roads within the Imperial City should proceed in a prudent way with conservation as the priority. Traffic volume should be gradually reduced. In Principle, the major urban roads in the Imperial City will remain unchanged in width.

Conservation ordnance for the Imperial City conservation district must be drawn up as soon as possible and management of the conservation and construction of this area must be strengthened.

Further deepen the work of the conservation district of the Imperial City, and hammer out detailed working plan.

Work out regulations of protection, maintenance and use of typical residential houses, which, once passed as law, will be seriously implemented by each household.

For special areas undergoing redevelopment, various designs for renovation and population resettlement should be produced for government to act upon and for inhabitants to choose from.

A platform of open communication should be set up, so as to solicit opinions from experts and residents of the conservation districts concerning the ways of preserving historical values and traditional features.

As a government act, conservation of the Imperial City is a public welfare activity, which should not seek economic profits. Instead, the planning and rectification can only be conducted under the policy of the government.

10. Conclusion

On October 15 and 17, 2002, the 142nd mayoral working conference of the municipal government and the 22nd plenary session of the Capital Urban Planning Commission examined and approved in principle the Conservation Plan of the Imperial City in Beijing. In April 2003, the Beijing people's government issued a document (no. 27, 2003), formally approving the above plan. It is our wish that guided by the Conservation Plan, the conservation and planning management of the Imperial City will be integrated in a legal system, so that the historical features of the Imperial City will be everlasting.

April 2003

Bibliography

- Chen Zongfan, *A Collection of Studies of the Capital of Yan*,
  Beijing Ancient Books Press, 1991.
- Dong Guangqi, *Reflections on Strategic Planning of Beijing*,
  China Architectural Industry Press, 1998.
- Compiled by Deng Hui, *Features of Imperial City of Beijing*,
  Publishing House of China Geology University, November 1997.
- Compiled by Fu Gongyue, *Old pictures of Beijing*,
  People's Art Press, 1989.
- Fang Ke, *Investigation, Research and Exploration of the renovation of Contemporary Beijing*,
  China Architectural Industry Press, 2000.
- Edited by Hou Renzhi, *A Collection of Historical Atlas of Beijing (Vol,1)*,
  Beijing Publishing House, 1997.
- Liang Sicheng, *Completed Works of Liang Sicheng, Vol, 4*,
  China Architectural Industry Press, 2001.
- *Painted* by Sheng Xishan, *Customs and Ways of Old Peking*,
  Foreign Language Publishing House, 1999.
- Wang Jinghui, etc. *Theory and Planning for Conservation of Historic Cities*,
  Publsihing House of Tongji University, 1999.
- Written by Wu Jianyong,etc. *Urban Life History of Beijing*,
  Kaiming Publishing House, 1997.
- Wang Qiming, *Quodrangles in Beijing*,
  China Book Store, 1999.
- Wang Tongzhen, *Old Beijing City*,
  Beijing Yanshan Publishing House, 1997.
- (Ming Dynasty) Zhang Jue, *A Collection of hutongs in Beijing*,
  Beijing Ancient Books Press, 1982.
- (Qing Dynasty) Zhu Yixin, *Manuscript of Streets in Beijing*,
  Beijing Ancient Books Press, 1982.
- Trans. Zhang Zongping and Lu Yonghe, *Materials from Records of Beijing at the end of the Qing Dynasty*,
  Beijing Yanshan Publishing House, 1994.
- *Conservation Planning of 25 Historic Areas in Beijing Old City*
  —8 Streets in Jingshan Area:
    Beijing Municipal City Planning Commission China Academy of Urban Planning and Design
  —Beichizi Area:
    Beijing Municipal City Planning Commission Architecture Department,Beijing Polytechnic University
  —Nanchizi and Donghuamen Area:
    Beijing Municipal City Planning Commission School of Architecture, Tsinghua University
  —South and North Changjie Streets and Xihuamen Street:
    Beijing Municipal City Planning Commission Beijing Municipal Institute of City Planning and Design

皇城历史文化保护区文物保护单位名单

名 称	故 宫
级 别	国家级文物
年 代	明、清
地 址	东城区景山前街4号
文物价值	明清皇宫，明永乐四年（公元1406年）创建，明清两代迭有修缮改建。宏伟壮丽，具有极高的艺术价值，充分表现了明清宫廷建筑的特征。
保存情况	现为故宫博物馆。保存较好。现管理开放。
规划措施	保存现状
备 注	

文物保护范围、建设控制地带图及全景鸟瞰照片（由南向北）

名 称	社稷坛
级 别	国家级文物
年 代	明、清
地 址	东城区天安门西
文物价值	明清皇宫，明永乐十九年（公元1421年）创建，后迭经重修和扩建，社稷坛明代规制，拜殿为原来建筑，是北京保存最古老的建筑之一。
保存情况	现为中山公园。保存较好。管理开放。
规划措施	保存现状
备 注	管理使用单位：中山公园管理处。

文物保护范围、建设控制地带图、全景航拍及新旧照片

名 称	太 庙
级 别	国家级文物
年 代	明、清
地 址	东城区天安门东
文物价值	明清皇室的祖庙，明永乐十八年（公元1420年）创建，嘉靖二十四年（公元1545年）重修，清代重修，现为保存最完整的明代建筑群。
保存情况	现为劳动人民文化宫。保存较好。现管理开放。
规划措施	保存现状
备 注	现管理使用单位：北京市劳动人民文化宫。

文物保护范围、建设控制地带图、全景航拍及现状太庙照片

保护单位名单

名 称	天安门	文物保护范围，建设控制地带图及现状照片
级 别	国家级文物	
年 代	明	
地 址	天安门广场	
文物价值	建于明永乐十五年（公元1417年），原名承天门，是旧皇城的正门。天安门的布局是一个完整的艺术杰作。	
保存情况	现为开放空间，保存较好。现管理开放。	
规划措施	保存现状	
备 注	现管理使用单位：天安门管理处	

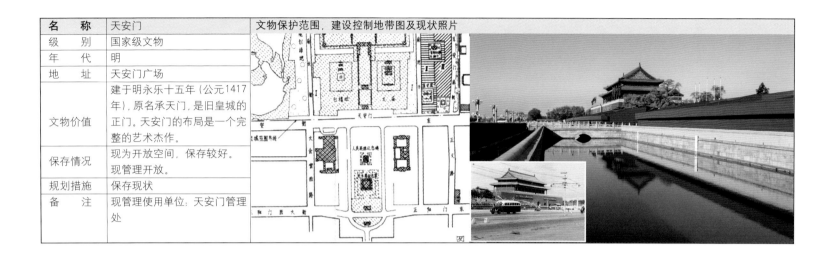

名 称	北海、团城	文物保护范围，建设控制地带图及现状照片
级 别	国家级文物	
年 代	明、清	
地 址	西城区文津街1号	
文物价值	辽、金（11世纪）开辟，现存元、明建筑遗制及清代建筑物，殿宇崇宏，为历代帝王别苑，名胜古迹很多。	
保存情况	现为公园及办公，保存较好。现管理开放，部分不开放。	
规划措施	保存现状	
备 注	现管理使用单位：北海景山公园管理处，北京市文物研究所、北海幼儿园等。	

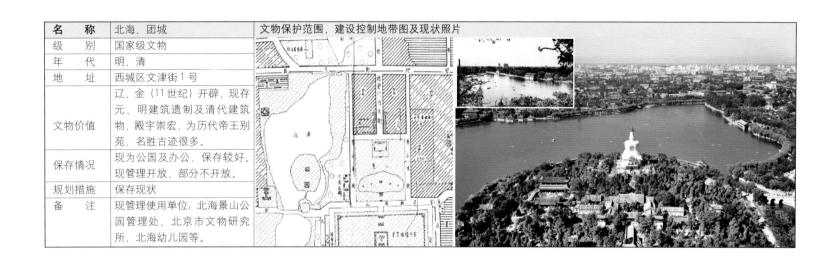

名 称	皇史宬	文物保护范围，建设控制地带图及现状照片
级 别	国家级文物	
年 代	明	
地 址	东城区南池子大街136号	
文物价值	明清宫廷的档案储藏所，明嘉靖十三年（公元1534年）建，清代重修。	
保存情况	大部分保存较好，部分被居民占用。管理开放。	
规划措施	保存现状	
备 注	现管理使用单位：皇史宬文保公司皇史宬管理部。	

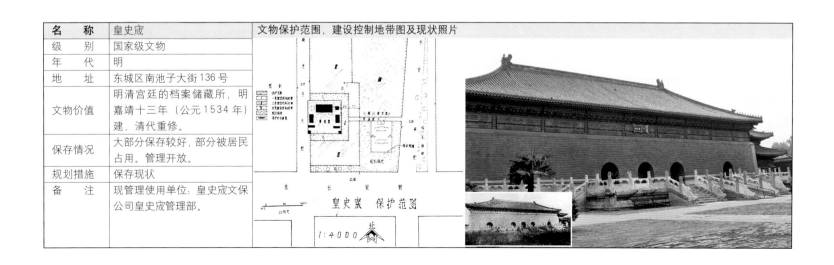

北京皇城保护规划·保护单位名单

名　　称	大高玄殿	文物保护范围，建设控制地带图及现状照片
级　　别	国家级文物	
年　　代	明	
地　　址	景山西街21、23号	
文物价值	明嘉靖二十一年（公元1542年）建，是目前仅存的一处明清皇家道观。主体建筑乾元阁风格独特，为两层阁楼式建筑，其是为圆顶代表天，下为方形的坤贞宇代表地。	
保存情况	年久失修，不开放。	
规划措施	归还文物部门，整修后开放。	
备　　注		

名　　称	北京大学红楼	文物保护范围，建设控制地带图及现状照片
级　　别	国家级文物	
年　　代	民初	
地　　址	东城区五四大街	
文物价值	始建于1916年，1918年落成。通体用红砖砌成，故名红楼。楼呈工字型，为砖木结构，连半地下室共五层。	
保存情况	现为文物局所属单位办公用，保存较好。一层部分已经对外开放。	
规划措施	保存现状	
备　　注	现管理使用单位：国家文物局，文物出版社。	

名　　称	景山	文物保护范围，建设控制地带图及现状照片
级　　别	国家级文物	
年　　代	明、清	
地　　址	西城区景山前街	
文物价值	明永乐年间（15世纪初）建置，清扩建。景山是明初皇宫里的镇山，所有建筑物和附属文物，是故宫体系内的一部分。	
保存情况	保存较好，东北角有市少年宫，存在一些隐患，现管理开放。	
规划措施	保存现状	
备　　注	现管理使用单位：北海景山公园管理处，北京市少年宫。	

保护单位名单

名 称	中南海	文物保护范围，建设控制地带图及现状照片
级 别	国家级文物	
年 代	辽～清	
地 址	西城区西长安街	
文物价值	中海为金、元（13世纪）开辟，南海为明初（15世纪）开辟，楼台殿阁在建筑上具有特殊风格。	
保存情况	保存较好。不开放。	
规划措施	保存现状	
备 注		

名 称	毛主席故居	文物保护范围，建设控制地带图及现状照片
级 别	市级文物	
年 代	1918年	
地 址	景山东街三眼井吉安所左巷8号	
文物价值	毛主席1918年第一次来京的住处，在此居住期间，毛主席组织了湖南留法勤工俭学并开始研读马列主义。	
保存情况	居民用房，建筑急需维修，搭建混乱，不开放。	
规划措施	搬迁居民，拆除搭建，整修住房，对外开放。	
备 注	现管理使用单位：东城区房管局景山房管所。	

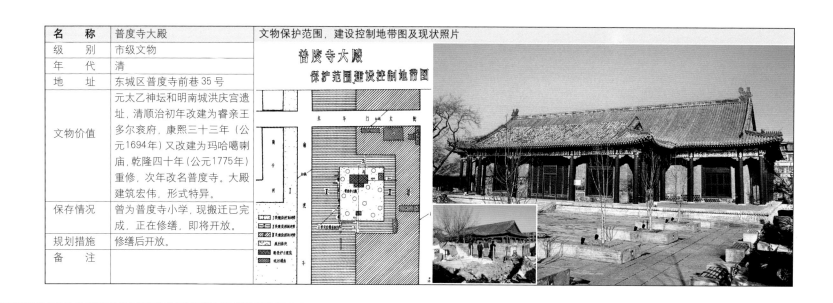

名 称	普度寺大殿	文物保护范围，建设控制地带图及现状照片
级 别	市级文物	
年 代	清	
地 址	东城区普度寺前巷35号	
文物价值	元太乙神坛和明南城洪庆宫遗址，清顺治初年改建为睿亲王多尔衮府，康熙三十三年（公元1694年）又改建为玛哈噶喇庙，乾隆四十年（公元1775年）重修，次年改名普度寺，大殿建筑宏伟，形式特异。	
保存情况	曾为普度寺小学，现搬迁已完成，正在修缮，即将开放。	
规划措施	修缮后开放。	
备 注		

北京皇城保护规划·保护单位名单

北京历史文化名城北京皇城保护规划

名　　称	嵩祝寺及智珠寺
级　　别	市级文物
年　　代	清
地　　址	东城区普度寺前巷35号
文物价值	清雍正十一年（公元1733年），皇帝为蒙古活佛章嘉胡图克图在京修建的宗教场所，名嵩祝寺。寺院规模宏伟，建筑轩昂。其西部与智珠寺毗邻，原有古建大部保存。
保存情况	嵩祝寺保存较好，为宾馆用房，智珠寺被工厂所占，用作库房，破坏严重。南侧高楼影响环境。不开放。
规划措施	搬迁工厂，整修后开放。
备　　注	现管理使用单位：北京林达房地产公司，北京市文体百货工业联合公司，北京装潢设计研究所。

文物保护范围，建设控制地带图及现状照片

名　　称	凝和庙
级　　别	市级文物
年　　代	清
地　　址	东城区北池子46号
文物价值	清雍正八年（公元1730年）建，祭祀云神的地方。原建筑基本完整，现状保存。
保存情况	现为北池子小学，中轴线建筑保存良好。即将开始修缮。不开放。
规划措施	保存现状
备　　注	现管理使用单位：北池子小学

文物保护范围，建设控制地带图及现状照片

名　　称	宣仁庙
级　　别	市级文物
年　　代	清
地　　址	东城区北池子2、4号
文物价值	清雍正六年（公元1728年）建，祭祀风神的地方。原建筑基本完整。
保存情况	北部为单位办公用房，保存尚可，南部为居民住房，已部分搬迁，失修严重，不开放。
规划措施	搬迁单位和居民，修缮维护，争取管理开放。
备　　注	现管理使用单位：北京市中医药管理局，市卫生局宿舍。

文物保护范围，建设控制地带图及现状照片

保护单位名单

名　称	京师大学堂建筑遗存
级　别	市级文物
年　代	清、民初
地　址	东城区沙滩后街55号、57号
文物价值	创办于清光绪二十四年（公元1898年），原为乾隆帝四女和嘉公主府空闲府第。光绪帝实行新政，开办京师大学堂为新政措施之一。民国成立后改为北京大学。现有建筑为原和嘉公主府的正殿、公主院等清式建筑和民国年间建成的数学系楼及"西斋"十四排中式平房，是我国近代成立的第一座最高学府。"五·四"运动前，共产主义读书会小组在此成立。
保存情况	出版社占用，新建办公楼和住宅楼，严重破坏文物保护单位的原有格局和历史文化环境，高大体量严重影响从景山观赏皇城的整体景观。
规划措施	拆除办公楼和住宅楼，整修后作为革命教育基地对外开放。
备　注	

名　称	北京大学地质馆旧址
级　别	市级文物
年　代	1931年
地　址	东城区沙滩北街15号
文物价值	建于1931年，是我国著名建筑学家梁思成、林徽音少数设计作品之一。平面、立面均为不对称式，体形随功能要求变化，总体造型明快简洁，是我国最早引进西方"现代建筑"的优秀作品之一，在中国近代建筑史上具有重要地位。
保存情况	中国社会科学院法学研究所办公用。不开放。
规划措施	保存现状
备　注	现管理使用单位：中国社科院法学所、政治学研究所。

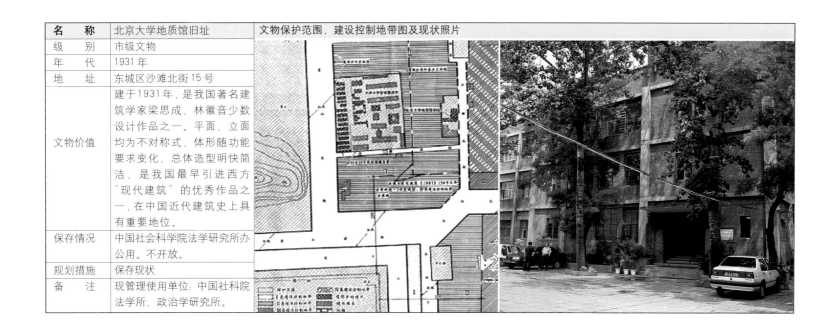

文物保护范围，建设控制地带图及现状照片

北京皇城保护规划·保护单位名单

名　称	军调部1946年中共代表团驻地
级　别	市级文物
年　代	1931年
地　址	东城区南河沿大街1号
文物价值	建于20世纪30年代初，原为日本侵华战争时华北驻军司令宅邸。1946年1月10日，周恩来代表中国共产党和国民党政府签订"停战协定"时，该处曾是中共代表团驻地。现存主楼是20世纪50年代按原状重新修复的，建筑平面呈"T"形，主体二层，中部三层，仿古式建筑。
保存情况	现为翠明庄客房，已翻建。不开放。
规划措施	保存现状
备　注	现管理使用单位：翠明庄宾馆。

位置图、全景航拍及现状照片

名　称	子民堂
级　别	市级文物
年　代	1947年
地　址	东城区北河沿大街77号，嵩祝寺北巷4、6号
文物价值	原为清乾隆大学士傅恒之宅第旧址，子民堂全称子民纪念堂，系北京大学为纪念蔡元培先生而建。两进院落，南有垂花门，正殿五间；有月台，东西配殿各五间；后院七间后堂，东西有配廊。
保存情况	现为文化部办公用房，保存较好，不开放。
规划措施	保存现状
备　注	现管理使用单位：文化部。

位置图及现状照片

保护单位名单

名　　称	福佑寺
级　　别	市级文物
年　　代	清
地　　址	西城区北长街20号
文物价值	清圣祖玄烨在紫禁城外的避痘处。雍正元年（公元1723年）建正殿，后改为庙宇。1918年毛泽东同志来京时，曾在此居住。后为班禅驻北平办事处。解放后为班禅驻京办事处。
保存情况	办公用房。已重建修复，保存较好。不开放。
规划措施	保存现状
备　　注	现管理使用单位：中国民族博物馆筹备处。

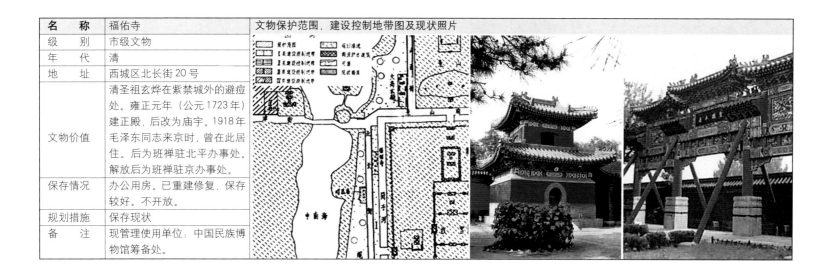

文物保护范围、建设控制地带图及现状照片

名　　称	昭显庙
级　　别	市级文物
年　　代	清
地　　址	西城区北长街71号
文物价值	清雍正十年（公元1732年）建，是皇家昔日祭祀雷神之所。
保存情况	今北长街小学。原有建筑仅余大殿及南侧一琉璃影壁。不开放。
规划措施	加强对文物建筑的维护，避免不合理使用。
备　　注	现管理使用单位：北长街小学。

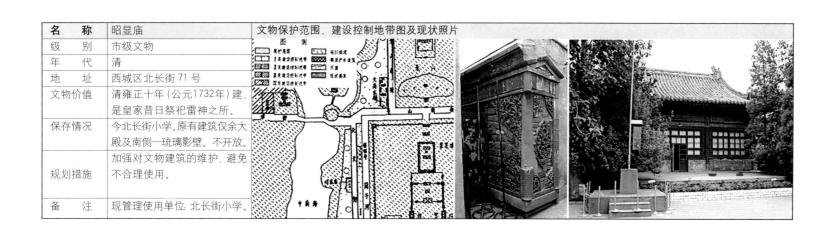

文物保护范围、建设控制地带图及现状照片

名　　称	西什库教堂
级　　别	市级文物
年　　代	清
地　　址	西城区西什库大街2号
文物价值	曾是北京最大的天主教堂，原在府右街北口（蚕池口），清光绪十六年（公元1890年）迁建于明代的西什库旧址，俗称"西什库教堂"。
保存情况	保存较好。管理开放。
规划措施	保存现状
备　　注	现管理使用单位：市天主教爱国会。

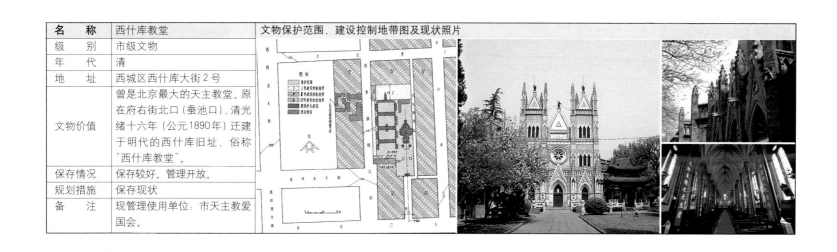

文物保护范围、建设控制地带图及现状照片

北京皇城保护规划·保护单位名单

147

名 称	升平署戏楼
级 别	市级文物
年 代	清
地 址	西城区西长安街1号
文物价值	现存的清代戏楼及四合院，保存完整。
保存情况	现为长安中学一部分，正在修复，不开放。
规划措施	保存现状
备 注	现管理使用单位：长安中学。

文物保护范围、建设控制地带图及现状照片

名 称	北京图书馆主楼
级 别	市级文物
年 代	1931年
地 址	西城区文津街7号
文物价值	1931年建成，同年，国立北平图书馆迁于此。
保存情况	现为北京图书馆，已经修复，保存较好，管理开放。
规划措施	保存现状
备 注	现管理使用单位：北京图书馆善本部。

文物保护范围、建设控制地带图及现状照片

名 称	盛新中学与佑真女中旧址
级 别	市级文物
年 代	1928年
地 址	西城区教场胡同2、4号
文物价值	原为法国天主教仁爱遣使会所属两所中学。现存教室楼二栋，礼堂一座，砖木结构，欧洲折衷主义建筑风格，立面三段划分，红瓦坡屋顶，红砖清水墙嵌以石料装饰。比例严谨，工艺精致，是20～30年代北京城内典型的教会学校建筑。
保存情况	现为北海中学和北京艺术与体育职业学校，保存较好，不开放。
规划措施	保存现状
备 注	现管理使用单位：北海中学、四十中。

文物保护范围、建设控制地带图及现状照片

保护单位名单

名　　称	北京水准原点旧址	位置图及现状照片
级　　别	市级文物	
年　　代	1915年	
地　　址	西城区西安门大街1号	
文物价值	始建于1915年。由花岗岩石砌筑，平面为正方形单层建筑，仿古希腊神庙式造型，是北京及至华北地区惟一一处建筑历史最早的水准点。建筑结构严谨，造型独特，对于科研、工交、文化、军事、建筑、地震、水文考察都具重要的意义和价值。	
保存情况	现位于北大医院妇产儿童医院内，保存良好，唯环境局促，不开放。	
规划措施	保存现状，为将来的环境整治留有余地。	
备　　注	现管理使用单位：北大医院妇产儿童医院。	

名　　称	陈独秀旧居	位置图及现状照片
级　　别	市级文物	
年　　代	民初	
地　　址	东城区箭杆胡同20号	
文物价值	陈独秀（1879～1942年）是中国共产党的创建人和早期领导人之一。1917年至1919年，陈独秀在此居住，该院坐北朝南，如意大门，正房三间，南房三间，硬山合瓦过垅脊。此处为宣传民主和科学的重要场所，具有重要历史价值。现位于北大医院妇产儿童医院内，保存良好，唯环境局促，不开放。	
保存情况	现为居住杂院，失修严重，不开放。	
规划措施	搬迁居民，修复整治，辟为展览。	
备　　注		

北京皇城保护规划·保护单位名单

北京历史文化名城北京皇城保护规划 / Conservation plan for the Historic City of Beijing and Imperial City of Beijing / 北京皇城

名　称	欧美同学会（原普胜寺）	文物保护范围、建设控制地带图、全景航拍及现状照片
级　别	区级文物	
年　代	清	
地　址	南河沿大街111号	
文物价值		
保存情况	保存良好，不开放。	
规划措施		
备　注	维护现状，争取对外开放。	

名　称	吉安所	位置图及现状照片
级　别	区级文物	
年　代	清	
地　址	吉安所右巷10号	
文物价值	是清代宫眷死后运棺柩之所，建筑风格独特。现存建筑有仪门三间，西有顺山房五间，东有假山，另有东西配房各五间，南北为侯棂大殿，绿琉璃瓦，黄卷边，歇山顶。	
保存情况	现为首长住宅。不开放。	
规划措施	现状保存	
备　注		

名　称	皇城墙	文物保护范围，建设控制地带图及现状照片
级　别	区级文物	
年　代	明	
地　址	东长安街，景山东街	
文物价值	始建于明永乐五年，由城砖砌筑。上涂朱红，顶覆黄琉璃瓦，高6m，墙厚2m。现除长安街外，石景山东街，地安门内大街，黄化门街尚保存部分皇城墙。	
保存情况	局部地段保存较好。	
规划措施	对黄瓦墙进行保护和周围环境整治。	
备　注		

保护单位名单

名　称	张自忠故居	位置图及现状照片
级　别	区级文物	
年　代	民国	
地　址	府右街椅子胡同15号	
文物价值	为爱国将领张自忠故居。张自忠1935年至1937年在此居住。中院北房东屋为卧室，东西厢房为客厅。1988年在故居院内立纪念碑，局部地段保存较好。	
保存情况	现为自忠小学，不开放。	
规划措施	加强对新增建筑造型、高度、色彩等方面的控制，使其与文物建筑的历史环境更好地融合。	
备　注		

名　称	万寿兴隆寺	位置图及现状照片
级　别	区级文物	
年　代	清代	
地　址	西城区北长街39号	
文物价值	明代为兵仗局佛堂，康熙三十九年敕改为万寿兴隆寺。后为内监修道养老之地。	
保存情况	现为数个居住杂院，原有建筑格局仍在，但被几个居住院落自行隔开。内部建筑质量尚可，加建极其混乱，亟需整修维护。	
规划措施	迁出居民，对文物建筑及院落环境进行整治。	
备　注		

名　称	永佑庙	位置图及现状照片
级　别	区级文物	
年　代	清雍正九年（公元1731年）建	
地　址	府右街1~3号	
文物价值	为祭祀皇城城隍之所。外垣门东向，主要建筑坐北朝南，依次有小门三间，屏墙一座，钟鼓楼各一间，前殿三间，中殿三间，后殿五间。	
保存情况	现为府右街小学，不开放。	
规划措施	根据西城区教育资源整合的需要，迁出小学。恢复用地为文化娱乐用地。	
备　注		

北京皇城保护规划・保护单位名单

北京普查登记在册文物

序号	名称	地址	公布批次(分类)	年代	保存情况	保护理由(说明)
1	四合院	南池子大街32号	东城区 V	民初		
2	四合院	北池子23号	东城区 V	清	保存较好，正房、耳房、东西厢房、大门、影壁等存在，垂花门有遗迹	传为某太监住宅，典型而完备的四合院格局，砖雕精美。
3	庄士敦故居	油漆作1、3号				庄士敦为英国人，末代皇帝溥仪的老师。坐北朝南，为大型四合院，现为单位宿舍。现存大门1个，倒座房6间，垂花门1间，二门1间，北房3间，东西厢房各3间，西院北5间，北房5间，为晚清中式建筑，硬山灰瓦屋脊。
4	外交学会	南池子大街71号	东城区 V	民国		
5	原丹麦使馆	南池子大街25号	东城区 V	近代		
6	暗八仙门墩	南河沿大街109号	东城区 VI	民国	保存较好	
7	门楼砖雕	南湾子15号	东城区 VI	民国	已拆	
8	砖雕和象眼刻画	南池子49号	东城区 VI	民国		
9	如意门砖雕	嵩祝寺西巷4～6号	东城区	民国	封闭作为房间外墙使用，局部损坏	砖雕极尽装饰之能事。
10	雕花松鼠门墩	三眼井93号	东城区	民国	局部损坏	形态生动、栩栩如生。
11	门墩	慈惠胡同11号	东城区	民国		
12	门楼象眼刻画	箭杆胡同15号	东城区	民国		
13	静生生物调研所	文津街3号	西城区 V	民国	1931年落成，已拆！	三层楼房。1928年，为纪念热衷研究生物学的范静生成立"静生生物调查所"。原址位于其住宅(石驸马大街83号，今新文化街86号)。后因房舍不够，在文津街3号设新所址，在学术上贡献很大。1941年12月，被侵华日军强占用作细菌站研究。
14	天主教共教进行会	西安门大街101号	西城区	民国	保存较好	西安门大街101号，建于20世纪20～30年代。建筑为欧式风格，南立面采用三分法，有水泥做的花苇，花柱，圆拱形窗，中式墙基，青石砌筑的须弥座。坐北朝南，三层楼房。
15	西什库小学	刘兰塑胡同14号	西城区 V	清光绪二十五年公元(1899年)	现为西什库小学，保存较好，已翻修。	原名圣心小学，后改为盛新小学，1952年改为现名。由法籍人士林(JARLIN)主教创办。自建校以来，百年来为国家共培养近2万名毕业生，其中不乏社会精英。教学楼坐北朝南呈倒F形。为仿欧式教堂建筑，砖木结构，灰砖红瓦(现改为水泥灰瓦)，阳台式连通走廊，原有楼板楼梯均为松木制成。1997年实施保护性大修，在楼内墙体和楼板实施钢筋水泥结构全面加固，收到了"整旧如旧"的效果。
16	若瑟会修女院	西什库大街76号	西城区 V	清		教会建筑，为天主教北京教区若瑟总院。院内建成2排东西向平房，北为17间，南为18间，中间为南北向房3排，每排各5间。
17	二层住宅楼	景山前街27号	西城区	民国	保存较好	近代中西结合式建筑，坐北朝南，主楼为二层。现为故宫宿舍。
18	大马关帝庙	恭俭胡同18号	西城区 III	清	现为居住大院，破坏严重。	始建于明万历时期，清光绪年间重修。现仅存大殿3间，硬山黄琉璃瓦顶，现为民居。
19	三官庙	恭俭胡同43号	西城区 III	清	居住院落	现存正殿3间，东西配殿各3间，硬山灰筒瓦顶，始建于明万历年，清光绪时重修。
20	冰窖	恭俭胡同5巷5号	西城区 III	清	现为北京宏远助邦商贸有限公司仓库	清代为宫廷冰窖，坐西朝东，勾连搭建筑，筒瓦顶，现为某单位仓库。

保护单位名单

21	庆云寺	景山后街10号	西城区 III	不详	居住院落	坐北朝南,现存正殿3间,后殿3间,两侧耳房各1间,东西配殿各3间,均为硬山筒瓦顶,保存基本完整。现为民居。
22	雪地冰窖（石刻1）	雪池胡同10号	西城区	清	一个现为北海公园职工自行车库,一个闲置。	是供大内御用的冰窖,现为某单位仓库,现存2栋建筑,绿琉璃瓦顶。
23	清内务府御使衙门	陟山门街5号	西城区	清	居住院落	清代建筑,坐北朝南,二进院落,大门3间,西侧转角房各5间,东西厢房各3间,后院北房5间,东西厢房各3间,现为故宫博物院宿舍。
24	祉园寺	西楼巷19号	西城区	清	居住院落	始建于清乾隆年间,现存山门1间,大殿3间,东耳房2间,东西配殿各3间,均为硬山筒瓦顶。现为民居。
25	西楼巷真武庙	西楼巷7号	西城区	清		
26	关帝庙	勤劳胡同18号				
27	三时学会旧址	北长街27号	西城区 III	不详		坐北朝南,大门东向,三排北房,原为佛教三时学会。现存前排房7间,中排7间,后排5间,两侧各1间,现《中国佛教》杂志编辑部。
28	会计司旧址	北长街113号	西城区 III	清		原为清代会计司。坐北朝南,现存官衙的中启门3间,前排北房3间,西排北房5间,西房9间,均为硬山筒瓦顶。
29	静默寺	北长街81、83号	西城区 III	清		始建于明崇祯年间,清康熙年重建。现存山门3间,前中后殿各3间,中院东西配殿各3间,前东西配殿各7间,后东西配殿各5间。
30	升平署	大晏乐胡同11号	西城区	清		为清代承应奏乐、演戏事务的机构。坐北朝南,分东、西两路,大门在东路,往北进二门有正殿,配殿,后殿,后罩房。
31	清内务府铁库	西华门大街4号	西城区 III	清		坐南朝北,现存大门3间库房3间。均为硬山、脊筒瓦顶。
32	明清西皇城墙遗址	西黄城根	西城区 III	清		
33	四合院	大红罗厂南巷1号	西城区	清		
34	双吉寺	双吉胡同	西城区	清	保存较好。现为西城区环卫局。	坐北朝南,现存二进大殿,正殿3间,东西配殿各3间,后殿3间,均为筒瓦顶,正带廊,砖雕精美。

北京皇城保护规划·保护单位名单

胡同名称列表

地区	编号	名称	建设朝代	名称来源	地区	编号	名称	建设朝代	名称来源
南池子地区	1	普度寺西巷	清	寺庙		18	火药局五条	清	局
	2	普度寺北巷	清	寺庙		19	火药局六条	清	局
	3	普度寺东巷	清	寺庙		20	东安门北街	清	城门
	4	普度寺前巷	清	寺庙		21	织染局胡同	清	局
	5	瓷器库北巷	清	库		22	水簸箕胡同	清	
	6	瓷器库胡同	清	库		23	后局大院	清	局
	7	瓷器库南巷	清	库		24	蜡库胡同	清	库
	8	灯笼巷胡同	清	库		25	钟鼓胡同	清	钟鼓寺
	9	缎库胡同	清	库		26	纳福胡同	明	内府库
	10	北湾子胡同	清	河		27	吉安所北巷	清	吉安所
	11	南湾子胡同	清	河		28	吉安所右巷	明	吉安所
	12	菖蒲河沿	清	河		29	吉安所左巷	清	吉安所
	13	东银丝胡同	清	东银丝沟		30	碾子胡同	清	
	14	西银丝胡同	民国	西银丝沟		31	三眼井胡同	清	井
	15	小苏州胡同	民国			32	横栅栏胡同	清	
	16	大苏州胡同	民国			33	嵩祝院西巷	清	寺
	17	飞龙桥胡同	清	桥		34	嵩祝院北巷	清	寺
						35	东高房胡同	清	
北池子地区	1	银闸胡同	清	闸		36	大学夹道	清	
	2	沙滩南巷	民国			37	沙滩北街	清	
	3	草垛胡同	明			38	沙滩后街	明	
	4	骑河楼北巷	明	骑河楼		39	中老胡同	清	
	5	骑河楼街	明	骑河楼		40	西老胡同	清	
	6	福禄巷	明	寺	景山西区	1	西楼巷	清	
	7	骑河楼南巷	明	骑河楼		2	油漆作胡同	清	作
	8	智德北巷	明			3	磨盘院胡同	清	
	9	箭杆胡同	清			4	米粮库胡同	民国	库
	10	智德西巷	清			5	恭俭胡同	清	内官监
	11	智德东巷	清			6	恭俭一巷	清	内官监
	12	文书馆巷	清			7	恭俭二巷	清	内官监
	13	北池子头条	清			8	恭俭三巷	清	内官监
	14	北池子二条	清			9	恭俭四巷	清	内官监
	15	北池子三条	清			10	恭俭五巷	清	内官监
景山东区	1	北月牙胡同	清			11	北海北夹道	清	
	2	慈慧胡同	清	慈慧寺		12	高卧胡同	清	
	3	南月牙胡同	清			13	房钱库胡同	清	库
	4	西吉祥胡同	清			14	园景胡同	清	
	5	东吉祥胡同	明			15	雪池胡同	清	冰窖
	6	帘子库胡同	明	库		16	陟山门街	明	
	7	黄化门街	清	黄化门		17	大石作胡同	清	作
	8	锥把胡同	清		南北长街	1	后宅胡同	清	
	9	东板桥街	清	桥		2	前宅胡同	清	
	10	东板桥西巷	清	桥		3	教育夹道	清	
	11	东板桥东巷	清	桥		4	道义巷	清	
	12	北河胡同	明	河		5	会计司胡同	清	司
	13	焕新胡同	清	火神庙		6	西小巷	清	
	14	火药局头条	清	局		7	养廉胡同	清	
	15	火药局二条	清	局		8	勤劳胡同	清	原老爷庙
	16	火药局三条	清	局		9	西苑门夹道	清	门
	17	火药局四条	清	局					

保护单位名单

地区	编号	名称	建设朝代	名称来源
	10	南长街西巷	清	
	11	百代胡同	清	
	12	水井胡同	清	苦水井
	13	小桥北河沿	清	
	14	后铁门	清	
	15	大宴乐胡同	清	署
	16	织女河东河沿	清	河
	17	玉钵胡同	清	玉钵庙
府右街	1	西岔胡同	清	
	2	双吉胡同	清	双吉寺
	3	惜薪胡同	清	惜薪司
	4	西红门胡同	民国	
	5	东红门胡同	民国	
	6	南红门胡同	民国	
	7	光明胡同	明	光明殿
	8	西椅子胡同	清	
	9	图样胡同	清	兔儿山
	10	石板房胡同	清	
	11	后达里	民国	
	12	后达里东巷	民国	
	13	后达里西巷	民国	
	14	互助巷	清	原鞑子营
	15	博学胡同	清	饽饽房
	16	枣林大院	清	
	17	石板房头条	清	
	18	石板房二条	清	
	19	石板房三条	清	
西什库	1	爱民一巷	清	
	2	爱民二巷	清	
	3	爱民三巷	清	
	4	爱民街	清	原旃檀寺西街
	5	天庆胡同	清	天庆宫
	6	草岚子胡同	清	
	7	刘兰塑胡同	明	人名
	8	真如镜胡同	清	
	9	酒醋局胡同	清	局
	10	扁担胡同	清	
	11	教场胡同	民国后	教场
	12	西什库大街	明	库
总计	137			

北京皇城保护规划·保护单位名单

北京历史文化名城北京皇城保护规划
Conservation plan for the Historic City of Beijing and Imperial City of Beijing

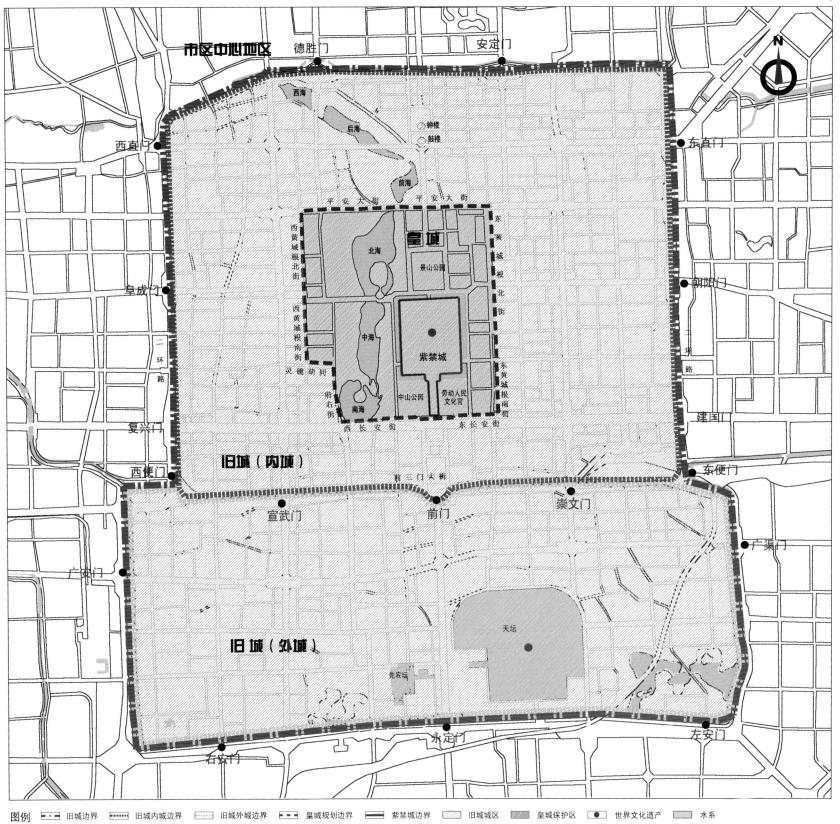

规划范围图
Scope of Planning

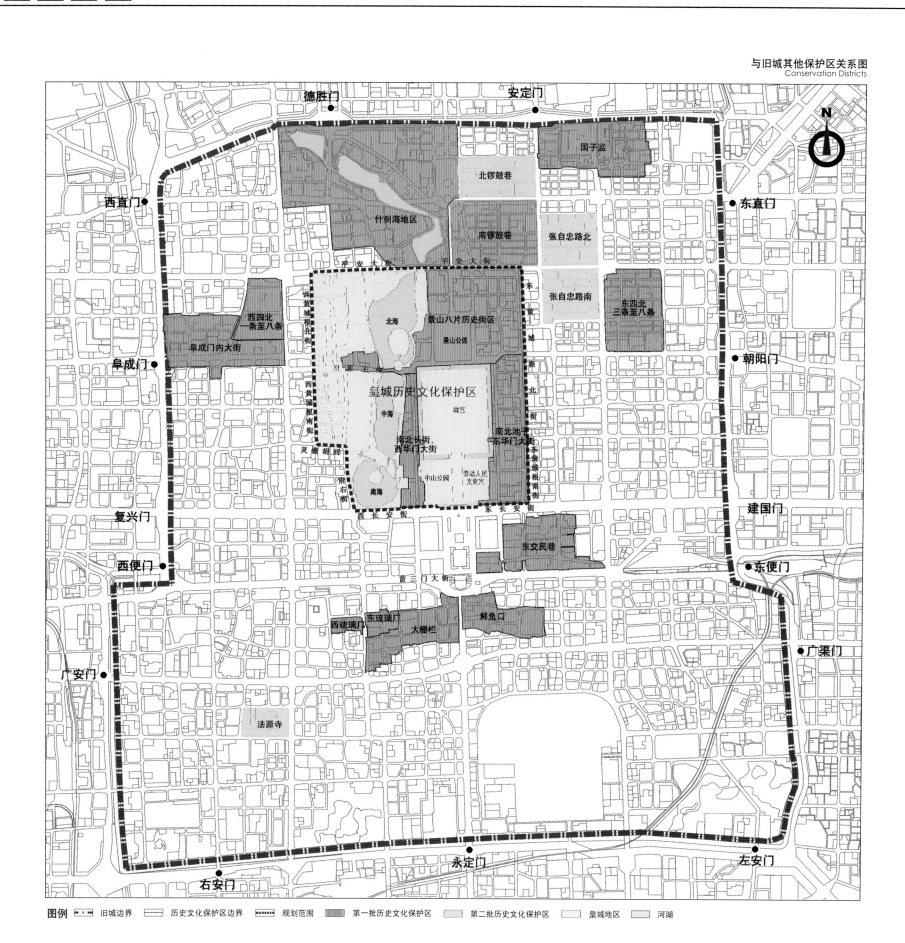

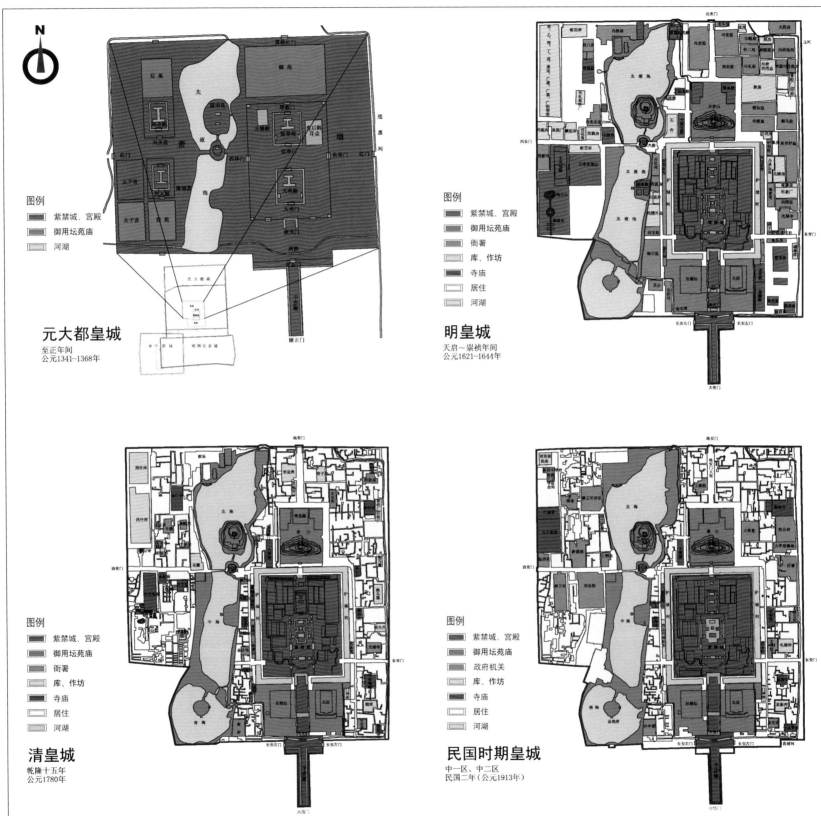

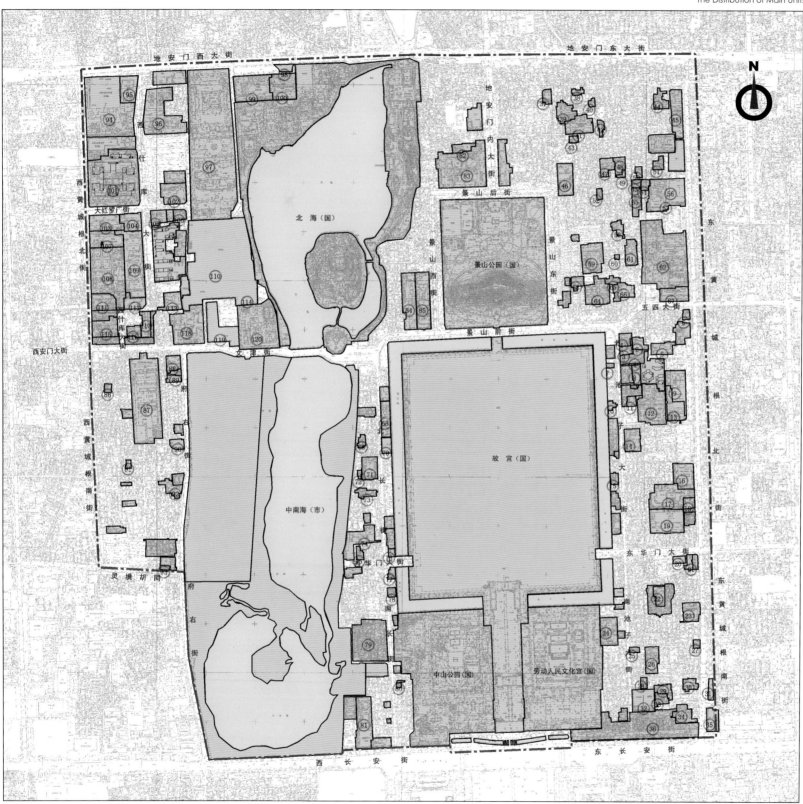

北京皇城

北京历史文化名城北京皇城保护规划
Conservation plan for the Historic City of Beijing and Imperial City of Beijing

土地使用功能现状图
Land Use

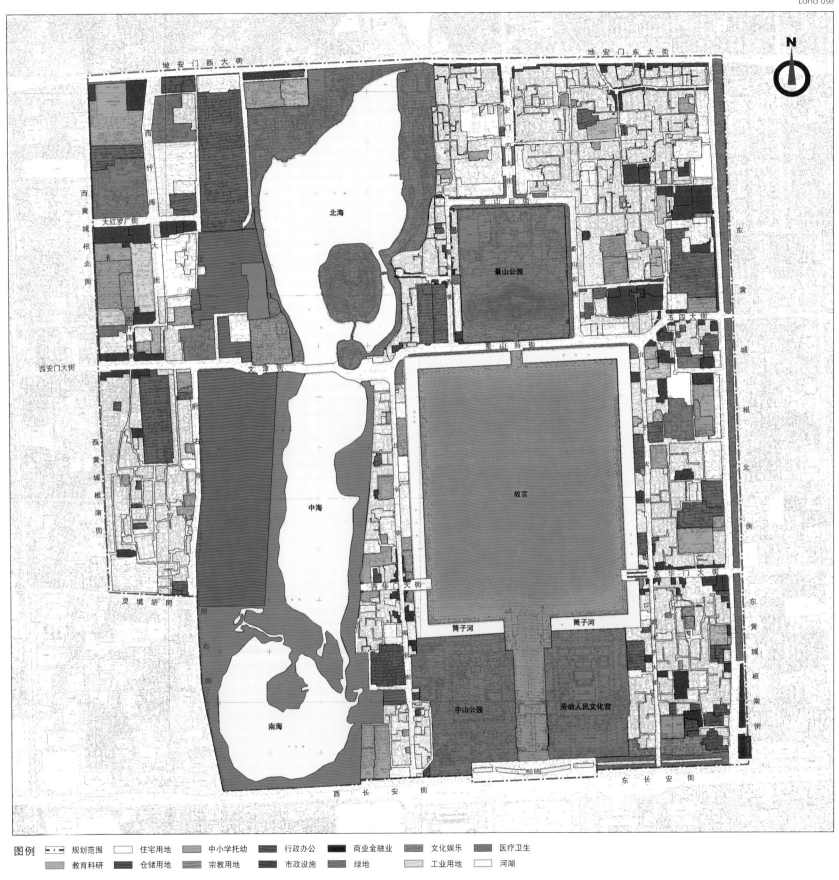

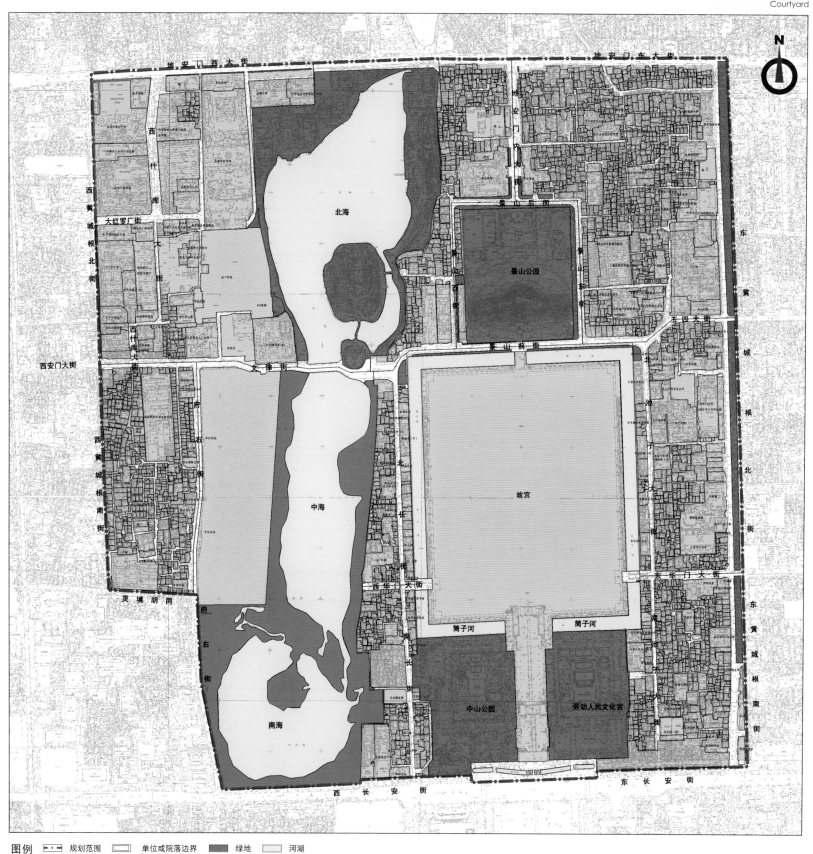

现状行政院落划分图
Courtyard

北京皇城保护规划·规划图纸

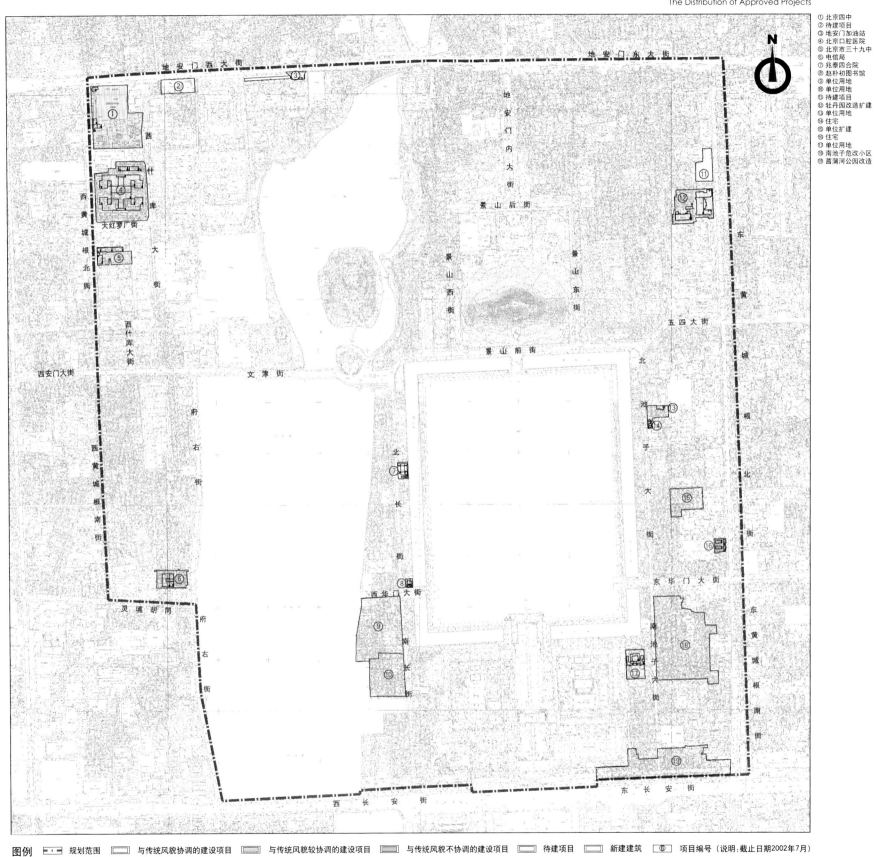

规 划 图 纸

现状居住人口分布图
Population

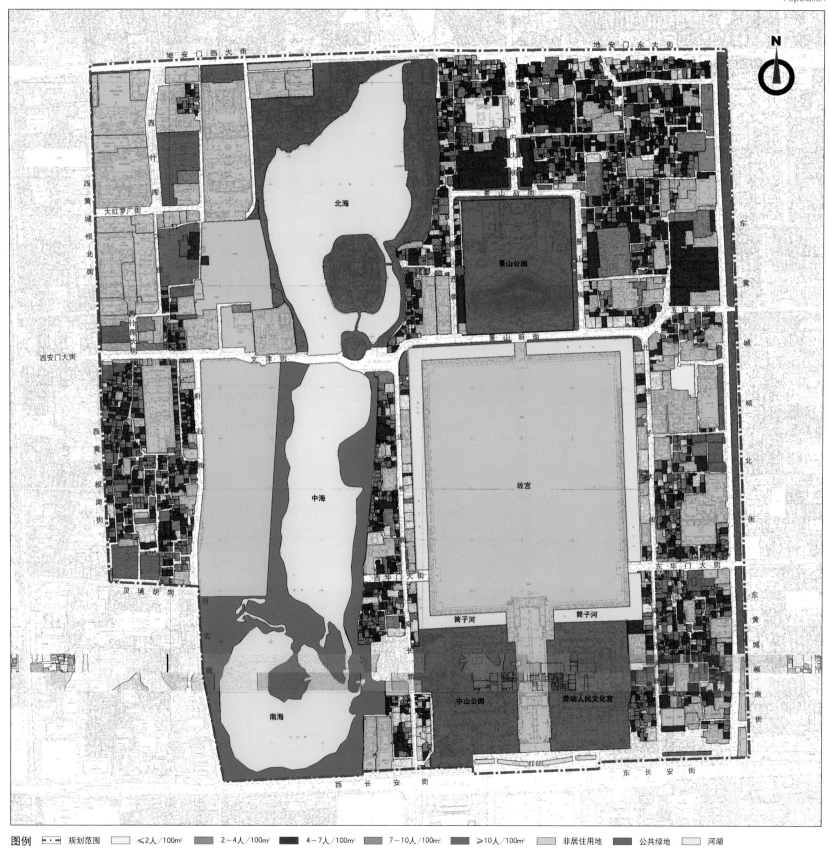

北京皇城保护规划·规划图纸

图例 ---规划范围 ≤2人/100m² 2~4人/100m² 4~7人/100m² 7~10人/100m² ≥10人/100m² 非居住用地 公共绿地 河湖

北京历史文化名城北京皇城保护规划

Conservation plan for the Historic City of Beijing and Imperial City of Beijing

现状街巷胡同分布图
Roads and Hutongs

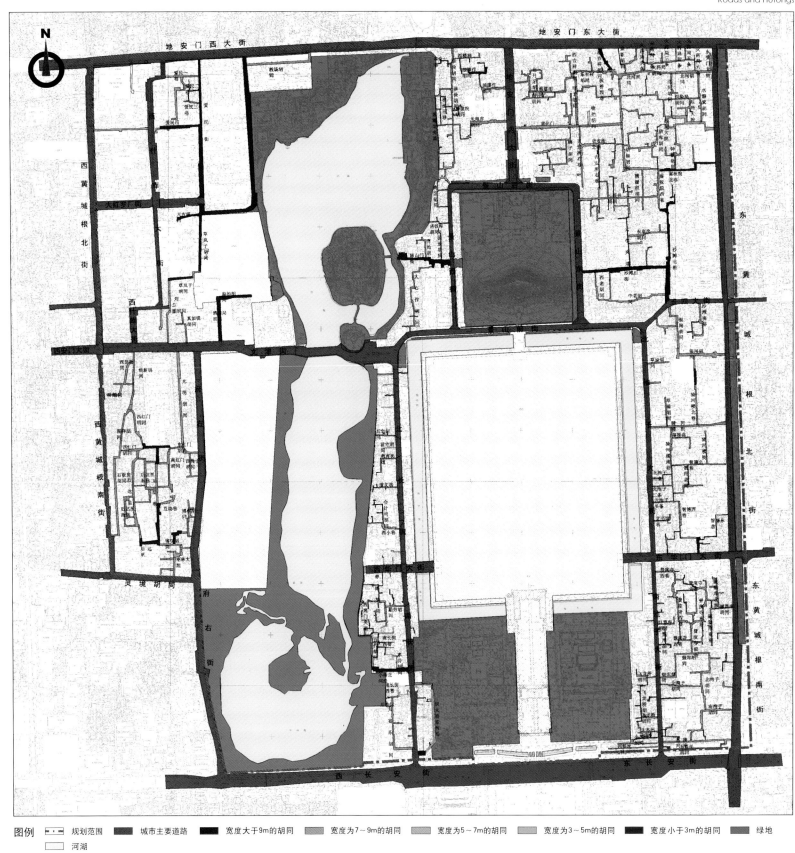

图例　规划范围　城市主要道路　宽度大于9m的胡同　宽度为7~9m的胡同　宽度为5~7m的胡同　宽度为3~5m的胡同　宽度小于3m的胡同　绿地　河湖

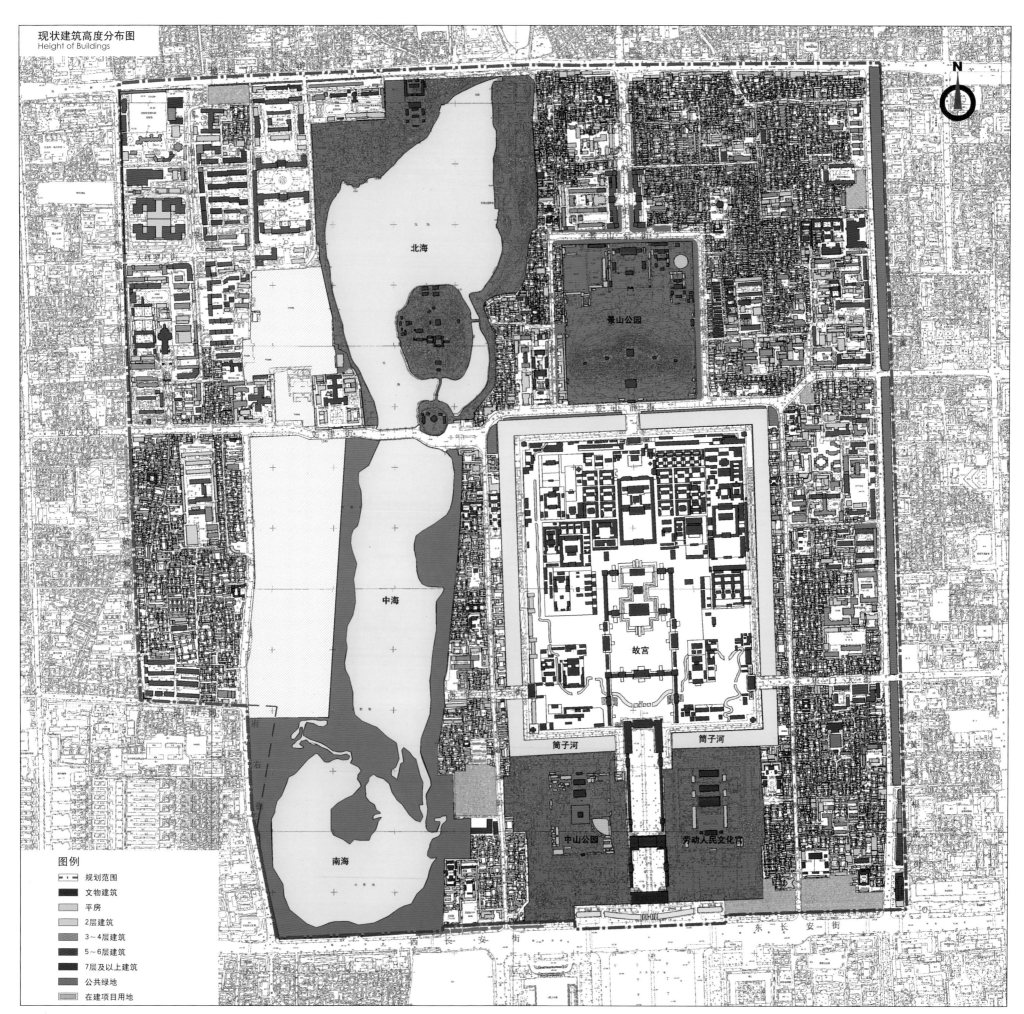

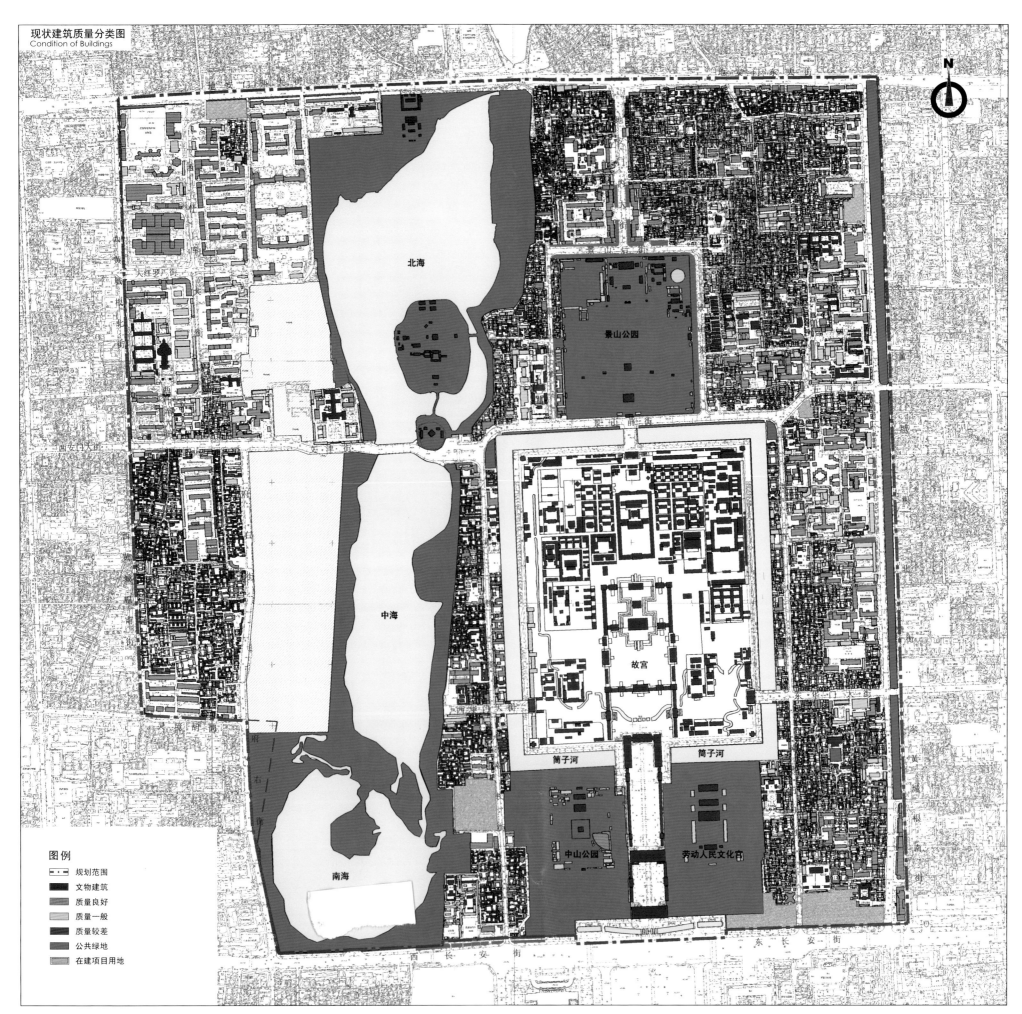

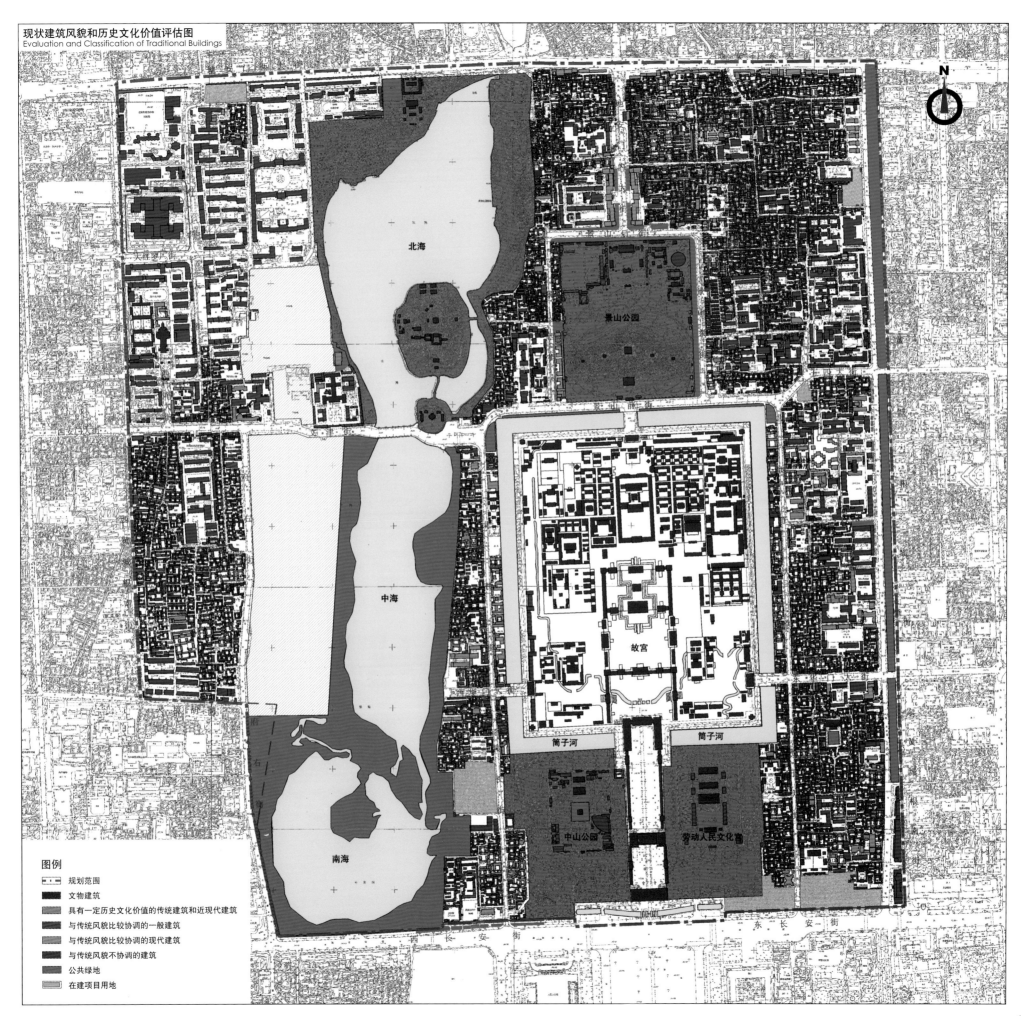

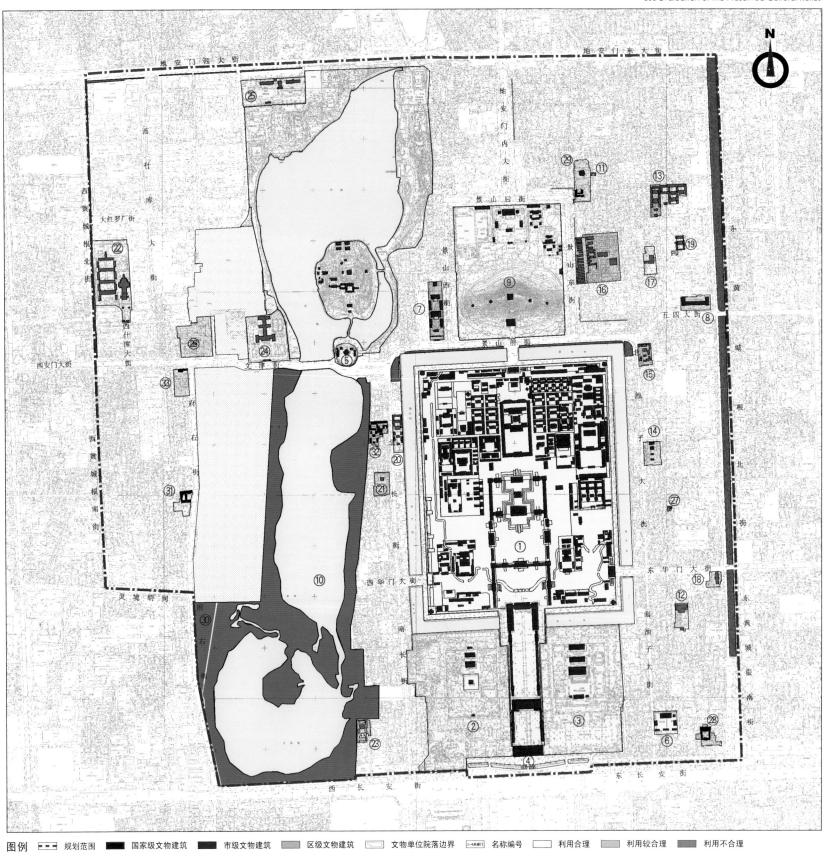

规划图纸

现状街区保护树木分布图
The Distribution of Preserved Trees

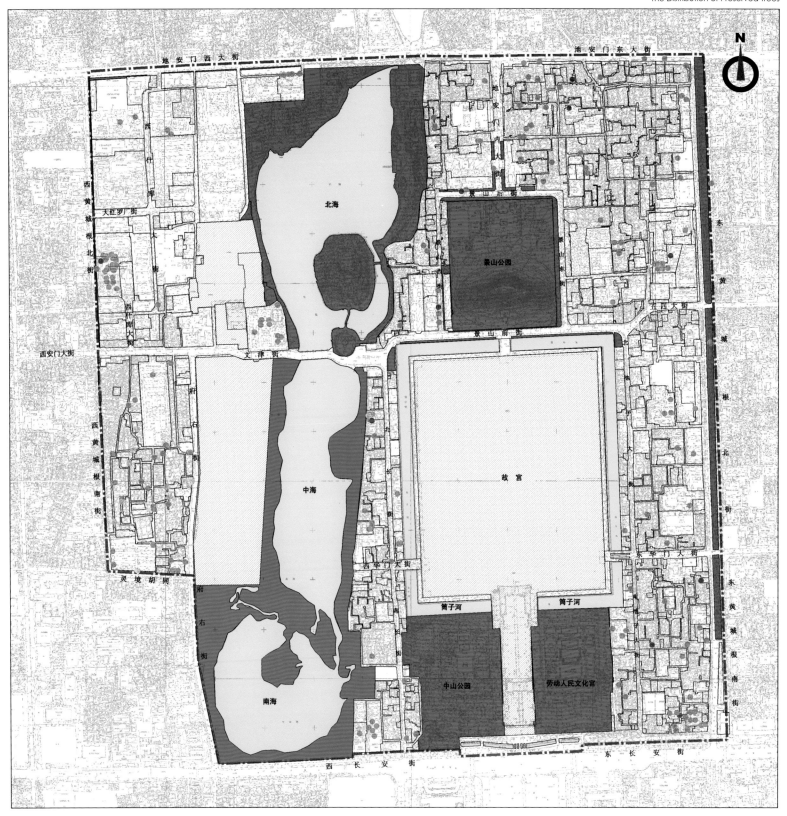

图例：规划范围　A类保护树木　B类保护树木　未挂牌保护树木　绿地　水体

说明：A类保护树木10棵，B类保护树木186棵，未挂牌保护树木15棵，合计211棵。（故宫、中山公园、劳动人民文化宫、景山公园、北海、中海、南海内的保护树木未统计在内）

北京皇城保护规划·规划图纸

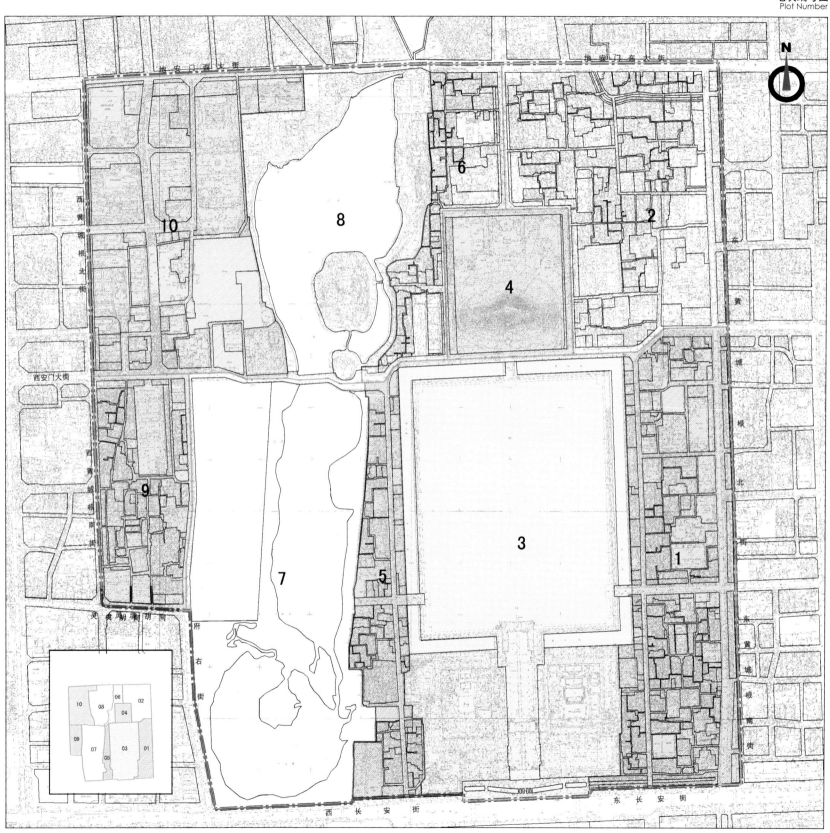

地块编号图
Plot Number

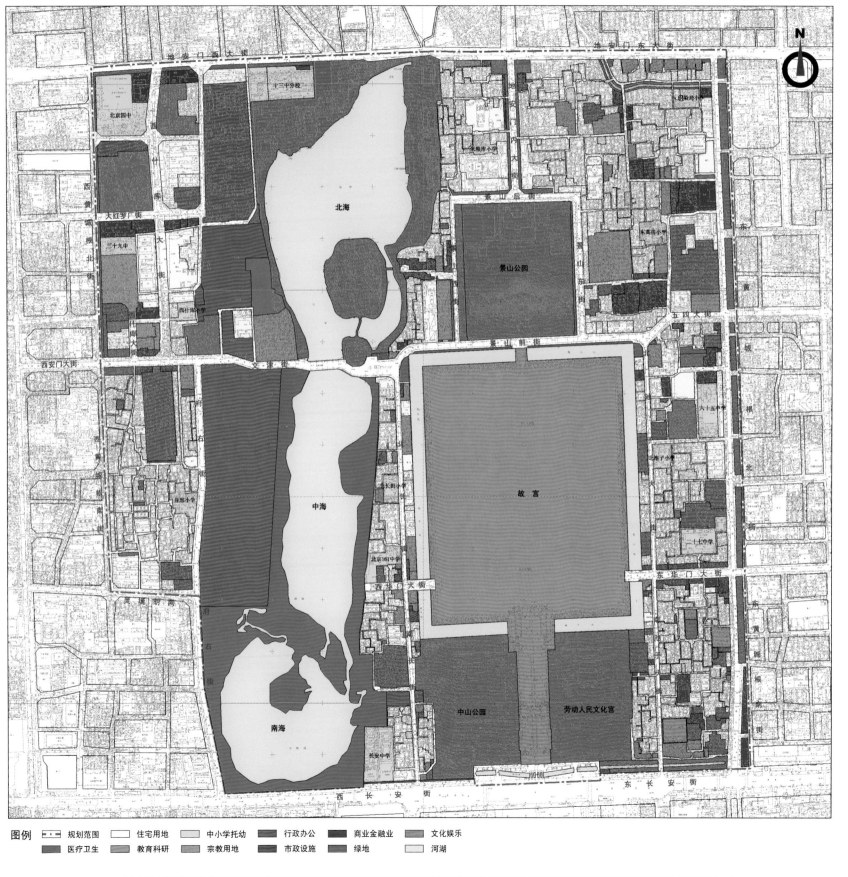

北京历史文化名城北京皇城保护规划
Conservation plan for the Historic City of Beijing and Imperial City of Beijing

绿化系统规划图
Planning of Green Network

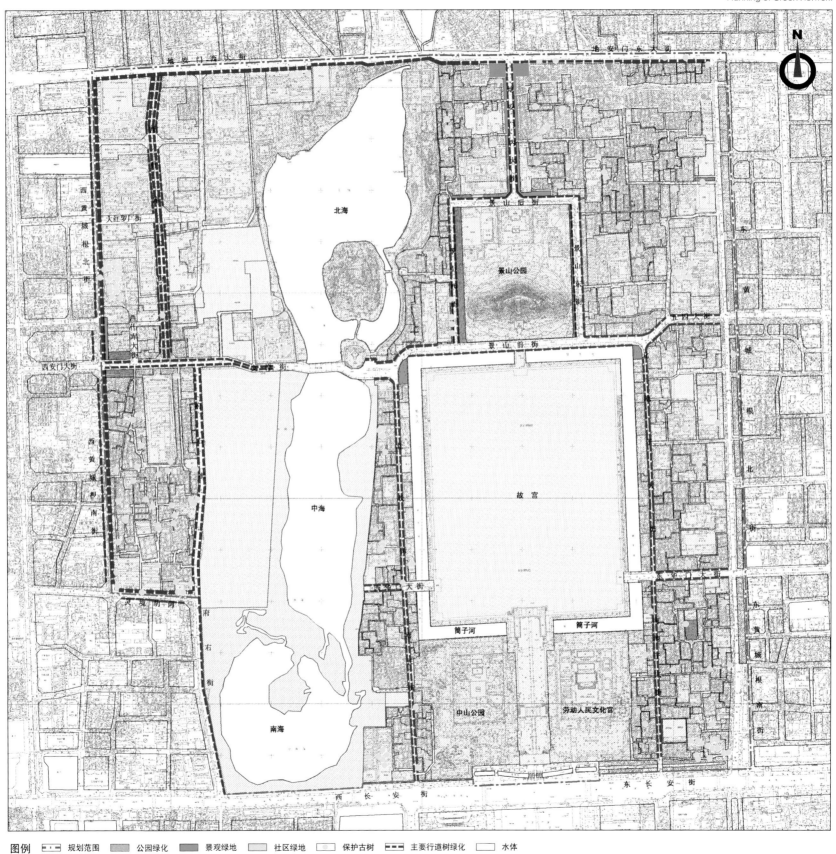

规划图纸

街巷胡同体系规划图
Planning of Hutongs System

北京皇城保护规划·规划图纸

图例: 规划范围 | 皇城周边城市干道 | 皇城内部城市主要道路 | 皇城内部街区主要胡同 | 皇城内部街区次要胡同 | 绿地 | 河湖

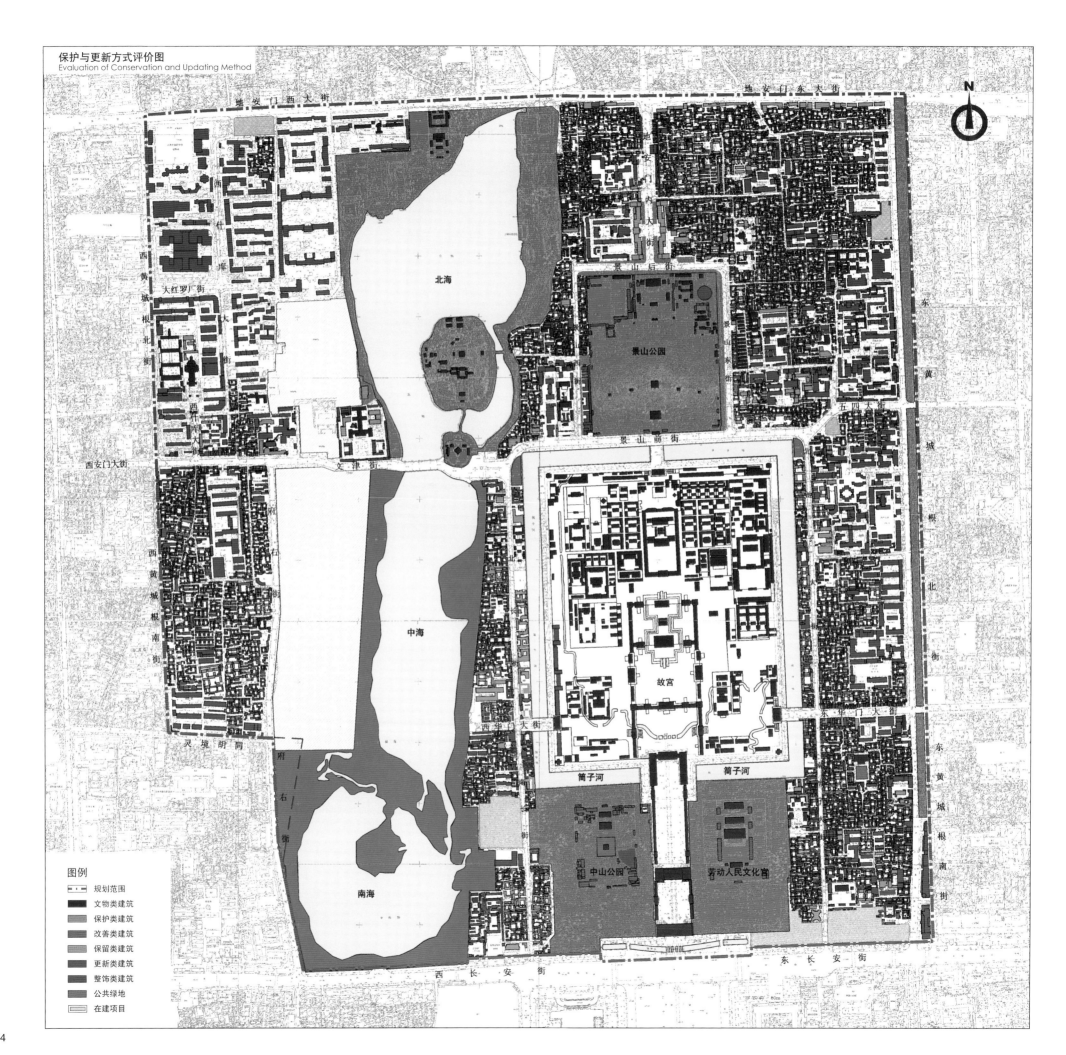

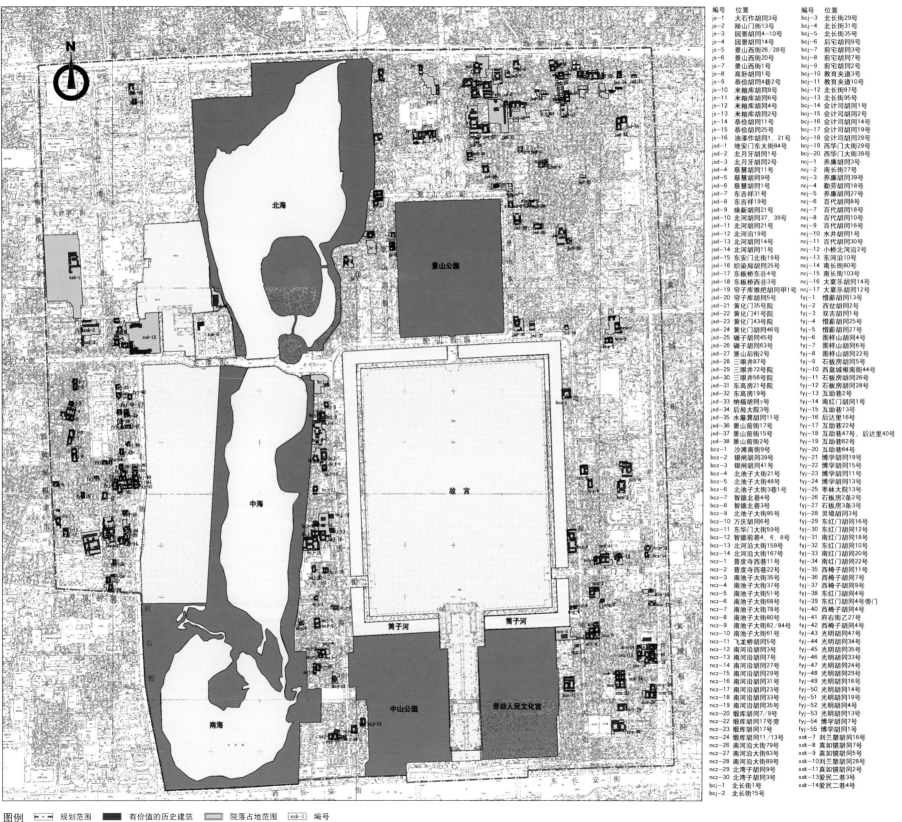

规划图纸

保护体系规划图（城墙、坛墙和水系）
Planning of Conservation System (City and Altar-Walls, Rivers System)

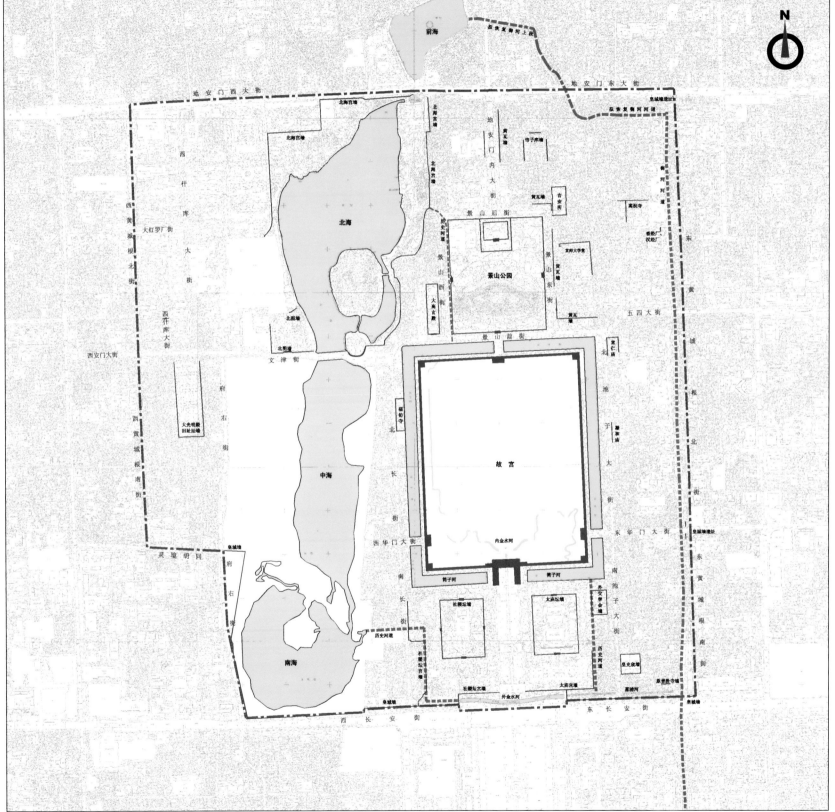

图例 ▬▬ 规划范围　▬▬ 城墙、坛墙遗存　▨ 现状水体　┈┈ 历史河道

北京皇城保护规划·规划图纸

北京皇城

北京历史文化名城北京皇城保护规划
Conservation plan for the Historic City of Beijing and Imperial City of Beijing

保护体系规划图（胡同）
Planning of Conservation System (Hutongs)

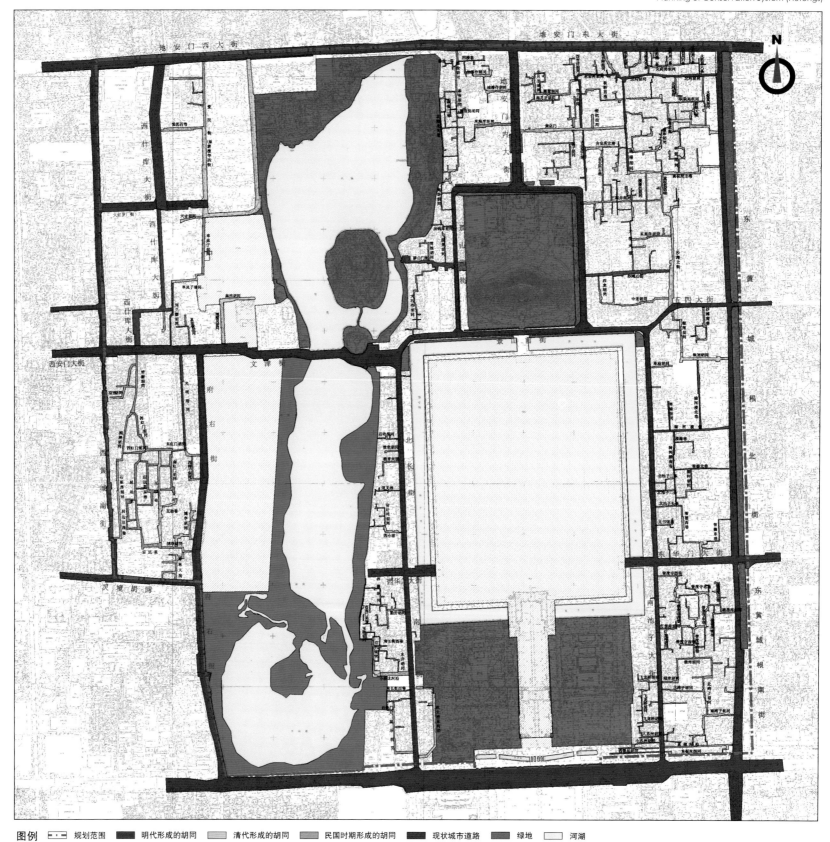

图例　规划范围　明代形成的胡同　清代形成的胡同　民国时期形成的胡同　现状城市道路　绿地　河湖

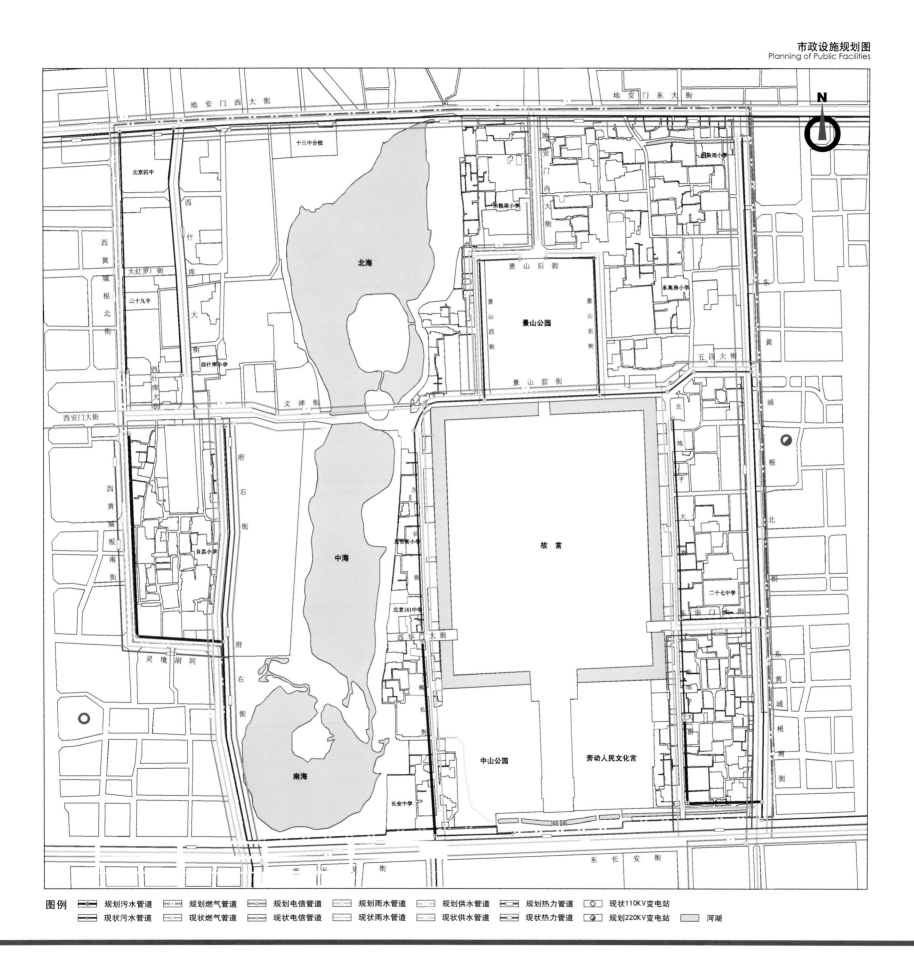

胡同内市政管道横断面图
Section of Hutongs

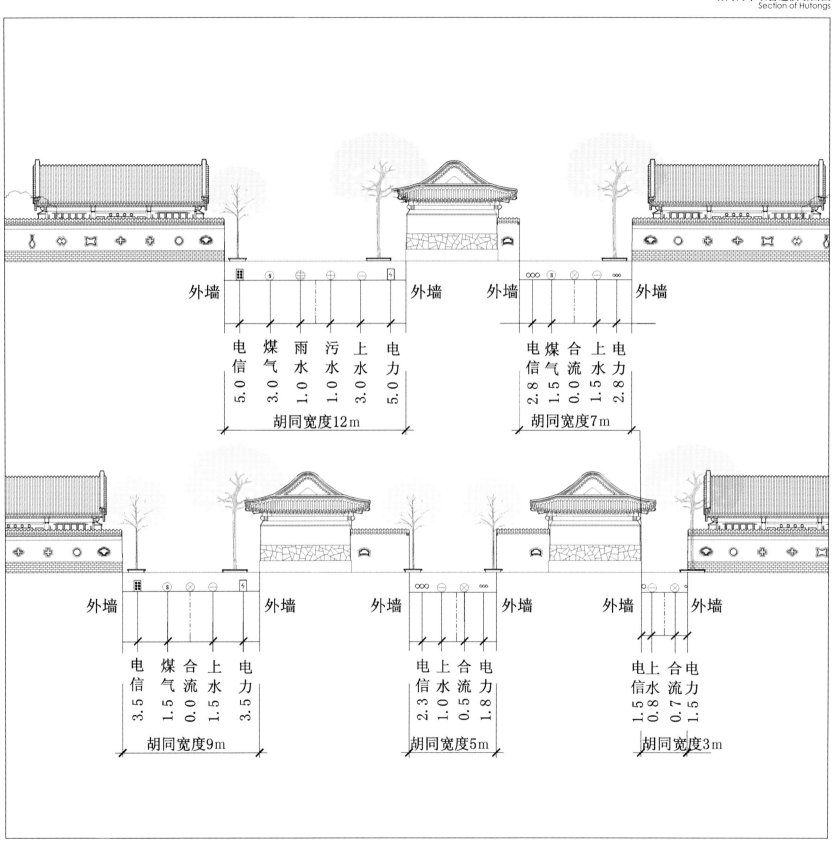

说明：图中数字标注为该管道中心线距道路中心线距离，单位为米(m)。

规划图纸

传统空间格局(特色)分析图
Analysis of Traditional Space Pattern

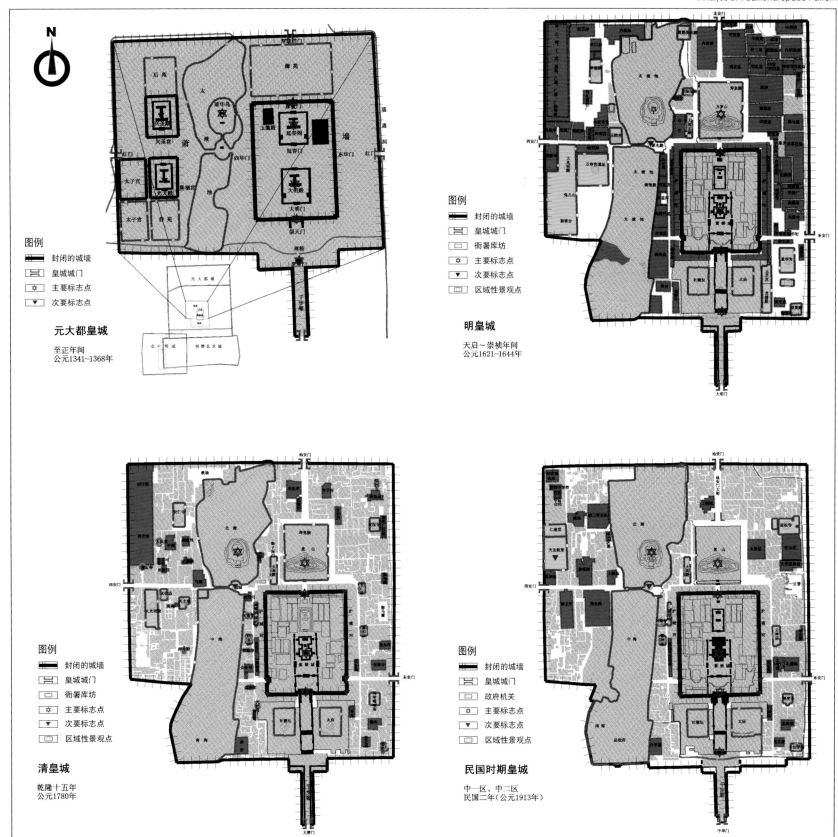

元大都皇城
至正年间
公元1341~1368年

明皇城
天启~崇祯年间
公元1621~1644年

清皇城
乾隆十五年
公元1780年

民国时期皇城
中一区、中二区
民国二年(公元1913年)

北京皇城保护规划·规划图纸

北京皇城

北京历史文化名城北京皇城保护规划
Conservation plan for the Historic City of Beijing and Imperial City of Beijing

现状空间格局分析图
Analysis of Current Space Pattern

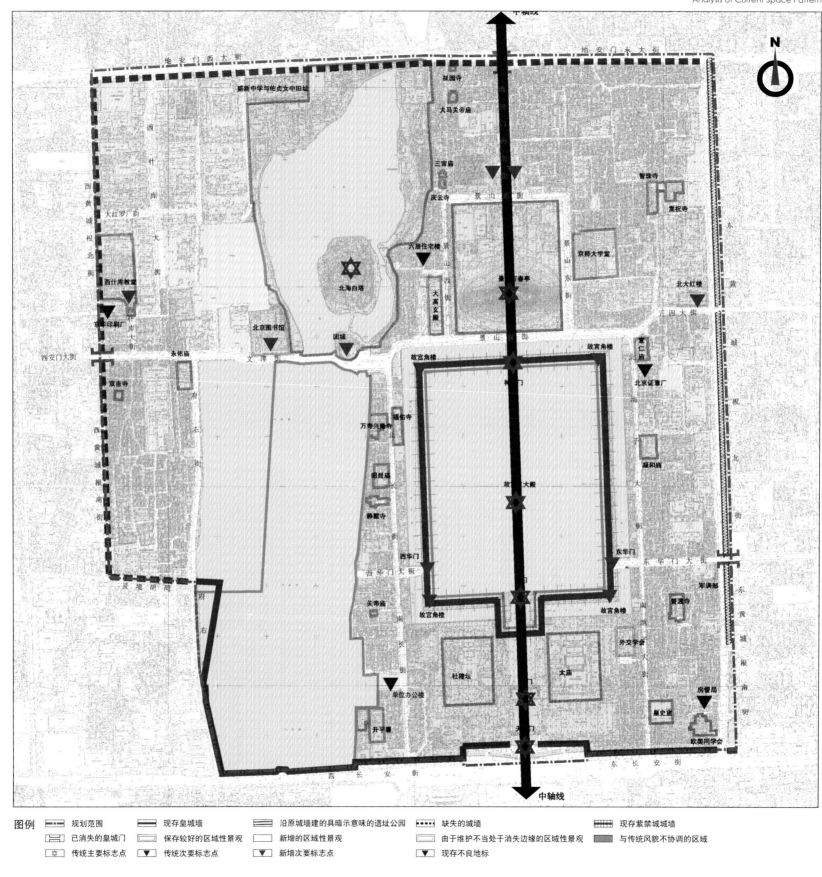

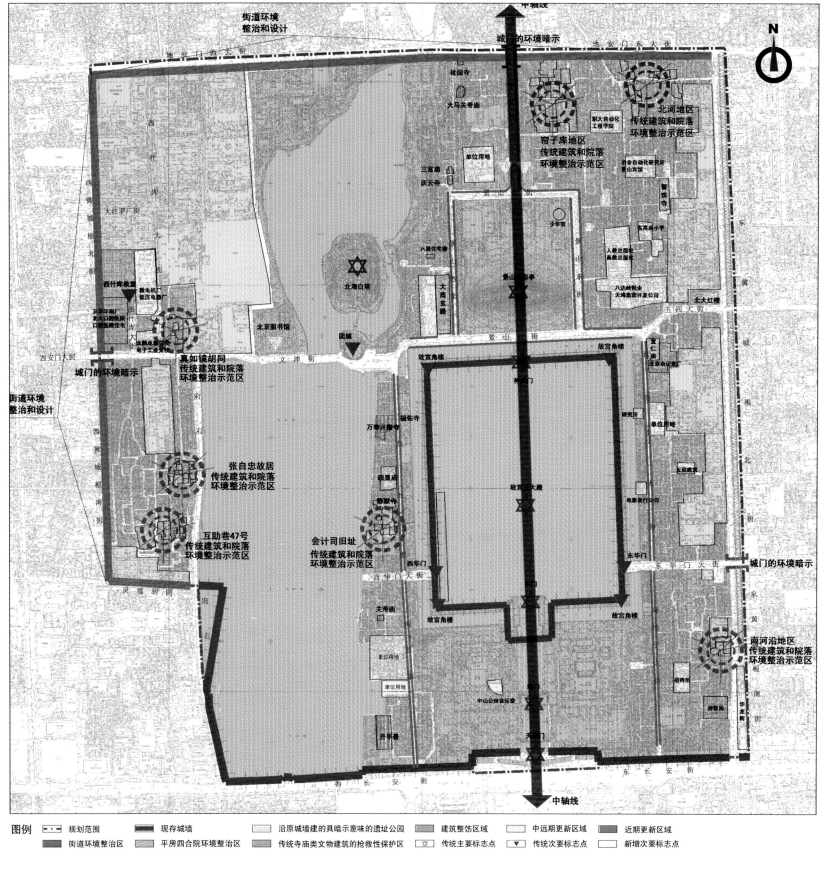

环境整治分析图
Analysis of Environment Rectification

北京皇城保护规划·规划图纸

北京历史文化名城北京皇城保护规划
Conservation plan for the Historic City of Beijing and Imperial City of Beijing

多层建筑整治更新计划图
Rectification of Multi-Storied Buildings

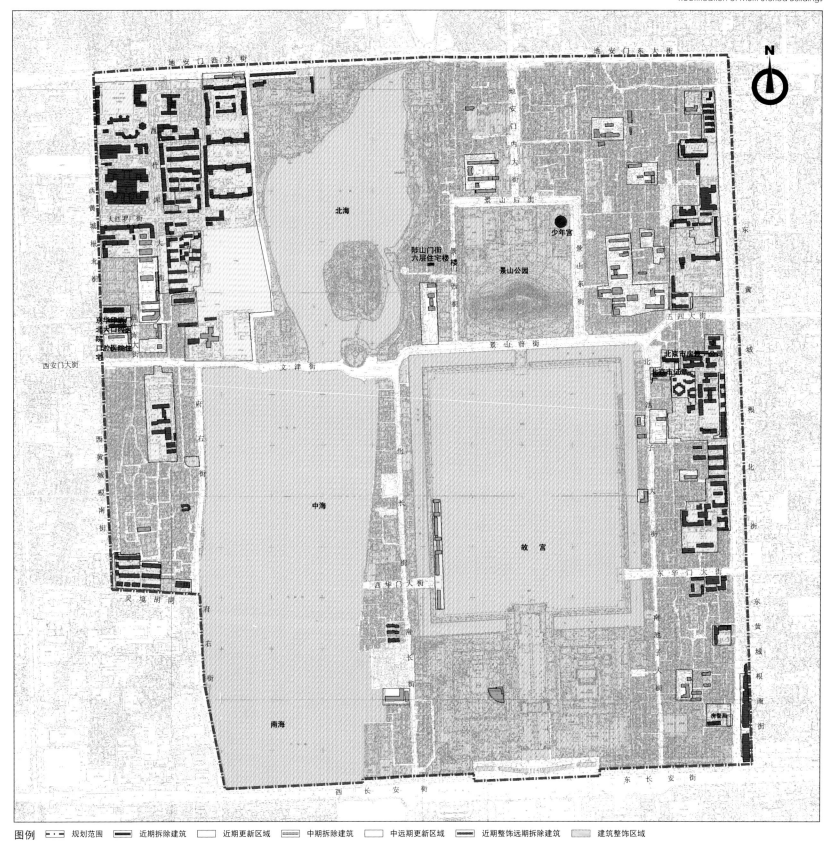

《北京皇城保护规划》专家评审会评审意见

2002年8月2日,市规划委、市文物局、市规划院联合召开了《北京皇城保护规划》(以下简称"保护规划")专家论证会,刘敬民副市长主持会议。

1. 对"保护规划"的总评价

专家们一致认为:由北京市规划院编制完成的《北京皇城保护规划》是一件非常重要的事情,对进一步加强北京历史文化名城的保护,加强对皇城的保护具有重要意义。这次"保护规划"内容翔实、具体,调查深入、细致,概念明了、清晰,在理论和实践上发展了对历史文化保护区保护的内涵,是一个令人兴奋的好规划。专家们建议对"保护规划"进行必要的修改后,尽快报请市政府审查批准,并在规划管理和实践中积极贯彻和组织实施。

2. 建议和意见

(1) 关于"保护规划"的总体要求

"保护规划"不同于一般规划,它没有规划期限,是一个逐步实施的过程,因此要一步到位,对保护、整治的内容尽可能提出较高的要求,不要太迁就近期实施的困难。(王景慧、董光器)

明朝,整个皇城就是紫禁城的后院,一个王府都没有,都是服务机构,在清朝却不一样,应把明朝和清朝时期主要的宫廷服务机构在图上标示出来。(王世仁)

皇城是一个历史概念,主要指明清皇城,应强调它的整体性,除各级文物、第一批历史文化保护区中的14片街区外,皇城西部地区(府右街、西什库地区)也是皇城的有机组成部分,应提出严格的保护控制要求。(王景慧)

对保护的内容要尽量明确,尤其作为立法和宣传,到底保什么不保什么,以什么方式保护,这样可以避免出现理解上的差异。(王景慧)

(2) 关于保护范围

由于大部分皇城城墙没有了,大家对皇城的范围认识不明确。加强皇城景观范围可视性的宣传,假设用绿化和其他标识标示出来非常重要。(赵中枢)

(3) 关于建设控制地带

25片历史文化保护区有一个核心保护区和建设控制地带,皇城里是否也要有一点区别,如西什库周边几乎已成为多层建筑区,想恢复原貌已不可能。(王景慧)

皇城的外围环境应当加以控制,应该加一个皇城保护的建设控制范围或者叫建设控制地带,这样皇城以外的建设就能得到控制,如什刹海体校、中南海的南侧地块也应该控制。(李准、刘小石)

(4) 关于用地功能

皇城内应适当降低行政办公、医疗卫生、工业、仓储、商业金融、教育科研等用地规模和建筑规模的比例,大容量的内容不能再往皇城内放,需要做减法,一个是建筑量减,一个是用地面积减。原则上应当外迁与皇城性质不符合的各类土地。(李准)

(5) 关于建筑高度的控制

皇城是历史文化保护区,其建筑高度的控制应当严格,一般都是平房,个别建二层,不应允许建设三层以上的房屋。(李准、王景慧)

(6) 关于文物及有历史价值的建筑的保护

许多文物保护单位的使用情况并不合理,故宫也不是铁板一块,里面有许多不符合文物保护要求的内容,文物保护单位仍有很多工作需要做。如故宫里的一些宫殿当库房用,应该提出来加以改进。(董光器、王世仁、赵中枢)

对文物使用不合理的单位应予以腾退、外迁,如景山公园内的少年宫、大高玄殿、国防部大楼、欧美同学会。(王世仁、宣祥鎏)

要保护历史名城的风貌,最现实的就是保护四合院,这是数量最多的,也是最有价值的。(刘小石)

(7) 关于对不协调建筑的处理

对破坏皇城风貌的建筑应进一步细分,采取不同的政策进行控制和管理,区别对待,提出各自的处理方案来,有的搬迁、腾退、恢复原貌,有的保留、利用,有的改造或拆除。(董光器、宣祥鎏)

应明确提出对环境有破坏作用、未来需拆除的建筑,如故宫西华门内加建的第一历史档案馆、中山音乐堂、南长街的七层办公楼等。(王世仁、傅熹年、董光器、李准)

对有些近期无法解决的问题,如某些多、高层建筑的拆除,只能从长远角度出发,留给后人去解决,但"保护规划"必须将需要更新的建筑明确提出来,逐步解决问题,逐步恢复皇城的风貌。(傅熹年)

(8) 关于水系

皇城内有4条历史河道,应尽可能创造条件加以恢复。(王世仁、董光器)

(9) 关于人口的疏解

皇城的居住人口密度较大,逐步向外疏散人口是改善环境的必要条件,应该计算皇城人口的合理容量。(董光器、王景慧、刘小石)

我赞成居民外迁,但不要强制,不要动不动就变成政府行为。应该说危改不要以区里为主,不要以开发公司为主,应该以居民为主,居民有权选择自建或者合作方式,或者居民委托开发公司来改造。(刘小石)

通过对合理的人口和交通容量进行研究以后,最好对危房改造有一个实施建议,有一个内外对应解决的办法。(赵中枢)

(10) 关于道路交通

道路交通问题需要专题研究,如南北长街、南北池子的宽度问题等,这些道路宽度基本上维持现状,有利于保护,但应和旧城整体交

通系统保持衔接。历史地段的道路不要拓宽，这是国际通行的办法。（刘小石、宣祥鎏、李准）

应当保护现存的胡同肌理，哪怕盖楼房，胡同也要留住，北京老城里边剩下的一些四合院，更重要的应属胡同的肌理和格局，希望在规划当中能重点提出。（王景慧）

根据提高居民生活质量和改善街区市政基础条件的要求，允许适当打通或拓宽个别胡同，但是总体上要满足保护的要求。（李准）

（11）关于皇城环境和景观的整治

皇城整体在景观上的目标是作为故宫和景山的背景，这一点规划比较明确。（王世仁）

皇城保护对历史真实性要求更高，对于历史的环境应采取恢复、展示、提示、标志等手段，突出皇城的文化内涵。（王世仁、傅熹年、王景慧）

西北角地区已成为多层建筑区，对皇城风貌破坏较大，应提出拆改对策，力争远期逐步恢复传统风貌。（宣祥鎏、董光器、傅熹年）

非常同意"保护规划"把不良标示点、不合理使用的文物、与传统风貌不协调的区域标示在图上。这三个方面让居民参与了解，群众就会拥护我们的规划。规划跟群众见面，这一点非常好。（赵中枢）

（12）关于危旧房改造

南池子改造社会争论较大，应对改造的方式和控制的程度提出一些细致的规则，避免争论。（董光器）

传统四合院是做政府官员的官邸，还是贵族化卖给谁，它的出路需要慢慢来解决。今天的看法和明天的看法不一样，不妨留给后代。只要这些四合院还留着就好说，如果拆了就再也没有了。（王景慧）

对危旧房屋的改造方式，不必急于下结论，应留给后人去解决，眼前重要的是改善一下特别差的环境，改善一下市政基础设施，改善一下危房，不是先将传统房子拆掉。希望这个规划要考虑一下操作的问题。总之，用开发的办法解决古城保护肯定是不可能的。（王景慧、李准）

皇城内不得进行"开发"式的建设，或者说不应该由区里掌握开发权，或让开发公司主导，这个机制必须改，否则必然使城市变丑。我赞成不通过开发而采取"微循环"式的建设方式，就是对危旧房慢慢改，不能推平头。（李准、刘小石）

（13）关于实施建议

"保护规划"审批通过后，应尽快编制一个详细的管理法规和详细规划，严格控制各类建设。（董光器、傅熹年）

在加强皇城保护的同时，应积极筹备申报世界文化遗产的事宜。（傅熹年、宣祥鎏）

应根据历史记载将一些重要的历史遗迹标示出来，让大家了解这个地方原来是什么样，原来是怎么布置的，原来是什么东西，使人们在游览中有所启迪，增强皇城的文化积淀。（傅熹年）

"保护规划"应该有一个近期实施行动的规划，在三年或五年之内要干什么，干到什么程度，建筑、市政、道路要怎么样，要有一个说法。（宣祥鎏）

多层建筑平改坡可以，但希望能再砍掉一层，而不是在这一层加坡顶，否则又高出半层。（王景慧）

原则上应停止审批楼房和与传统建筑不协调的建筑，一般不要批楼房。（李准）

应该一个点一个点地研究多、高层建筑对皇城的影响，应该拿出一个单子，应该具体化、明确化。（宣祥鎏）

附：出席会议专家名单

傅熹年	中国建筑设计研究院	工程院院士
宣祥鎏	首都规划建设委员会	原副主任兼秘书长
王景慧	中国城市规划设计研究院	总工程师
	中国历史文化名城保护专家委员会	秘书长
李 准	原北京市城市规划管理局	总工程师
王世仁	北京市文物局	首席研究员
刘小石	原北京市城市规划管理局	总工程师
董光器	原北京市城市规划设计研究院	副院长
赵中枢	中国城市规划设计研究院	教授级高级工程师

附件

北京市人民政府市长办公会议纪要

2002年10月16日下午,刘淇市长主持召开第142次市长办公会议。会议讨论了：

市规划委、市文物局、市规划院关于报审《北京皇城保护规划》的请示

会议讨论结果如下：

1. 关于北京皇城保护规划

会议原则同意市规划委、市文物局、市规划院提出的《北京皇城保护规划》(以下简称"规划"),由市规划委、市文物局和市规划院根据会议所提意见对其进行修改完善,报首都规划委员会审议后组织实施。

2. 会议决定：

(1) 皇城内人口应逐步向外疏散。要进一步研究皇城内人口降低比例问题。

(2) 关于皇城西、北边界区域规划一条5m左右绿化带问题,需进一步研究后确定。

(3) 皇城内严重影响皇城整体空间景观的多、高层建筑必须拆除,与皇城不协调的多层建筑,要加强整饰工作,涉及到的市政府有关部门和市属企事业单位要带头做好有关工作。

(4) 由市规划委、市文物局、市规划院负责制订落实《规划》的行动计划,根据实际情况分步实施。

(5) 关于皇城内市政设施建设问题,需根据南池子地区试点情况确定方案。

2002年10月20日

首都规划建设委员会第22次全体会议纪要

2002年10月17日,首都规划建设委员会召开第22次全体会议。会议由首规委副主任刘淇同志主持,首规委主任汪光焘、马凯、刘敬民、周友良以及首规委委员等出席会议,首规委主任贾庆林同志出席会议并作了重要讲话。

会议审议了《北京皇城保护规划》,现将会议讨论和议定的事项纪要如下：

关于《北京皇城保护规划》

(1) 北京皇城保护是北京历史文化名城保护的核心,对全国乃至世界都有重要影响。制定《北京皇城保护规划》非常重要,原则同意规划内容,要进一步深化。《北京皇城保护规划》的实施难度大,关键要统一认识。既要强调发展服从保护的要求,以保护为主,修缮并合理利用文物、修饰现有建筑、整治环境；还要为党、政、军首脑机关办公创造良好的环境。

(2) 进一步完善规划,把保护、修缮、整治的目标落实到具体的建筑、文物、院落、街道和胡同中,制定实施方案。

(3) 在保护风貌的前提下,可适当增加一些中央办公用地。

(4) 有关皇城申报世界文化遗产问题,涉及的方面和内容很复杂,建议再研究。

(5)《北京皇城保护规划》修改后,组织首规委专家论证,然后由市政府报国务院审批。

关于报请批复《北京皇城保护规划》的请示

市政府：

北京皇城总用地约6.8km²，是北京历史文化名城的重要组成部分，具有极高的历史文化价值。为加强皇城的整体保护，深入落实《北京历史文化名城保护规划》，根据市领导的指示，市规划委、市文物局组织编制了《北京皇城保护规划》。

此次编制《北京皇城保护规划》的主要依据是《北京城市总体规划》、《北京历史文化名城保护规划》、《北京旧城25片历史文化保护区保护规划》等，指导思想是把皇城作为一个整体加以保护，正确处理皇城的保护与建设、发展与变化的关系，并贯彻"以人为本"和"可持续发展"的思想。规划的主要内容包括：规划范围、皇城性质和历史文化价值、土地分区和特征、土地使用功能、文物保护单位及有历史文化价值的建筑的保护和利用、城廓、城墙、坛墙、水系、绿地、道路和胡同体系、建筑高度控制、保护与更新方式评价、市政设施规划及皇城整体环境整治等等，并提出实施建议。

2002年10月16日，经第142次市长办公会研究，原则同意《北京皇城保护规划》，并要求根据会议所提意见修改完善，报首都规划建设委员会审议。10月17日，经首都规划建设委员会第22次全体会议讨论，原则同意该规划的内容，并提出了完善规划的意见。2003年1月，我委将修改后的规划上报市政府审批。2003年2月24日，根据刘淇书记的指示精神，由市政府组织召开了《北京皇城保护规划》专家论证会，邀请吴良镛、周干峙等11位专家及人大代表、政协委员，对《北京皇城保护规划》作进一步的论证。会议由孟学农市长主持，有关领导出席，刘淇书记做了重要指示。现已根据专家意见和会议要求，对"规划依据"、"规划指导思想"、"规划范围"、"皇城的性质和历史文化价值"、"城廓、城墙、坛墙"、"水系"、"绿化"、"保护与更新方式评价"、"环境整治的实施建议"等方面内容进行了研究、修改、充实和完善，已基本达到上报审批的要求。因此，建议市政府对《北京皇城保护规划》予以批复。

妥否，请示。

北京市规划委员会
2003年3月20日

北京市人民政府关于《北京皇城保护规划》的批复

市规划委：

你委《关于报请批复〈北京皇城保护规划〉的请示》（京规文【2003】356号）收悉。经研究，现批复如下：

1. 原则同意《北京皇城保护规划》（以下简称《规划》）。皇城是北京历史文化名城的重要组成部分，《规划》是《北京历史文化名城保护规划》的深化，是加强皇城整体保护，正确处理皇城保护与城市现代化建设关系的重要保障。各有关区政府和市政府各有关部门要认真贯彻执行。

2. 皇城的规划范围为东至东黄（皇）城根，南至东、西长安街，西至黄（皇）城根、灵境胡同、府右街，北至平安大街。规划占地面积约6.8km²。

3. 同意《规划》的指导思想，即把皇城作为一个整体加以保护，正确处理皇城保护与城市现代化建设的关系，建设要服从保护要求，保证皇城整体风貌与空间格局的延续，坚持"以人为本、保护为主"的方针，保持可持续发展。

4. 加强对紫禁城等重点文物保护单位的保护，搞好文物的普查和升级工作。对尚未列入文物保护单位但具有一定历史文化价值的建筑或院落，要按照文物保护要求进行保护，并改善其使用环境。

5. 维护皇城城市形态和生态环境，强化皇城历史边界的可识别性，对城廓、城墙、坛墙等加以保护；严格保护皇城内的水系、绿地和现有道路、胡同，保持皇城传统空间尺度和肌理，保留传统地名和胡同名。

6. 严格控制皇城内建筑高度，保护其传统的平缓、开阔的空间形态。在对皇城内现状建筑质量和建筑风貌充分评估的基础上，分类确定保护和更新方式。

7. 以历史文化保护为前提，加快研究制订房屋产权改革、房屋管理和修缮、人口外迁和疏导的相关政策，因地制宜地规划建设市政基础设施，逐步改善工作和生活条件。

8. 市计划、规划、财政、文物、园林、国土房管、市政管理等有关部门要建立相应的工作制度，明确管理责任，落实保护资金，规范建设和管理行为。东城区政府和西城区政府要严格执行《规划》，切实做好相关工作。

北京市人民政府
2003年4月7日

附件

有关历史文化名城保护的国内、国际法案摘编

1. 国际古迹保护与修复宪章
(第二届历史古迹建筑师及技师国际会议
于1964年5月25~31日在威尼斯通过)

世世代代人民的历史古迹,饱含着过去岁月的信息留存至今,成为人们古老的活的见证。人们越来越意识到人类价值的统一性,并把古代遗迹看作共同的遗产,认识到为后代保护这些古迹的共同责任。将它们真实地、完整地传下去是我们的职责。

古代建筑的保护与修复指导原则应在国际上得到公认并作出规定,这一点至关重要。各国在各自的文化和传统范畴内负责实施这一规划。

1931年的雅典宪章第一次规定了这些基本原则,为一个国际运动的广泛发展做出了贡献,这一运动所采取的具体形式体现在各国的文件之中,体现在国际博物馆协会和联合国教育、科学及文化组织的工作之中,以及在由后者建立的国际文化财产保护与修复研究中心之中。一些已经并在继续变得更为复杂和多样化的问题已越来越受到注意,并展开了紧急研究。现在,重新审阅宪章的时候已经来临,以便对其所含原则进行彻底研究,并在一份新文件中扩大其范围。

为此,1964年5月25~31日在威尼斯召开了第二届历史古迹建筑师及技师国际会议,通过了以下文本:

定义

第一条:历史古迹的概念不仅包括单个建筑物,而且包括能从中找出一种独特的文明、一种有意义的发展或一个历史事件见证的城市或乡村环境。这不仅适用于伟大的艺术作品,而且亦适用于随时光流逝而获得文化意义的过去一些较为朴实的艺术品。

第二条:古迹的保护与修复必须求助于对研究和保护考古遗产有利的一切科学技术。

宗旨

第三条:保护与修复古迹的目的旨在把它们既作为历史见证,又作为艺术品予以保护。

保护

第四条:古迹的保护至关重要的一点在于日常的维护。

第五条:为社会公用之目的使用古迹永远有利于古迹的保护。因此,这种使用合乎需要,但决不能改变该建筑的布局或装饰。只有在此限度内才可考虑或允许因功能改变而需做的改动。

第六条:古迹的保护包含着对一定规模环境的保护。凡传统环境存在的地方必须予以保存,决不允许任何导致改变主体和颜色关系的新建、拆除或改动。

第七条:古迹不能与其所见证的历史和其产生的环境分离。除非出于保护古迹之需要,或因国家或国际之极为重要利益而证明其必要,否则不得全部或局部搬迁该古迹。

第八条:作为构成古迹整体一部分的雕塑、绘画或装饰品,只有在非移动而不能确保其保存的惟一办法时方可进行移动。

修复

第九条:修复过程是一个高度专业性的工作,其目的旨在保存和展示古迹的美学与历史价值,并以尊重原始材料和确凿文献为依据。一旦出现臆测,必须立即予以停止。此外,即使如此,任何不可避免的添加都必须与该建筑的构成有所区别,并且必须要有现代标记。无论在任何情况下,修复之前及之后必须对古迹进行考古及历史研究。

第十条:当传统技术被证明为不适用时,可采用任何经科学数据和经验证明为有效的现代建筑及保护技术来加固古迹。

第十一条:各个时代为一古迹之建筑物所做的正当贡献必须予以尊重,因为修复的目的不是追求风格的统一。当一座建筑物含有不同时期的重叠作品时,揭示底层只有在特殊情况下,在被去掉的东西价值甚微,而被显示的东西具有很高的历史、考古或美学价值,并且保存完好足以说明这么做的理由时才能证明其具有正当理由。评估由此涉及的各部分的重要性以及决定毁掉什么内容不能仅仅依赖于负责此项工作的个人。

第十二条:缺失部分的修补必须与整体保持和谐,但同时须区别于原作,以使修复不歪曲其艺术或历史见证。

第十三条:任何添加均不允许,除非它们不致于贬低该建筑物的有趣部分、传统环境、布局平衡及其与周围环境的关系。

第十四条:古迹遗址必须成为专门照管对象,以保护其完整性,并确保用恰当的方式进行清理和开放。在这类地点开展的保护与修复工作应得到上述条款所规定之原则的鼓励。

发掘

第十五条:发掘应按照科学标准和联合国教育、科学及文化组织1956年通过的适用于考古发掘国际原则的建议予以进行。

遗址必须予以保存,并且必须采取必要措施,永久地保存和保护建筑风貌及其所发现的物品。此外,必须采取一切方法促进对古迹的了解,使它得以再现而不曲解其意。

然而对任何重建都应事先予以制止,只允许重修,也就是说,把现存但已解体的部分重新组合。所用粘结材料应永远可以辨别,并应尽量少用,只须确保古迹的保护和其形状的恢复之用便可。

出版

第十六条:一切保护,修复或发掘工作永远应有用配以插图和照片的分析及评论报告这一形式所做的准确的记录。

清理、加固、重新整理与组合的每一阶段,以及工作过程中所确认的技术及形态特征均应包括在内,这一记录应存放于一公共机构的档案馆内,使研究人员都能查到,该记录应建议出版。

2. 关于历史地区的保护及其当代作用的建议
(联合国教育、科学及文化组织大会第十九届会议
于1976年11月26日在内罗毕通过)

联合国教育、科学及文化组织大会于1976年10月26日至11月30日在内罗毕举行第十九届会议。

考虑到历史地区是各地人类日常环境的组成部分，它们代表着形成其过去的生动见证，提供了与社会多样化相对应所需的生活背景的多样化，并且基于以上各点，它们获得了自身的价值，又得到了人性的一面；

考虑到自古以来，历史地区为文化、宗教及社会活动的多样化和财富提供了最确切的见证，保护历史地区并使它们与现代社会生活相结合是城市规划和土地开发的基本因素；

考虑到面对因循守旧和非个性化的危险，这些昔日的生动见证对于人类和对那些从中找到其生活方式缩影及其某一基本特征的民族，是至关重要的；

注意到整个世界在扩展或现代化的借口之下，拆毁（却不知道拆毁的是什么）和不合理不适当重建工程正给这一历史遗产带来严重的损害；

考虑到历史地区是不可移动的遗产，其损坏即使不会导致经济损失，也常常会带来社会动乱；

考虑到这种情况使每个公民承担责任，并赋予公共当局只有他们才能履行的义务；

考虑到为了使这些不可替代的财产免受它们所面临的退化甚至全部毁坏的危险，各成员国当务之急是采取全面而有力的政策，把保护和复原历史地区及其周围环境作为国家、地区或地方规划的组成部分；

注意到在许多情况下缺乏一套有关建筑遗产及其与城市规划、领土、地区或地方规划相互联系的相当有效而灵活的立法；

注意到大会已通过了保护文化和自然遗产的国际文件，如：《关于适用于考古发掘的国际原则的建议》(1956年)、《关于保护景观和遗址的风貌与特征的建议》(1962年)、《关于保护受到公共或私人工程威胁的文化财产的建议》(1972年)；

希望补充并扩大这些国际文件所确定的标准和原则的适用范围；

收到关于历史地区的保护及其当代作用的建议，该问题作为本届会议第27项议程；

第十八次会议决定该问题应采取向各成员国的建议的形式。于1976年11月26日通过本建议。

大会建议各成员国应通过国家法律或其他方式制定使本建议所规定的原则和准则在其所管辖的领土上生效的措施，以适用以上规定。

大会建议各成员国应将本建议提请与保护历史地区及其周围环境有关的国家、地区和地方当局、事业单位、行政部门或机构以及各种协会的注意。

大会建议各成员国应按大会决定的日期和形式向大会提交有关本建议执行情况的报告。

2.1 定义

(1) 为本建议之目的

1) "历史和建筑(包括本地的)地区"系指包含考古和古生物遗址的任何建筑群、结构和空旷地，它们构成城乡环境中的人类居住地，从考古、建筑、史前史、历史、艺术和社会文化的角度看，其凝聚力和价值已得到认可。

在这些性质各异的地区中，可特别划分为以下各类：史前遗址、历史城镇、老城区、老村庄、老村落以及相似的古迹群。不言而喻，后者通常应予以精心保存，维持不变。

2) "环境"系指影响观察这些地区的动态、静态方法的、自然或人工的环境。

3) "保护"系指对历史或传统地区及其环境的鉴定、保护、修复、修缮、维修和复原。

2.2 总则

(2) 历史地区及其环境应被视为不可替代的世界遗产的组成部分。其所在国政府和公民应把保护该遗产并使之与我们时代的社会生活融为一体作为自己的义务。国家、地区或地方当局应根据各成员国关于权限划分的情况，为全体公民和国际社会的利益，负责履行这一义务。

(3) 每一历史地区及其周围环境应从整体上视为一个相互联系的统一体，其协调及特性取决于它的各组成部分的联合，这些组成部分包括人类活动、建筑物、空间结构及周围环境。因此一切有效的组成部分，包括人类活动，无论多么微不足道，都对整体具有不可忽视的意义。

(4) 历史地区及其周围环境应得到积极保护，使之免受各种损坏，特别是由于不适当的利用、不必要的添建和诸如将会损坏其真实性的错误的或愚蠢的改变而带来的损害，以及由于各种形式的污染而带来的损害。任何修复工程的进行应以科学原则为基础，同样，也应十分注意组成建筑群并赋予各建筑群以自身特征的各个部分之间的联系与对比所产生的和谐与美感。

(5) 在导致建筑物的规模和密度大量增加的现代城市化的情况下，历史地区除了遭受直接破坏的危险外，还存在一个真正的危险：新开发的地区会毁坏临近的历史地区的环境和特征。建筑师和城市规划者应谨慎从事，以确保古迹和历史地区的景色不致遭到破坏，并确保历史地区与当代生活和谐一致。

(6) 当存在建筑技术和建筑形式的日益普遍化可能造成整个世界的环境单一化的危险时，保护历史地区能对维护和发展每个国家的文化和社会价值作出突出贡献，这也有助于从建筑上丰富世界文化遗产。

2.3 国家、地区和地方政策

(7) 各成员国应根据各国关于权限划分的情况制定国家、地区和地方政策，以便使国家、地区和地方当局能够采取法律、技术、经济和社会措施，保护历史地区及其周围环境，并使之适应于现代生活的需要。由此制定的政策应对国家、地区或地方各级的规划产生影响，并为各级城市规划，以及地区和农村发展规划，为由此而产生的并构成制定目标和计划重要组成部分的活动、责任分配以及实施行为提供指导。在执行保护政策时，应寻求个人和私人协会的合作。

2.4 保护措施

(8) 历史地区及其周围环境应按照上述原则和以下措施予以保护，

附件

具体措施应根据各国立法和宪法权限以及各国组织和经济结构来决定。

立法及行政措施

（9）保护历史地区及其周围环境的总政策之适用应基于对各国整体有效的原则。各成员国应修改现有规定，或必要时，制定新的法律和规章以便参照本章及下列章节所述之规定，确保对历史地区及其周围环境的保护。它们应鼓励修改或采取地区或地方措施以确保此种保护。有关城镇和地区规划以及住宅政策的法律应予以审议，以便使它们与有关保护建筑遗产的法律相协调、相结合。

（10）关于保护历史地区的制度的规定应确立关于制定必要的计划和文件的一般原则，特别是：

适用于保护地区及其周围环境的一般条件和限制；

关于为保护和提供公共服务而制定的计划和行动说明；

将要进行的维护工作并为此指派负责人；

适用于城市规划、再开发以及农村土地管理的区域；

指派负责审批任何在保护范围内的修复、改动、新建或拆除的机构；

保护计划得到资金并得以实施的方式。

（11）保护计划和文件应确定：

被保护的区域和项目；

对其适用的具体条件和限制；

在维护、修复和改进工作中所应遵守的标准；

关于建立城市或农村生活所需的服务和供应系统的一般条件；

关于新建项目的条件。

（12）原则上，这些法律也应包括旨在防止违反保护法的规定，以及防止在保护地区内财产价值的投机性上涨的规定，这一上涨可能危及为整个社会利益而计划的保护和维修。这些规定可以包括提供影响建筑用地价格之方法的城市规划措施，例如：设立邻里区或制定较小型的开发计划，授予公共机构优先购买权、在所有人不采取行动的情况下，为了保护、修复或自动干预之目的实行强制购买。这些规定可以确定有效的惩罚，如：暂停活动、强制修复和适当的罚款。

（13）个人和公共当局有义务遵守保护措施。然而，也应对武断的或不公正的决定提供上诉的机制。

（14）有关建立公共和私人机构以及公共和私人工程项目的规定应与保护历史地区及其周围环境的规定相适应。

（15）有关贫民区的房产和街区以及有补贴住宅之建设的规定，尤其应本着符合并有助于保护政策的目的予以制订或修改。因此，应拟定并调整已付补贴的计划，以便专门通过修复古建筑推动有补贴的住宅建筑和公共建设的发展。在任何情况下，一切拆除应仅限于没有历史或建筑价值的建筑物，并对所涉及的补贴应谨慎予以控制。另外，应将专用于补贴住宅建设的基金拨出一部分，用于旧建筑的修复。

（16）有关建筑物和土地的保护措施的法律后果应予以公开并由主管官方机构作出记录。

（17）考虑到各国的具体条件以及各个国家、地区和地方当局的责任划分，下列原则应构成保护机制运行的基础：

1）应设有一个负责确保长期协调一切有关部门，如国家、地区和地方公共部门或私人团体的权力机构；

2）跨学科小组一旦完成了事先一切必要的科学研究后，应立即制订保护计划和文件，这些跨学科小组特别应由以下人员组成：

保护和修复专家，包括艺术史学家；

建筑师和城市规划师；

社会学家和经济学家；

生态学家和风景园林师；

公共卫生和社会福利的专家；

并且更广泛地说，所有涉及历史地区保护和发展学科方面的专家；

3）这些机构应在传播有关民众的意见和组织他们积极参与方面起带头作用；

4）保护计划和文件应由法定机构批准；

5）负责实施保护规定和规划的国家、地区和地方各级公共当局应配有必要的工作人员和充分的技术、行政管理和财政来源。

技术、经济和社会措施

（18）应在国家、地区或地方一级制订保护历史地区及其周围环境的清单。该清单应确定重点，以使可用于保护的有限资源能够得到合理的分配。需要采取的任何紧急保护措施，不论其性质如何，均不应等到制定保护计划和文件后再采取。

（19）应对整个地区进行一次全面的研究，其中包括对其空间演变的分析，它还应包括考古、历史、建筑、技术和经济方面的数据。应制订一份分析性文件，以便确定哪些建筑物或建筑群应予以精心保护、哪些应在某种条件下予以保存，哪些应在极例外的情况下经全面记录后予以拆毁。这将能使有关当局下令停止任何与本建议不相符合的工程。此外，出于同样目的，还应制订一份公共或私人开阔地及其植被情况的清单。

（20）除了这种建筑方面的研究外，也有必要对社会、经济、文化和技术数据与结构以及更广泛的城市或地区联系进行全面的研究。如有可能，研究应包括人口统计数据以及对经济、社会和文化活动的分析、生活方式和社会关系、土地使用问题、城市基础设施、道路系统、通讯网络以及保护区域与其周围地区的相互联系。有关当局应高度重视这些研究并应牢记没有这些研究，就不可能制定出有效的保护计划。

（21）在完成上述研究之后，并在保护计划和详细说明制定之前，原则上应有一个实施计划，其中既要考虑城市规划、建筑、经济和社会问题，又要考虑城乡机构吸收与其具体特点相适应的功能的能力。实施计划应使居住密度达到理想水平，并应规定分期进行的工作及其进行中所需的临时住宅，以及为那些无法重返先前住所的居民提供永久性的住房。该实施计划应由有关的社区和人民团体密切参与制订。由于历史地区及其周围环境的社会、经济及自然状态方面会随时间流逝而不断变化，因此，对其研究和分析应是一个连续不断的过程。所以，至关重要的是在能够进行研究的基础上制定保护计划并加以实施，而不是由于推敲计划过程而予以拖延。

（22）一旦制订出保护计划和详细说明并获得有关公共当局批准，最好由制订者本人或在其指导下予以实施。

（23）在具有几个不同时期特征的历史地区，保护应考虑到所有这些时期的表现形式。

(24) 在有保护计划的情况下，只有根据该计划方可批准涉及拆除既无建筑价值和历史价值且结构又极不稳固、无法保存的建筑物的城市发展或贫民区治理计划，以及拆除无价值的延伸部分或附加楼层，乃至拆除有时破坏历史地区整体感的新建筑。

(25) 保护计划未涉及地区的城市发展或贫民区治理计划应尊重具有建筑或历史价值的建筑物和其他组成部分及其附属建筑物。如果这类组成部分可能受到该计划的不利影响，应在拆除之前制定上述保护计划。

(26) 为确保这些计划的实施不致有利于牟取暴利或与计划的目标相悖，有必要经常进行监督。

(27) 任何影响历史地区的城市发展或贫民区治理计划应遵守适用于防止火灾和自然灾害的通用安全标准，只要这与适用于保护文化遗产的标准相符合。如果确实出现了不符的情况，各有关部门应通力合作，找出特别的解决方法，以便在不损坏文化遗产的同时，提供最大的安全保障。

(28) 应特别注意对新建筑物制订规章并加以控制，以确保该建筑能与历史建筑群的空间结构和环境协调一致。为此，在任何新建项目之前，应对城市的来龙去脉进行分析，其目的不仅在于确定该建筑群的一般特征，而且在于分析其主要特征，如：高度、色彩、材料及造型之间的和谐，建筑物正面和屋顶建造方式的衡量、建筑面积与空间体积之间的关系及其平均比例和位置。特别应注意基址的面积，因为存在着这样一个危险，即基址的任何改动都可能带来整体的变化，均对整体的和谐不利。

(29) 除非在极个别情况下并出于不可避免的原因，一般不应批准破坏古迹周围环境而使其处于孤立状态，也不应将其迁移它处。

(30) 历史地区及其周围环境应得到保护，避免因架设电杆、高塔、电线或电话线、安置电视天线及大型广告牌而带来的外观损坏。在已经设置这些装置的地方，应采取适当措施予以拆除。张贴广告、霓虹灯和其他各种广告、商业招牌及人行道与各种街道设备应精心规划并加以控制，以使它们与整体相协调。应特别注意防止各种形式的破坏活动。

(31) 各成员国及有关团体应通过禁止在历史地区附近建立有害工业，并通过采取预防措施消除由机器和车辆所带来的噪音、振动和颤动的破坏性影响，保护历史地区及其周围环境免受由于某种技术发展，特别是各种形式的污染所造成的日益严重的环境损害。另外还应做出规定，采取措施消除因旅游业的过分开发而造成的危害。

(32) 各成员国应鼓励并帮助地方当局寻求解决大多数历史建筑群中所存在的一方面机动交通，另一方面建筑规模以及建筑质量之间的矛盾的方法。为了解决这一矛盾并鼓励步行，应特别重视设置和开放既便于步行、服务通行又便于公共交通的外围乃至中央停车场和道路系统。许多诸如在地下铺设电线和其他电缆的修复工程，如果单独实施耗资过大，可以简单而经济地与道路系统的发展相结合。

(33) 保护和修复工作应与振兴活动齐头并进。因此，适当保持现有的适当作用，特别是贸易和手工艺，并增加新的作用是非常重要的，这些新作用从长远来看，如果具有生命力，应与其所在的城镇、地区或国家的经济和社会状态相符合。保护工作的费用不仅应根据建筑物的文化价值而且应根据其经使用获得的价值进行估算。只有参照了这两方面的价值尺度，才能正确看待保护的社会问题。这些作用应满足居民的社会、文化和经济需要，而又不损坏有关地区的具体特征。文化振兴政策应使历史地区成为文化活动的中心并使其在周围社区的文化发展中发挥中心作用。

(34) 在农村地区，所有引起干扰的工程和经济、社会结构的所有变化应严加控制，以使具有历史意义的农村社区保持其在自然环境中的完整性。

(35) 保护活动应把公共当局的贡献同个人或集体所有者、居民和使用者单独或共同作出的贡献联系起来，应鼓励他们提出建议并充分发挥其积极作用。因此，特别应通过以下方法在社区和个人之间建立各种层次的经常性的合作：适合于某类人的信息资料，适合于有关人员的综合研究，建立附属于计划小组的顾问团体，所有者、居民和使用者在对公共企业机构发挥咨询作用方面的代表性。这些机构负责有关保护计划的决策、管理和组织实施的机构或负责创建参与实施计划。

(36) 应鼓励建立自愿保护团体和非营利性协会以及设立荣誉或物质奖励，以使保护领域中各方面卓有成效的工作能得到认可。

(37) 应通过中央、地区和地方当局足够的预算拨款，确保得到保护历史地区及其环境计划中所规定的用于公共投资的必要资金。所有这些资金应由受委托协调国家、地区或地方各级一切形式的财政援助，并根据全面行动计划发放资金的公共、私人或半公半私的机构集中管理。

(38) 下述形式的公共援助应基于这样的原则：在适当和必要的情况下，有关当局采取的措施，应考虑到修复中的额外开支，即与建筑物新的市场价格或租金相比，强加给所有者的附加开支。

(39) 一般来说，这类公共资金应主要用于保护现有建筑，特别包括低租金的住宅建筑，而不应划拨给新建筑的建设，除非后者不损害现有建筑物的使用和作用。

(40) 赠款、补贴、低息贷款或税收减免应提供给按保护计划所规定的标准进行保护计划所规定的工程的私人所有者和使用者。这些税收减免、赠款和贷款可首先提供给拥有住房和商业财产的所有者或使用者团体，因为联合施工比单独行动更加节省。给予私人所有者和使用者的财政特许权，在适当情况下，应取决于要求遵守为公共利益而规定的某些条件的契约，并确保建筑物的完整，例如：允许参观建筑物，允许进入公园、花园或遗址，允许拍照等。

(41) 应在公共或私人团体的预算中，拨出一笔特别资金，用于保护受到大规模公共工程和污染危害的历史建筑群。公共当局也应拨出专款，用于修复由于自然灾害所造成的损坏。

(42) 另外，一切活跃于公共工程领域的政府部门和机构应通过既符合自己目的，又符合保护计划目标的融资，安排其计划与预算，以便为历史建筑群的修复作出贡献。

(43) 为了增加可资利用的财政资源，各成员国应鼓励建立保护历史地区及其周围环境的公共或私人金融机构，这些机构应有法人地位，并有权接受来自个人、基金会以及有关工业和商业方面的赠款。对捐赠人可给予特别的税收减免。

(44）通过建立借贷机构为保护历史地区及其周围环境所进行的各种工程的融资工作，可由公共机构和私人信贷机构提供便利，这些机构将负责向所有者提供低息长期贷款。

(45）各成员国和其他有关各级政府部门可促进非赢利组织的建立。这些组织负责以周转资金购买，或如果合适在修复后出售建筑物。这笔资金是为了使那些希望保护历史建筑物、维护其特色的所有人能够在其中继续居住而专门设立的。

(46）保护措施不应导致社会结构的崩溃，这一点尤为重要。为了避免因翻修给不得不从建筑物或建筑群迁出的最贫穷的居民所带来的艰辛，补偿上涨的租金能使他们得以维持家庭住房、商业用房、作坊以及他们传统的生活方式和职业，特别是农村手工业、小型农业、渔业等。这项与收入挂钩的补偿，将会帮助有关人员偿付由于进行工程而导致的租金上涨。

2.5 研究、教育和信息

(47）为了提高所需技术工人和手工艺者的工作水平，并鼓励全体民众认识到保护的必要性并参与保护工作，各成员国应根据其立法和宪法权限，采取以下措施。

(48）各成员国和有关团体应鼓励系统地学习和研究：
城市规划中有关历史地区及其环境方面；
各级保护和规划之间的相互联系；
适用于历史地区的保护方法；
材料的改变；
现代技术在保护工作中的运用；
与保护不可分割的工艺技术。

(49）应采用与上述问题有关的包括实习培训期的专门教育。另外，至关重要的是鼓励培养专门从事保护历史地区，包括其周围的空间地带的专业技术工人和手工艺者。此外，还有必要振兴受工业化进程破坏的工艺本身。在这方面有关机构有必要与专门的国际机构进行合作，如在罗马的文化财产保护与修复研究中心、国际古迹遗址理事会和国际博物馆协会。

(50）对地方在历史地区保护方面发展中所需行政人员的教育，应根据实际需要，按照长远计划由有关当局提供资金并进行指导。

(51）应通过校外和大学教育，以及通过诸如书籍、报刊、电视、广播、电影和巡回展览等信息媒介增强对保护工作必要性的认识。还应提供不仅有关美学而且有关社会和经济得益于进展良好的保护历史地区及其周围环境的政策方面的、全面明确的信息。这种信息应在私人和政府专门机构以及一般民众中广为传播，以使他们知道为什么以及怎样才能按此方法改善他们的环境。

(52）对历史地区的研究应包括在各级教育之中，特别是在历史教学中，以便反复向青年人灌输理解和尊重昔日成就，并说明这些遗产在现代生活中的作用。这种教育应广泛利用视听媒介及参观历史建筑群的方法。

(53）为了帮助那些想了解历史地区的青年人和成年人，应加强教师和导游的进修课程以及对教师的培训。

2.6 国际合作

(54）各成员国应在历史地区及其周围环境的保护方面进行合作，如有必要，寻求政府间的和非政府间的国际组织的援助，特别是联合国教育、科学及文化组织——国际博物馆协会——国际古迹遗址理事会文献中心的援助。此种多边或双边合作应认真予以协调，并应采取诸如下列形式的措施：

1）交流各种形式的信息及科技出版物；
2）组织专题研讨会或工作会；
3）提供研究或旅行基金，派遣科技和行政工作人员并发送有关设备；
4）采取共同行动以对付各种污染；
5）实施大规模保护、修复与复原历史地区的项目，并公布已取得的经验。在边境地区，如果发展和保护历史地区及其周围的环境导致影响边境两边的成员国的共同问题，双方应协调其政策和行动，以确保文化遗产以尽可能的最佳方法得到利用和保护；
6）邻国之间在保护共同感兴趣并具有本地区历史和文化发展特点的地区方面应互相协助。

(55）根据本建议的精神和原则，一成员国不应采取任何行动拆除或改变其所占领土之上的历史区段、城镇和遗址的特征。

以上乃1976年11月30日在内罗毕召开的联合国教育、科学及文化组织大会第十九届会议正式通过之公约的作准文本。

特此签字，以昭信守。

3. 保护历史城镇与城区宪章

(国际古迹遗址理事会全体大会第八届会议
于1987年10月在华盛顿通过)

3.1 序言与定义

(1）所有城市社区，不论是长期逐渐发展起来的，还是有意创建的，都是历史上各种各样的社会的表现。

(2）本宪章涉及历史城区，不论大小，其中包括城市、城镇以及历史中心或居住区，也包括其自然的和人造的环境。除了它们的历史文献作用之外，这些地区体现着传统的城市文化的价值。今天，由于社会到处实行工业化而导致城镇发展的结果，许多这类地区正面临着威胁，遭到物理退化、破坏甚至毁灭。

(3）面对这种经常导致不可改变的文化、社会甚至经济损失的惹人注目的状况，国际古迹遗址理事会认为有必要为历史城镇和城区起草一国际宪章，作为"国际古迹保护与修复宪章"（通常称之为"威尼斯宪章"）的补充。这个新文本规定了保护历史城镇和城区的原则、目标和方法。它也寻求促进这一地区私人生活和社会生活的协调方法，并鼓励对这些文化财产的保护。这些文化财产无论其等级多低，均构成人类的记忆。

(4）正如联合国教育、科学及文化组织1976年华沙——内罗毕会议"关于历史地区保护及其当代作用的建议"以及其他一些文件所规定的，"保护历史城镇与城区"意味着这些城镇和城区的保护、保存和修复及其发展并和谐地适应现代生活所需的各种步骤。

3.2 原则和目标

（1）为了更加卓有成效，对历史城镇和其他历史城区的保护应成为经济与社会发展政策的完整组成部分，并应当列入各级城市和地区规划。

（2）所要保存的特性包括历史城镇和城区的特征以及表明这种特征的一切物质的和精神的组成部分，特别是：

1）用地段和街道说明的城市的形制；
2）建筑物与绿地和空地的关系；
3）用规模、大小、风格、建筑、材料、色彩以及装饰说明的建筑物的外貌，包括内部的和外部的；
4）该城镇和城区与周围环境的关系，包括自然的和人工的；
5）长期以来该城镇和城区所获得的各种作用。

任何危及上述特性的威胁，都将损害历史城镇和城区的真实性。

（3）居民的参与对保护计划的成功起着重大的作用，应加以鼓励。历史城镇和城区的保护首先涉及它们周围的居民。

（4）历史城镇和城区的保护需要认真、谨慎以及系统的方法和学科，必须避免僵化，因为，个别情况会产生特定问题。

3.3 方法和手段

（5）在作出保护历史城镇和城区规划之前必须进行多学科的研究。保护规划必须反映所有相关因素，包括考古学、历史学、建筑学、工艺学、社会学以及经济学。

保护规划的主要目标应该明确说明达到上述目标所需的法律、行政和财政手段。

保护规划的目的应旨在确保历史城镇和城区作为一个整体的和谐关系。

保护规划应该决定哪些建筑物必须保存，哪些在一定条件下应该保存以及哪些在极其例外的情况下可以拆毁。在进行任何治理之前，应对该地区的现状作出全面的记录。

保护规划应得到该历史地区居民的支持。

（6）在采纳任何保护规划之前，应根据本宪章和威尼斯宪章的原则和目的开展必要的保护活动。

（7）日常维护对有效地保护历史城镇和城区至关重要。

（8）新的作用和活动应该与历史城镇和城区的特征相适应。使这些地区适应现代生活需要认真仔细地安装或改进公共服务设施。

（9）房屋的改进应是保存的基本目标之一。

（10）当需要修建新建筑物或对现有建筑物改建时，应该尊重现有的空间布局，特别是在规模和地段大小方面。

与周围环境和谐的现代因素的引入行为应受到鼓励，因为，这些特征能为这一地区增添光彩。

（11）通过考古调查和适当展出考古发掘物，应使一历史城镇和城区的历史知识得到拓展。

（12）历史城镇和城区内的交通必须加以控制，必须划定停车场，以免损坏其历史建筑物及其环境。

（13）城市或区域规划中作出修建主要公路的规定时，这些公路不得穿过历史城镇或城区，但应改进接近它们的交通。

（14）为了保护这一遗产并为了居民的安全与安居乐业，应保护历史城镇免受自然灾害、污染和噪音的危害。

不管影响历史城镇或城区的灾害的性质如何，必须针对有关财产的具体特性采取预防和维修措施。

（15）为了鼓励全体居民参与保护，应为他们制定一项普通信息计划，从学龄儿童开始。

与遗产保护相关的行为亦应得到鼓励，并应采取有利于保护和修复的财政措施。

（16）对一切与保护有关的专业应提供专门培训。

4.昆明宣言

中国历史文化名城委员会1999年昆明年会

作为国家历史文化名城的代表，欣逢世纪之交，本着对历史负责、对祖先和子孙负责的精神，热切期盼国家尽快颁布保护历史文化名城的专项法规，并就以下九条达成共识，共同遵守，欢迎社会各界和广大人民群众的监督。

1.历史文化名城所在地政府必须严格执行《中华人民共和国文物保护法》和《中华人民共和国城市规划法》及当地制定的名城保护条例、规章，设立或指定负责名城保护工作的综合协调机构，每年检查执法保护工作情况，报告当地党委和人民代表大会，同时通过传媒公之于众，请各界人士监督、帮助。

2.历史文化名城必须根本改变"以旧城为中心""大规模改造旧城"的开发思路，削减旧城区内的开发建设强度，促进旧城区城市功能的合理调整疏散，扩大旧城内"历史文化保护区"的面积，认真编制、修改好历史文化名城保护规划。其中，重点保护的街区和项目，要进行城市设计，编制控制性详细规划。保护规划按照法定程序批准后，必须广为宣传，严格按规划办事。

3.历史文化名城的保护必须特别注意分析研究不同城市的特点，按照保护文化古迹，保护历史地段，保护和延续古城风貌特色等不同层次的要求进行保护，继承和发扬优秀的传统文化。

历史文化名城要千方百计保留和保护一批具有不同民族、地方和时代特色的传统民居与名人故居及其整体空间环境，为当地客居异国他乡的儿女寻根访旧、重温乡情，留下一片人们记忆中的故土家园。

4.历史文化名城的旧城和新区，要区别对待，采取不同的方针。对于旧城的保护、整治，尊重历史，慎之又慎，切忌急于求成，大拆大建，要抢救濒临破坏的文物古迹、历史街区，注意保护历史环境，严格控制建筑高度，延续传统风貌。同时积极改善基础设施，改善生活条件，促进经济繁荣。

对于新区的开发、建设，要从当地实际出发，继承和发扬优秀历史文化传统，切忌盲目追求"大、高、洋"，切实加强基础设施建设，使之逐步成为融贯古今，富有当地特色的城市新区。

5.历史文化名城要尽心竭力保护文物古迹的真实性、完整性和相关历史环境，从设计、构造、材料、样式、色彩、体量及空间位置等多方面千方百计保存历史原物、原状。对不复存在、仅存残部或遗迹的文物古迹，应着重保存遗址现状，除保护遗址所需的保护性措施和修复外，不提倡重建，尤其反对无根据、未经严谨论证和严格审批的重建。

6.历史文化名城必须从不同城市的历史实际出发，努力发掘城市的传统文脉，以及由于不同的历史、地理环境形成的固有模式与显著特色和世代传承的传统艺术、民间艺术、民俗精华、传统产业、名人轶事等有形或无形的优秀历史文化，用以激发当地人民的凝聚力和自豪感，促进名城的有效保护、合理建设和科学利用。

7.历史文化名城要努力创造条件，根据不同保护项目的性质、规划和特点，采取勒石纪事、参观券署名等鼓励措施和不同的优惠政策，鼓励和吸引国内外企业、社会团体和友好人士的投资、捐款，广开筹资渠道，逐步建立起具有一定规划的历史文化名城保护基金，同时当地政府每年从地方财政中单列一定数额的名城保护资金，为名城保护的持续、健康发展提供比较稳定的资金保证。

8.历史文化名城要坚持不懈地开展宣传工作，在全体市民中普及名城知识，开展"保护名城，人人有责"、"建设名城，振兴中华"等群众性活动。同时充分利用名城的历史文化优势，发展具有当地特色的旅游事业，使历史文化名城既成为吸引国内外游客的旅游胜地，又成为向世界展现中华优秀文化的窗口。

9.国家历史文化名城要继续发展与外国历史城市和友好城市的友好往来，促进经济、文化交流，同时要加强与国际文物、古迹保护等机构的交流合作，将中国的名城保护融入国际共同事业，为人类的文明进步作出应有贡献。

5.北京共识

中国文化遗产保护和城市发展国际会议

联合国教科文组织、世界银行和中国国家文物局、建设部于2000年7月5日至7日在北京召开"文化遗产保护和城市发展"会议。

考虑到历史文化遗产在现代化城市建设中的重要意义；

考虑到保护世界文化遗产有助于各国人民相互了解，并有利于促进世界和平事业；

考虑到当前文化遗产的保护正受到来自城市建设方面的冲击；

同时，也考虑到文化遗产的保护和现代化城市建设的根本利益趋向一致，是制定城市发展建设规划的基本目标，来自世界的与会代表就以下问题达成共识：

保存在城市中的文化遗产不仅是历史上不同传统和精神成就的载体和见证，同时也体现了全世界各民族的基本特征，构成了各个城市面貌和特点的基本要素。它凝聚了数千年人类辛勤的劳动和无穷的智慧，沉淀了人类文明世代相传的宝贵精神资源和物资财富，并作为一种精神动力支撑着城市居民构筑21世纪美好家园的信心和理念。

长期以来世界各国为保护城市中的历史文化遗产所付出的努力，不仅有效地延续了城市的历史文化，同时极大地促进了当地旅游业乃至经济的发展。在人类即将进入21世纪的今天，我们更加认识到保护城市中的历史文化遗产的重要性，同时，随着广大民众文明程度的不断提高和对精神文化生活的迫切要求，城市化进程中的文化遗产保护必将拥有十分美好的前景。

在经济快速发展的21世纪，许多历史城市中的文化遗产遭受冲击，甚至面临着遭受破坏的危险。城市人口的增加，城市向大型化、现代化、经济化的发展，正日益侵蚀着历史文化遗产赖以生存的环境，许多具有历史意义的传统文化街区的历史真实性正在消失。

要妥善保护城市中的历史文化遗产，迎接来自各个方面的挑战，必须采取相应的保护措施。首先需要制定一个完整的保护法规体系，无论国际组织，还是各个国际乃至地方，建立一个更加完善、更加丰富、更加具体的法规体系是历史文化遗产保护的基本前提。其次需要一个与城市建设相吻合的切合实际的历史文化遗产保护规划，并且严格按照规划进行各项城市建设。第三需要一个城市的市长以及政府有关机构具有重视城市文化遗产保护的长远目光和胆识，需要广大公民，特别是城市中的居民能够充分认识到所负有的保护城市历史文化遗产的责任感和使命感，使保护历史文化遗产成为每一位公民的责任和义务。长期以来新闻媒体在保护历史文化中发挥着积极的作用，我们期待有更多的社会力量参与到历史文化的保护工作中来，以形成更为强大的社会舆论。

愿通过我们的共同努力，使历史文化遗产更多地保留在日益发展的现代化城市中，并得以永世流传。

后 记

北京历史文化名城保护是一项长期的、连续的工作，一直得到了北京市委、市政府的高度重视。继《北京旧城二十五片历史文化保护区保护规划》图集向社会公开发行以来，北京的名城保护得到了国内外社会各界的广泛关注。

目前，《北京历史文化名城保护规划》和《北京皇城保护规划》，又得到了北京市政府的正式批复，这极大地丰富了北京历史文化名城保护工作的内涵。我们将这两个规划编辑成册，推出《北京旧城二十五片历史文化保护区保护规划》图集的"续集"，以报社会各界对北京名城保护工作的关爱，同时为更多的专业人士提供宝贵的科学资料。当然，现在呈献给大家的这本图集内容仅仅是反映现阶段对名城保护的认识水平，随着时间的推移，认识肯定会不断地提高和深化，保护办法也会越来越多。

本图集的编辑出版，经过了上百个不眠之夜，这个过程，编辑组的人员完全利用工余时间完成了图集的采编、整理、校对工作；这个过程，凝聚了许多参与过此项工作的单位和个人的辛勤劳动和智慧结晶。当此即将付梓之时，一切辛苦化为欣喜，大家共同认为，这是为北京的名城保护工作又做了一件十分有意义的事儿。

现在，我们又开始了第二批历史文化保护区保护规划的编制工作，同时，旧城的第三批历史文化保护区的划定工作也在进行当中，名城保护图集很快就会再添"新传"。

由于编辑的水平和时间有限，本图集一定会有许多不尽人意之处，希望各位专家、同行以及广大读者多多赐教。

编者
2003年9月

Postscript

The Conservation of historic city of Beijing is a long and arduous task, which has always been attached great importance by the Beijing communist Party Committee and municipal government. Since the book Conservation Plan for the 25 Historic Areas in Beijing Old City was published, this issue has drawn the attention of people from walks of life.

The approval of the Conservation Plan of the Historic City of Beijing and the Conservation Plan of the Imperial City of Beijing has further enriched the implication of preservation work. In order to reciprocate the care shown by society for the conservation effort, as well as to provide professionals with more valuable references, we have combined the above-mentioned two plans together and published them as the sequence of the Conservation Plan for the 25 Historic Areas in Beijing Old City. We acknowledge that the new book is only a reflection of present knowledge towards the conservation campaign. There is no doubt that with the passage of time, new awareness and more approaches for the heritage preservation will inevitably emerge.

For the publication of this book, the editors endured and overcame countless difficulties including spending their spare time in collecting, editing and proofreading the contents. So this effort is virtually the crystallization of the hard-work and wisdom of many participating units and individuals. At the moment of sending the manuscript to be printed, all of us are feeling excited as our painstaking effort is turning into fruit, and we have reached the consensus that we have done another significant work for the conservation of Beijing.

Right now, the conservation plan for the second group of conservation areas of Beijing has started to be made and plan for determining the third group of conservation areas in old Beijing city is also going on. Therefore, new editions will be added to our series soon.

There are certainly much room for improvement as the compilation time is very limited, so we hope experts, colleagues and readers will be so kind as to give us timely advice and suggestions.

Editors
September 2003